RADICALS ON SURFACES

TOPICS IN MOLECULAR ORGANIZATION AND ENGINEERING

Volume 13

Honorary Chief Editor:

W. N. LIPSCOMB (*Harvard, U.S.A.*)

Executive Editor:

Jean MARUANI (*Paris, France*)

Editorial Board:

Henri ATLAN (*Jerusalem, Israel*)
Sir Derek BARTON (*Texas, U.S.A.*)
Christiane BONNELLE (*Paris, France*)
Paul CARO (*Meudon, France*)
Stefan CHRISTOV (*Sofia, Bulgaria*)
I. G. CSIZMADIA (*Toronto, Canada*)
P-G. DE GENNES (*Paris, France*)
J-E. DUBOIS (*Paris, France*)
Manfred EIGEN (*Göttingen, Germany*)
Kenishi FUKUI (*Kyoto, Japan*)
Gerhard HERZBERG (*Ottawa, Canada*)

Alexandre LAFORGUE (*Reims, France*)
J-M. LEHN (*Strasbourg, France*)
P-O. LÖWDIN (*Uppsala, Sweden*)
Patrick MacLEOD (*Massy, France*)
H. M. McCONNELL (*Stanford, U.S.A.*)
C. A. McDOWELL (*Vancouver, Canada*)
Roy McWEENY (*Pisa, Italy*)
Ilya PRIGOGINE (*Brussels, Belgium*)
Paul RIGNY (*Saclay, France*)
R. G. WOOLLEY (*Nottingham, U.K.*)

The titles published in this series are listed at the end of this volume.

RADICALS ON SURFACES

Edited by

Anders Lund

and

Christopher J. Rhodes

KLUWER ACADEMIC PUBLISHERS
DORDRECHT / BOSTON / LONDON

Library of Congress Cataloging-in-Publication Data

```
Radicals on surfaces / edited by Anders Lund and Christopher Rhodes.
      p.    cm. -- (Topics in molecular organization and engineering ;
   v. 13)
   Includes index.
   ISBN 0-7923-3108-7
   1. Surface chemistry.   2. Free radicals (Chemistry)   3. Catalysis.
 I. Lund, A. (Anders)   II. Rhodes, Christopher.   III. Series.
 QD506.R24  1995
 541.3'3--dc20                                                  94-31444
```

ISBN 0-7923-3108-7

Published by Kluwer Academic Publishers,
P.O. Box 17, 3300 AA Dordrecht, The Netherlands.

Kluwer Academic Publishers incorporates
the publishing programmes of
D. Reidel, Martinus Nijhoff, Dr W. Junk and MTP Press.

Sold and distributed in the U.S.A. and Canada
by Kluwer Academic Publishers,
101 Philip Drive, Norwell, MA 02061, U.S.A.

In all other countries, sold and distributed
by Kluwer Academic Publishers Group,
P.O. Box 322, 3300 AH Dordrecht, The Netherlands.

"The logo on the front cover represents the generative hyperstructure of alkanes", printed with permission from J.E. Dubois, Institut de Topologie et de Dynamique des Systèmes, Paris, France.

Printed on acid-free paper

All Rights Reserved
© 1995 by Kluwer Academic Publishers
No part of the material protected by this copyright notice may be reproduced or
utilized in any form or by any means, electronic or mechanical,
including photocopying, recording or by any information storage and
retrieval system, without written permission from the copyright owner.

Printed in the Netherlands

Introduction to the Series

The Series 'Topics in Molecular Organization and Engineering' was initiated by the Symposium 'Molecules in Physics, Chemistry, and Biology', which was held in Paris in 1986. Appropriately dedicated to Professor Raymond Daudel, the symposium was both broad in its scope and penetrating in its detail. The sections of the symposium were: 1. The Concept of a Molecule; 2. Statics and Dynamics of Isolated Molecules; 3. Molecular Interactions, Aggregates and Materials; 4. Molecules in the Biological Sciences, and 5. Molecules in Neurobiology and Sociobiology. There were invited lectures, poster sessions and, at the end, a wide-ranging general discussion, appropriate to Professor Daudel's long and distinguished career in science and his interests in philosophy and the arts.

These proceedings have been arranged into eighteen chapters which make up the first four volumes of this series: Volume I, 'General Introduction to Molecular Sciences'; Volume II, 'Physical Aspects of Molecular Systems'; Volume III, 'Electronic Structure and Chemical Reactivity'; and Volume IV, 'Molecular Phenomena in Biological Sciences'. The molecular concept includes the logical basis for geometrical and electronic structures, thermodynamic and kinetic properties, states of aggregation, physical and chemical transformations, specificity of biologically important interactions, and experimental and theoretical methods for studies of these properties. The scientific subjects range therefore through the fundamentals of physics, solid-state properties, all branches of chemistry, biochemistry, and molecular biology. In some of the essays, the authors consider relationships to more philosophic or artistic matters.

In Science, every concept, question, conclusion, experimental result, method, theory or relationship is always open to reexamination. Molecules do exist! Nevertheless, there are serious questions about precise definition. Some of these questions lie at the foundations of modern physics, and some involve states of aggregation or extreme conditions such as intense radiation fields or the region of the continuum. There are some molecular properties that are definable only within limits, for example, the geometrical structure of non-rigid molecules, properties consistent with the uncertainty principle, or those limited by the neglect of quantum-field, relativistic or other effects. And there are properties which depend specifically on a state of aggregation, such as superconductivity, ferroelectric (and anti), ferromagnetic (and anti), superfluidity, excitons, polarons, etc. Thus, any molecular definition may need to be extended in a more complex situation.

Chemistry, more than any other science, creates most of its new materials. At least so far, synthesis of new molecules is not represented in this series, although the principles of chemical reactivity and the statistical mechanical aspects are included. Similarly, it is the more physico-chemical aspects of biochemistry, molecular biology and biology itself that are addressed by the examination of questions related to molecular recognition, immunological specificity, molecular pathology, photochemical effects, and molecular communication within the living organism.

Many of these questions, and others, are to be considered in the Series 'Topics in Molecular Organization and Engineering'. In the first four volumes a central core is presented, partly with some emphasis on Theoretical and Physical Chemistry. In later volumes, sets of related papers as well as single monographs are to be expected; these may arise from proceedings of symposia, invitations for papers on specific topics, initiatives from authors, or translations. Given the very rapid development of the scope of molecular sciences, both within disciplines and across disciplinary lines, it will be interesting to see how the topics of later volumes of this series expand our knowledge and ideas.

WILLIAM N. LIPSCOMB

Table of Contents

ANDERS LUND AND CHRIS RHODES / Introduction — ix

Part I: Properties of Catalytic Surfaces — 1

I.1 CATHERINE LOUIS, CHRISTINE LEPETIT, AND MICHEL CHE / EPR Characterization of Oxide Supported Transition Metal Ions: Relevance to Catalysis — 3

I.2 R. I. SAMOILOVA, A. D. MILOV, AND YU. D. TSVETKOV / Study of Catalytic Site Structure and Diffusion of Radicals in Porous Heterogeneous Systems with ESR, ENDOR and ESE — 39

I.3 MASAHIDE OHNO / Theoretical Studies of Core Ionization, Excitation and De-excitation of Adsorbates — 61

Part II: Structure and Reactivity of Radicals on Surfaces — 87

II.1 R. B. CLARKSON, KAREN MATTSON, WENJUN SHI, WEI WANG, AND R. L. BELFORD / Electron Magnetic Resonance of Aromatic Radicals on Metal Oxide Surfaces — 89

II.2 CHRISTOPHER J. RHODES AND CHANTAL S. HINDS / ESR Studies of Organic Radical Cations in Zeolites — 119

II.3 ELIO GIAMELLO AND DAMIEN MURPHY / Surface Trapped Electrons on Metal Vapour Modified Magnesium Oxide. Nature of Surface Centres and Reactivity with Adsorbed Molecules — 147

II.4 MASARU SHIOTANI AND MIKAEL LINDGREN / Radicals on Surfaces Formed by Ionizing Radiation — 179

II.5 ALEXANDER M. VOLODIN, VADIM A. BOLSHOV, AND TATIANA A. KONOVALOVA / Photostimulated Formation of Radicals on Oxide Surfaces — 201

Part III: Trends in Modern Techniques — 227

III.1 HANS VAN WILLIGEN AND PATRICIA R. LEVSTEIN / Fourier Transform Electron Paramagnetic Resonance Studies of Photochemical Reactions in Heterogeneous Media — 229

III.2 EMIL RODUNER, MARTINA SCHWAGER, AND MEE SHELLEY / Muon Spin Resonance of Radicals on Surfaces — 259

III.3 RONALD L. BIRKE AND JOHN R. LOMBARDI / Investigation of Radical Ions with Time-Resolved Surface Enhanced Raman Spectroscopy   277

INDEX   311

Introduction

The entire area of 'catalysis' is one of immense current activity, as befits the intellectual and commercial challenges which it provides. A great deal is known regarding the physical properties of catalytic materials, particularly from techniques such as NMR, and studies of adsorbed molecules are the cornerstone of 'surface science'. What is less clear, however, are the mechanistic details by which molecular transformations occur on catalyst surfaces, but growing evidence tends to implicate organic free radicals as intermediates in a number of reactions – particularly those catalysed by zeolites and other metal oxides.

The present book addresses the broad aspect of radicals on surfaces, and concerns the properties of paramagnetic surface sites, the structure and reactivity of radicals on surfaces, and those trends and developments in spectroscopic techniques which are revealing new horizons of this topic.

We have invited specialists from around the world to contribute a chapter relating to one of the above topics, presented in the manner of a review: collectively, thus, we have an overview of the state of current knowledge, but we envisage that this will be added to as the various sub-areas progress. Since this is, predominantly, a book about radicals, we have quite deliberately avoided coverage of conventional surface science techniques, which are well dealt with elsewhere.

We are grateful to all our colleagues for their contributions, and to the editorial staff of Kluwer Academic Publishers, particularly Dr. David Larner, for their invaluable cooperation.

ANDERS LUND   CHRIS RHODES

Part I: Properties of Catalytic Surfaces

EPR Characterization of Oxide Supported Transition Metal Ions: Relevance to Catalysis

CATHERINE LOUIS, CHRISTINE LEPETIT and MICHEL CHE
Laboratoire de Réactivité de Surface, URA 1106 CNRS, Université P. et M. Curie, 4 place Jussieu, 75252 Paris Cedex 05, France

(Received: 15 December 1993; accepted: 5 July 1994)

Abstract. EPR spectroscopy is a powerful tool to identify at a molecular level, the different steps of catalyst preparation, and of catalytic reactions:
 (i) Deposition of paramagnetic transition metal ions onto a support is monitored, and the coordination sphere of the metallic center is characterized by EPR.
 (ii) The catalyst is also characterized after activation (thermal oxidation or reduction):
–the distribution among the different sites in zeolites can be determined;
–the dispersion of the active phase may be appreciated;
–the unsaturation degree of the active site may be evaluated using probe molecules such as water or ^{13}C enriched carbon monoxide.
 (iii) The catalytic mechanisms can be investigated by studying the elementary steps of the catalytic reaction, as illustrated for methanol oxidation over Mo/SiO_2 catalysts whose EPR results have extended the reaction mechanism proposed on the basis of kinetic data. In addition, reaction intermediates may be isolated in *quasi-in situ* conditions as in the case of olefin oligomerization catalyzed by Ni/SiO_2 systems.

Key words. EPR, coordination sphere, surface reactivity, catalysis.

1. Introduction

EPR spectroscopy has been extensively used to investigate surfaces of heterogeneous catalysts. Heterogeneous catalysts often consist of transition metal ions dispersed onto an oxide support of high surface area. These transition metal ions are the working sites in catalytic reactions (also called active sites or active phase). Because of its high sensitivity, EPR has been used in the field of catalysis to identify paramagnetic atoms, molecules or ions such as surface defects, adsorbed species or transition metal ions, sometimes stabilized by the support in unusual oxidation states, e.g. Ni^I. Since the publication of the first review by O'Reilly in 1960 [1], several other reviews have appeared [2].

The main challenges in the elaboration of an heterogeneous catalytic system involve the control of all the preparation steps, including the activation one (e.g. thermal treatment, reduction or oxidation, performed to make the catalyst active for the reaction). Spectroscopic characterization of the catalyst during its preparation is a useful tool to improve this control which should contribute subsequently to tailoring the catalytic properties.

The aim of this review is to show how the EPR techniques may help in determining at a molecular level, the state of paramagnetic species during the catalyst preparation and catalytic reactions.

During the deposition of transition metal ions onto supports and further activation processes of the catalysts, we will especially show how it is possible to determine from the EPR spectra:

- the oxidation state of paramagnetic ions
- the number of ligands in their coordination sphere and the symmetry of the surface complex
- the state of aggregation or dispersion of the ions onto support surface
- possibly, their migration in the bulk of the support or in the different cavities of zeolites and their location.

During catalytic reactions, we will show that using EPR, it may be possible:

- to study the adsorption of reactants, which may be of interest for the determination of the active site and the activation step
- to characterize intermediates species, which may be decisive for the determination of the mechanism of the catalytic reaction
- to study the desorption of products. It may help in the knowledge of the properties of poisoning or regeneration of the catalysts.

This paper does not intend to survey the literature in a comprehensive way but rather to give typical examples to illustrate the topics described above. The first section describes briefly how to relate the EPR parameters to the chemical states of paramagnetic transition metal ions.

2. General Background

A general presentation of EPR and its principles will not be given here since several books and reviews [2d, 3, 4] have already been published on this subject. The relation between the EPR parameters and the environment of the transition metal ions will be briefly described below.

2.1. g TENSOR

The general spin Hamiltonian for a paramagnetic species may be written as:

$$H = g\beta B \cdot S + \sum_{i}^{m} (AS \cdot I_i - g_n\beta_n B \cdot I_i) + H_{\text{quadrupole}} + H_{\text{ZFS}}$$

where β and β_n are Bohr and nuclear magneton, respectively, g and g_n are the electron and nuclear g factors, A is the hyperfine tensor, m is the number of different nuclei with non-zero nuclear spins, B is the magnetic field, and I and S are the nuclear and electron spin operators. The nuclear Zeeman term, $-g_n\beta_n B.I.$, is often neglected, $H_{\text{quadrupole}}$ describes the interaction of the electron spin with nuclear quadrupole moments present for nuclei with $I > 1/2$ and H_{ZFS} contains the zero-field splitting term for systems having $S > 1/2$. The goal of the

TABLE I
Relation between the relative order of g tensor components with the symmetry and ground state of a d^9 ion (from Ref. 40).

Symmetry	Ground state	Relative order
Distorted elongated tetrahedron	$d_{xy}, d_{x^2y^2}$	Temperature dependent-spectrum
Octahedron, ligands in the plane perpendicular to the symmetry axis dominate	$d_{x^2-y^2}$	$g_\| > g_\perp > g_e$
Trigonal bipyramid or distorted compressed tetrahedron	d_{z^2}	$g_\perp > g_\| \approx g_e$
Octahedron ligands along the symmetry axis dominate	d_{z^2}	$g_\perp > g_\| \approx g_e$

spectroscopist is to determine the spin Hamiltonian parameters as completely as possible.

A typical heterogeneous catalyst is generally a polycrystalline or amorphous powder rather than a single crystal. The spectrum of a powdered sample will be the envelope of spectra from all possible orientations of the paramagnetic species with respect to the applied magnetic field. Extraction of the spin Hamiltonian parameters is therefore more difficult than in the case of a single crystal. Usually, measurements of turning points (maxima, minima or points of inflexion) in the first derivative spectra of powders can give the magnitude of g and A tensor components.

The number of g tensor components is related to the symmetry of the paramagnetic species (isotropic, axial or orthorhombic) and/or its mobility.

The shift of each g component compared to g_e may be related to the strength of the crystal field surrounding the paramagnetic d elements (f transition elements are little affected by the crystal field).

The relative order of g components may be related to the symmetry of the paramagnetic ion as well as to its ground state [5]. This is illustrated in the case of d^9 ions in Table I.

2.2. HYPERFINE (hf), SUPERHYPERFINE (shf) STRUCTURES

If the central atom or surrounding ligands have non zero nuclear spins, the latter can interact with the electron spin leading to hf or shf structures, respectively. The analysis of the number and relative intensity of the resulting EPR lines allows one to determine the hf tensor A (or the shf tensor, called a) and the corresponding coupling constants. The hyperfine structure is a fingerprint of the central atom. This latter can therefore be unambiguously identified. The shf structure allows one to determine the number and equivalency of ligands in each typical class of positions, e.g., axial and equatorial, within the coordination sphere of the paramagnetic ion.

2.3. PROBE MOLECULES

Providing that there is no electron transfer involved, reaction of the paramagnetic ion with small probe molecules (isotopically labelled or not) may be used to evaluate its degree of unsaturation. The coordination number is increased without any modification of the number of spins. $^{13}CO/^{12}CO$ is one of the most commonly used probe molecules.

2.4. SIMULTANEOUS PRESENCE OF VARIOUS SPECIES

Some special techniques are useful in obtaining the EPR parameters of each species when the spectrum is intricate, e.g. when a mixture is present:
- Recording the spectra at several microwave powers or temperatures.
- third derivative spectra [6]: in the case of overlapping signals, the resolution may be enhanced on the third derivative spectrum.
- Use of several microwave frequencies [2c]: this permits a distinction to be made between the g tensor components and the hf(shf) lines. The separation between hf (shf) lines is not affected by the variation of the microwave frequency. In contrast, the separation between the g tensor components is a linear function of the microwave frequency, so recording the spectrum in the Q band ($\nu \approx 36$ GHz) rather than in the X band ($\nu \approx 9.5$ GHz) increases the field separation of the different species by a factor of about 4. However, signal resolution is observed to decrease with increasing microwave frequency due to an increase in line width [7].
- Simulation of spectra [3b]: the EPR parameters (g, A, a tensors, number of lines and relative intensity) obtained from the experimental spectrum, may be confirmed by computer simulation. This is particularly useful for getting accurate values in the case of mixtures of species and/or hf(shf) structures.
- Gas adsorption [3b]: only surface species are expected to react with the adsorbate. They can therefore be distinguished from bulk species.

2.5. QUANTITATIVE ANALYSIS [8]

There is a linear relationship between the EPR signal intensity and the number of paramagnetic species. The EPR signal intensity of the sample investigated is generally compared to a known standard. The spectra of both the investigated and standard samples have to be recorded in the very same experimental conditions. The standard should as much as possible be very similar to the investigated sample with respect to electronic structure, symmetry, nature of the ligands and concentration. Reference materials for quantitative measurements have been reviewed by Chang et al. [9].

3. Deposition of Transition Metal Ions on Oxide Surfaces

To our knowledge, there are very few examples in the literature where EPR spectroscopy has been used to provide information on the mechanisms occurring during deposition of transition metal ions onto the support. Two of them are described below.

3.1. PREPARATION OF Mo SUPPORTED ON SILICA BY THE GRAFTING METHOD (GAS-SOLID INTERFACE)

In the following example, the assignment of the spectrum of supported complexes of unknown structures is made by analogy of the magnetic parameters with those of well-defined complexes with known structures.

The grafting method used for the preparation of highly dispersed Mo supported on silica, involves a chemical reaction between surface hydroxyl groups of silica and $MoCl_5$, under air- and water-free conditions because of the high sensitivity of $MoCl_5$ toward the latter [10, 11]. In contrast to the impregnation method in which Mo is bonded to the support during calcination, the grafting reaction produces this bonding directly, which is expected to prevent Mo migration and aggregation during further calcination and to lead to a better Mo dispersion.

The choice of $MoCl_5$ is particularly attractive since it is paramagnetic ($4d^1$) and monomeric in the vapor phase and gives an isotropic signal at $g = 1.952$. The reactor used for the preparation was equipped with an EPR tube which permitted the monitoring of the grafting process by EPR.

When dehydrated silica was heated in the presence of gas phase $MoCl_5$ molecules at 200 °C, a Mo^{5+} EPR signal with an axial symmetry ($g_\perp = 1.952$, $g_\parallel = 1.968$, $A_\perp = 44$ G, $A_\parallel = 82$ G) appeared on silica and increased in intensity (Figure 1). The hyperfine pattern is due to the interaction between the unpaired electron with $^{95,97}Mo$ nuclei ($I = 5/2$) in ~25% natural abundance. The initial isotropic EPR signal of Mo^{5+} (monomeric $MoCl_5$) is modified upon grafting, suggesting that the Mo^{5+} coordination sphere is affected by the grafting process. The parameters of the EPR signal and of the electronic spectrum obtained by diffuse reflectance spectroscopy (DRS) after grafting were found to be very similar to those of the $[MoOCl_4]^-$ ion, suggesting the following grafting reaction [10, 11]:

$$MoCl_5 + \equiv SiOH \rightarrow \equiv SiOMoCl_4 + HCl$$

3.2. PREPARATION OF Cu SUPPORTED ON OXIDES BY CATION EXCHANGE (LIQUID-SOLID INTERFACE)

Cu^{2+} ions ($3d^9$, $I = 3/2$ for ^{63}Cu and ^{65}Cu in 69 and 31% natural abundance, respectively) have been extensively studied since they give rise to EPR spectra with resolved hyperfine structure and g anisotropy when they are isolated. The deposition of Cu^{2+} ions onto different supports has been monitored by EPR.

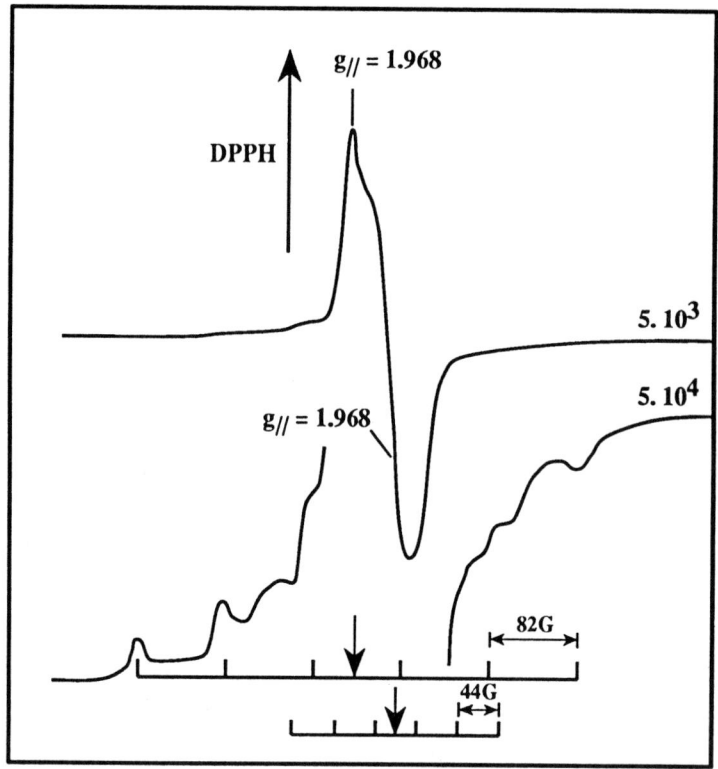

Fig. 1. EPR spectra (X band, 77 K) of a Mo/SiO$_2$ catalyst obtained after grafting with MoCl$_5$ vapor at 200 °C: (a) first derivative spectrum; (b) third derivative spectrum (from Ref. 11).

3.2.a. Cu Ammine Exchange on Silica

Cu^{2+} was introduced onto the silica support using the competitive cation-exchange method. Ammoniated silica was stirred at 25 °C with a solution of copper ammine complexes (pH = 11) for 48 h. The sample was then filtered, washed and dried at 80 °C overnight [12]. Analysis of DRS spectra shows that [Cu(NH$_3$)$_4$(H$_2$O)$_2$]$^{2+}$ in solution interacts with silica to give ($\equiv$SiO)$_2$Cu(NH$_3$)$_4$, and then transforms into ($\equiv$SiO)$_2$Cu(NH$_3$)$_2$(H$_2$O)$_2$ after drying. The EPR parameters of the dried complex are: $g_\perp = 2.025$, $g_\| = 2.280$, $A_\perp$ = unresolved, $A_\| = 170$ G. The relative order of the g tensor components, $g_\| > g_\perp > g_e$, is in agreement with a distorted octahedral symmetry and a $d_{x^2-y^2}$ ground state (Table I) and confirms the DRS analysis. The hyperfine splitting is due to the interaction between the unpaired electron and the spin $I = 3/2$ of the Cu nucleus.

3.2.b. Cu Ethylenediamine Exchange in X and Y Zeolites

In this example, the assignment of the EPR spectra of the prepared sample was made by comparison with those of similar complexes in aqueous solutions.

X and Y zeolites were exchanged with solutions of $[Cu(en)_2(H_2O)_2]^{2+}$ (pH ~ 11, stirring at 25 °C for 48 h) [13]. The spectrum of the exchanged Y-zeolite shows only one EPR signal ($g_\perp = 2.035$, $g_\parallel = 2.194$, $A_\perp = 29$ G, $A_\parallel = 189$ G) very close to that of the frozen $[Cu(en)_2(H_2O)_2]^{2+}$ solution ($g_\perp = 2.041$, $g_\parallel = 2.190$, $A_\perp = 30$ G, $A_\parallel = 203$ G), whether the sample was wet or dried at 25 °C, suggesting that $[Cu(en)_2(H_2O)_2]^{2+}$ retains its integrity upon exchange and drying. However, the analysis of the EPR and DRS spectra shows that the change of environment of $[Cu(en)_2(H_2O)_2]^{2+}$ from aqueous solution to the supercages of the zeolites induces a small change in the nature of the bonds perpendicular to the plane of the distorted octahedral complex, i.e., directed towards the surface. This may be accounted for by the replacement of axially coordinated water molecules by zeolite framework oxygens.

Similarly in wet X-zeolite, $[Cu(en)_2(H_2O)_2]^{2+}$ complexes were observed but they partially decomposed on the acidic surface sites upon drying, to form $[Cu(en)(H_2O)_5]^{2+}$ complexes ($g_\perp = 2.053$, $g_\parallel = 2.264$, $A_\perp = 12$ G, $A_\parallel = 169$ G). DRS also revealed the presence of $[Cu(H_2O)_6]^{2+}$.

4. Evaluation of the State of Dispersion of Supported Transition Metal Ions by Oxygen Adsorption

The dispersion of transition metals ions, i.e., the ratio of surface to total ion concentration is an important catalytic parameter since only surface ions are supposed to be catalytically active. The use of EPR to evaluate the dispersion is illustrated below for the case of molybdenum in the pentavalent state in Mo/SiO_2 catalysts. The distinction between bulk and surface ions is made by oxygen adsorption in such conditions that only the surface ions interact with O_2.

The measurement was performed by adsorption of oxygen at low pressure at 77 K on thermally reduced catalysts [14, 15]. Upon interaction with surface Mo^{5+} ions, oxygen is reduced into the paramagnetic radical anion O_2^- ($g_{xx} = 2.004$, $g_{yy} = 2.010$, $g_{zz} = 2.017$) via the following electron transfer reaction:

$$Mo^{5+} + O_2 \xrightarrow{77\ K} Mo^{6+} + O_2^-$$

The EPR signal intensity of O_2^- increases at the expense of the one of surface Mo^{5+} (Figure 2). The adsorption of oxygen must be performed at 77 K in order to limit the electron transfer to surface molybdenum ions and to avoid oxidation of the bulk [16]. In addition, the oxygen pressure must be low (less than 5 torr) in order to avoid dipolar broadening between the triplet state of O_2 and O_2^-.

From the comparison of the Mo^{5+} EPR signal intensity before and after oxygen chemisorption (Figure 2), the percentage of Mo^{5+} ions accessible to oxygen, i.e., the ratio of surface Mo^{5+} to total Mo^{5+} concentrations may be evaluated.

The Mo^{5+} dispersion has been found to be equal to 80 and 30% for grafted (see Section 3.1) and impregnated samples, respectively, for 1 wt.-% Mo content. Values as high as 100% have been obtained for grafted samples with molybdenum contents lower than 0.1 wt.-%, indicating that grafted Mo is much better dispersed than impregnated Mo. These results have been confirmed by other techniques (IR, UV-visible, photoluminescence, catalytic reaction) [10, 14, 15].

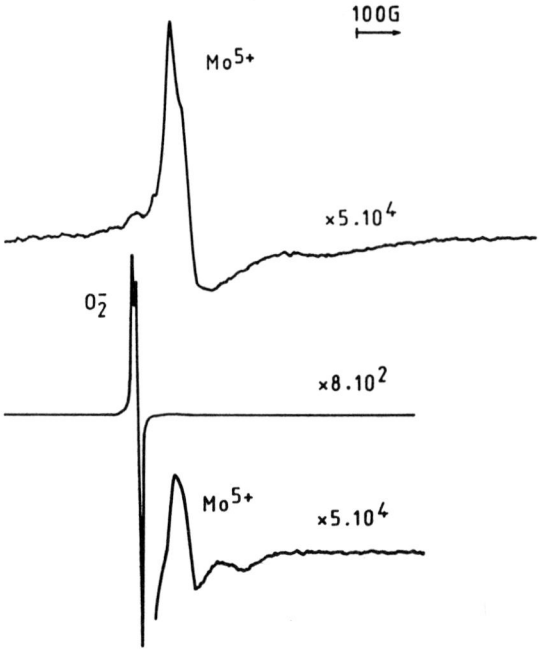

Fig. 2. EPR spectra (X band, 77 K) of a grafted Mo/SiO$_2$ catalyst (0.3 wt.-% Mo): (a) after reduction under H$_2$ at 600 °C; (b) after oxygen adsorption (~1 torr) at 77 K (from Ref. 15).

5. Coordination of Transition Metal Ions on Oxide Surfaces

The EPR technique, as well as a number of other techniques, has been employed to identify various types of metal ions on oxide matrices. Two types of transition metal ions will be described below: (1) those in extraframework positions on oxide surfaces; (2) those in the framework positions at the surface of oxide solid solutions or doped-oxides.

5.1. SUPPORTED TRANSITION METAL IONS IN EXTRAFRAMEWORK POSITION

The coordination sphere of supported paramagnetic transition metal ions may be studied by EPR using probe molecules such as H$_2$O, ^{12}CO, ^{13}CO, and deduced from a self-consistent analysis of:

(i) the relative order of the g tensor components which gives the probable ground state occupied by the single electron and the symmetry of the site;
(ii) the superhyperfine (shf) structure (number and relative intensities of shf lines), due to the interaction of the unpaired electron with the nuclear spin of ^{13}C ($I = 1/2$), which is related to the number and position (axial, equatorial) of ^{13}CO ligands.

This method is illustrated here with nickel and molybdenum complexes.

5.1.a. Nickel Coordination Chemistry

Coordination chemistry of supported Ni^I ions has been investigated by EPR using ^{12}CO and ^{13}CO as probe [17]. Depending on the CO pressure, complexes **1** to **4** have been obtained (Figure 3). **1** can reversibly be transformed into **2**, **3** and **4** by addition or evacuation of CO.

The g tensor of species **4** is axial and the relative order, $g_\perp > g_\parallel \approx g_e$, suggests a d_{z^2} ground state for the single electron, consistent either with a trigonal bipyramid or a distorted octahedron (Table I). The relative intensities of the shf lines (Figure 4), better resolved on the third derivative spectrum (Figure 4b), indicate that Ni^I is bound to four ^{13}CO ligands, three of them being equivalent. The trigonal bipyramidal structure drawn on Figure 3d is the only one consistent with the above analysis.

The EPR signal of complex **1** is pseudo-axial and the relative order of the g tensor components, $g_{xx} \approx g_{yy} \approx g_\perp > g_{zz} \approx g_\parallel \approx g_e$, suggests a d_{z^2} ground state for the single electron consistent with a slightly distorted compressed tetrahedron (Table I). Although no shf structure is observed in the presence of ^{13}CO, there is one and only one CO ligand in the coordination sphere of **1** as observed by IR spectroscopy [18]. The absence of shf structure and the distortion of the complex is explained by a tilted CO (angle α) with respect to the C_3 axis of the complex as presented below:

5.1.b. Molybdenum Coordination Chemistry

After thermal reduction under H_2 and further evacuation at 600 °C, the grafted Mo/SiO_2 samples (see Section 3.1) exhibit a complex asymmetric line (Figure 5a). The third derivative spectrum (Figure 5b) as well as EPR measurements at different temperatures and frequencies reveal the presence of three axial Mo^{5+} ions with $g_\perp > g_\parallel$ [19], referred to as **a**, **b**, and **c**.

The comparison of the above signals with those of the isopolyanion $[Mo_6O_{19}]^{3-}$ ($g_\perp = 1.930$, $g_\parallel = 1.919$), ^{95}Mo enriched Mo/SiO_2 catalysts and various molybdenyl ions [7, 19, 20] shows that ions **a**, **b**, and **c** possess oxygen ligands and a molybdenyl character.

Those assignments are in agreement with the behavior of **a**, **b**, and **c** upon adsorption of a probe molecule such as water [19]. A stepwise transformation of one Mo^{5+} ion into another without changing the total number of spins was observed: increasing the pressure of water adsorbed at 25 °C, leads to the disappearance of the signal **c** whereas that of **b** increases (Figure 6a), then to the disappearance of the **b** while that of **a** increases (Figure 6b) [19]. Water desorption at room temperature leads to the reappearance of **b** signal intensity, and at 500 °C, the

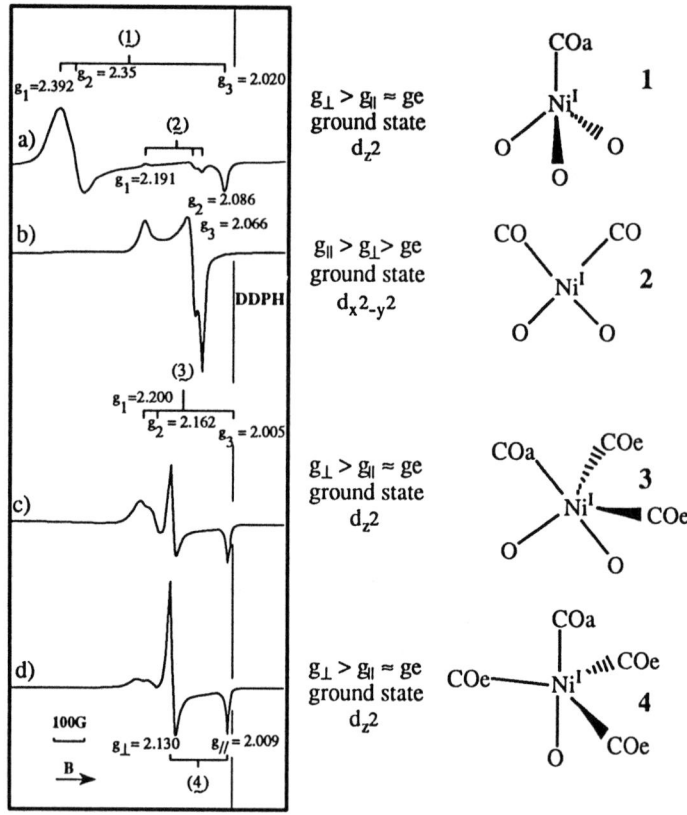

Fig. 3. EPR spectra (X band, 77 K) of $Ni^I(CO)_n$ ($n \leq 4$) complexes obtained upon adsorption at 293 K of ^{12}CO under a pressure of: (a) 10 torr followed by outgassing 15 min at 340 K; (b) 10 torr; (c) 100 torr; (d) 400 torr (from Ref. 17).

three initial Mo^{5+} signals are restored. Similar results are obtained with methanol instead of water [21].

These results suggest that the coordination sphere of **a** is saturated while those of **b** and **c** are unsaturated with one and two vacancies, respectively. They are reported in Table II as Mo^{5+}_{6c}, Mo^{5+}_{5c}, Mo^{5+}_{4c}, respectively where 4c, 5c, 6c stand for tetra, penta and hexacoordinated.

The mechanism of water adsorption occurs in two steps:

- admission of a first water molecule in the Mo^{5+}_{4c} coordination sphere.
- admission of a water molecule in the axial position of both the non hydrated Mo^{5+}_{5c} and the monohydrated Mo^{5+}_{4c} (Table II).

The previous scheme can be understood by considering, as is done in reference [20], that the g values of a molybdenyl ion in C_{4v} symmetry depend on the spin-orbit coupling of the ligands surrounding the central atom. As the water ligand is connected to the Mo^{5+} ion by the oxygen atom, the spin-orbit coupling constant is the same as that of O^{2-}. Hence, the Mo^{5+}_{4c} ion, hydrated by two molecules of

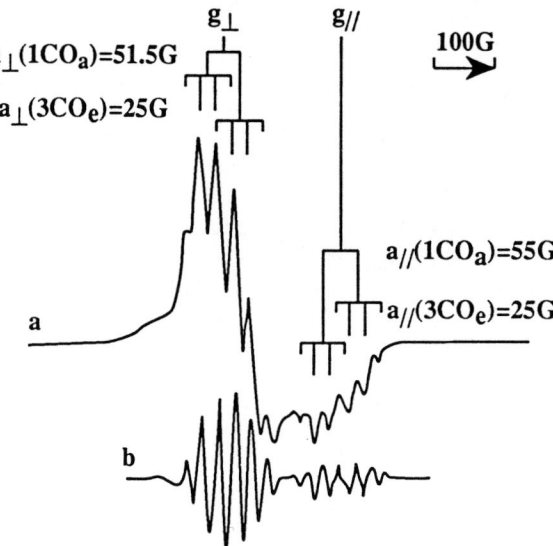

Fig. 4. EPR spectra (X band, 77 K) of complex **4** under a pressure of 400 torr of ^{13}CO: (a) first derivative spectrum; (b) third derivative spectrum (from Ref. 17).

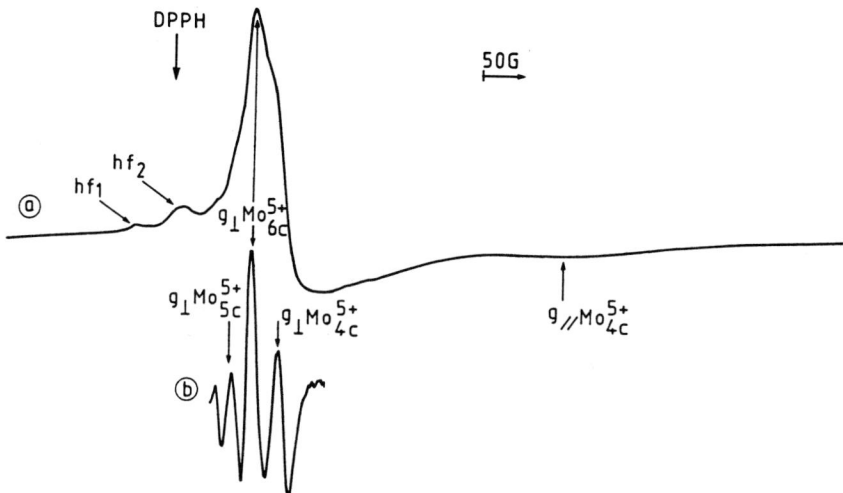

Fig. 5. EPR spectra (X band, 77 K) of a grafted Mo/SiO$_2$ catalyst (0.33 wt.-%) after reduction under H$_2$ at 600 °C: (a) first derivative spectrum; (b) third derivative spectrum (from Ref. 19).

water, gives the same signal as Mo^{5+}_{6c} with six O^{2-} ligands with one doubly bonded to Mo (Table II). These results have been confirmed using ^{13}CO as a probe molecule [19].

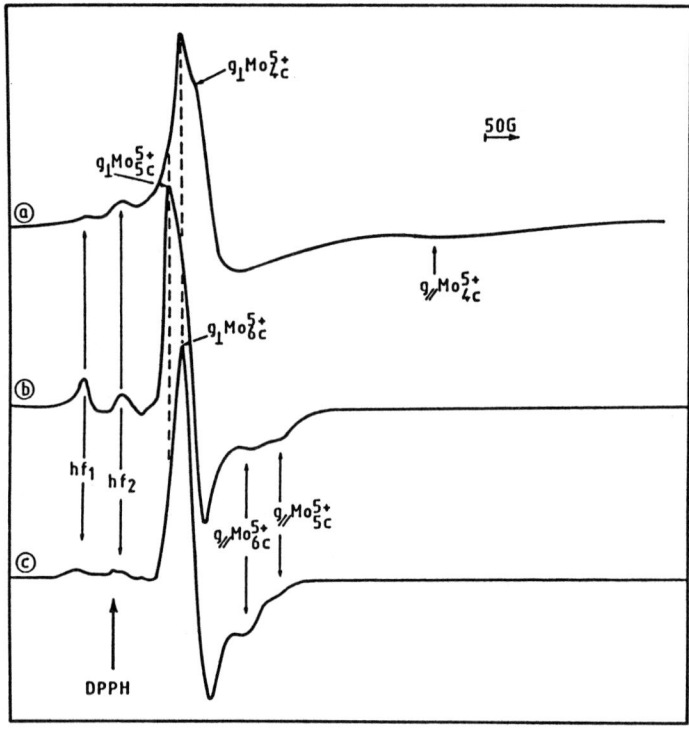

Fig. 6. EPR spectra (X band, 77 K) of a grafted Mo/SiO$_2$ catalyst (0.33 wt.-%): (a) after reduction under H$_2$ at 600 °C; (b) then after water adsorption at 25 °C (~1 torr); (c) then after water adsorption at 25 °C (~18 torr) (from Ref. 19).

TABLE II
Change of coordination upon water adsorption on surface Mo^{5+} ions (from Ref. 19).

Mo^{5+} species after reduction	Hydrated species	Notation	$g_\perp$	$g_\parallel$
		c: Mo$^{5+}_{4c}$	1.925	1.750
		b: Mo$^{5+}_{5c}$	1.957	1.866
		c: Mo$^{5+}_{6c}$	1.944	1.892

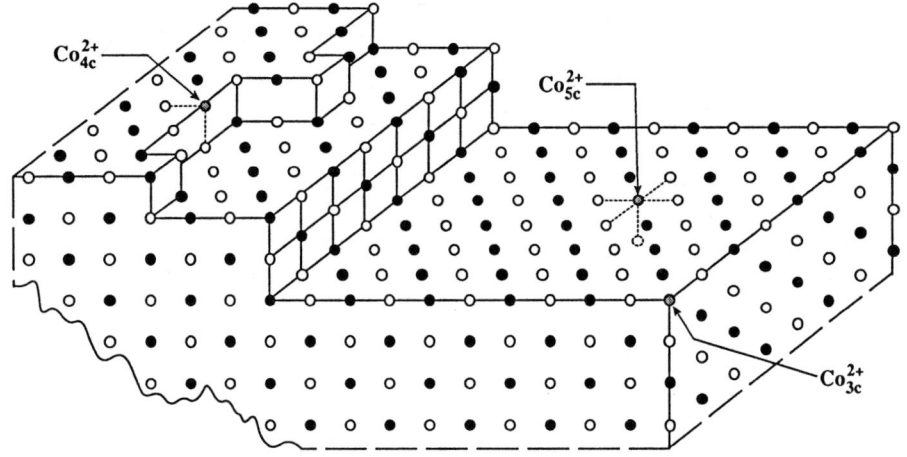

Fig. 7. Scheme of the (100) face of CoO–MgO showing the different types of coordination for Co^{2+} ions.

5.2. TRANSITION METAL IONS IN SURFACE FRAMEWORK POSITIONS: CoO–MgO SOLID SOLUTIONS

In the case of solid solutions of oxides or doped oxides, the transition metal ions are located both in the bulk and on the surface of the lattice. Only the ions located on the surface are of catalytic interest. In order to describe surface sites, the adsorption of two different probe molecules (O_2 isotopically labelled or not, and NO) was performed on a CoO–MgO solid solution (5 mol.-% of CoO in MgO) with nonzero nuclear spins ($I = 7/2$ for ^{59}Co, 100% natural abundance).

The CoO–MgO solid solution was obtained by thermal decomposition of the mixture of hydroxides [22], leading to cubelets with prevalently exposed (100) faces. Most of the surface Co^{2+} ions are five coordinated with C_{4v} symmetry (Co^{2+}_{5c}) but less coordinated Co^{2+} are present at edges, corners, kinks and steps (Figure 7).

Adsorption of O_2 at 77 K, followed by removal of excess oxygen, gives rise to a complex EPR spectrum made of two paramagnetic Co^{3+}–O_2^- species (**I** and **II**) characterized by axial g tensors and superhyperfine structures arising from the interaction between the unpaired electron of O_2^- and the nuclear spin $I = 7/2$ of ^{59}Co (Table III and Figure 8).

Upon raising the temperature up to 120–150 K, signals **I** and **II** disappear while a new Co^{3+}–O_2^- one (**III**) appears, superimposed on signal **IV** of Mg^{2+}–O_2^- (Figure 8). The features of signal **III** disappear upon evacuation at room temperature but can be restored by readmission of O_2 at the same temperature. Species **IV**, which is the most stable, is destroyed above 500 K.

The molecular nature of these various species has been confirmed using ^{17}O-enriched (58%) oxygen [23]. The analysis of the hyperfine structures due to the nuclear spin $I = 5/2$ of ^{17}O isotope, led us to conclude that species **I, II, III** exhibit nonequivalent oxygen nuclei corresponding to a bent Co^{3+}–O_2^- structure, while

TABLE III
Spin Hamiltonian parameters of oxygen species on CoO–MgO surface (from Ref. 22, 25).

Probe molecule	Species	g_1	g_2	g_3	A_1 (G)	A_2 (G)	A_3 (G)
O_2	I ($Co^{3+}O_2^-$)	2.120		1.983	38		17.5
	II ($Co^3 + O_2^-$)	2.113		1.983	38		17.5
	III ($Co^{3+}O_2^-$)	2.141	2.033	1.990	31	7	15
	IV (O_2^-/MgO)	2.077	2.009	2.002			
NO	I' (NO_2^{2-}/MgO)	2.0068		2.0025			43
	II' ($Co^0(NO^+)_2$)	2.11	2.06	2.00	38.5	37	93

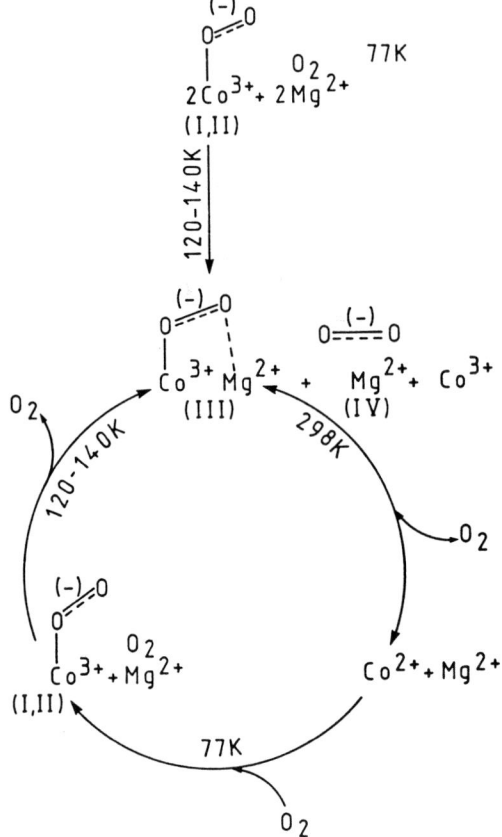

Fig. 8. Reaction for the interaction of oxygen with the CoO–MgO surface (from Ref. 22).

Mg^{2+}–O_2^- species exhibits equivalent oxygen nuclei equidistant to the Mg^{2+} adsorbing cation [24].

With the help of IR spectroscopy [22], the structures of these paramagnetic species have been fully determined. By analogy with Co complexes in solution, species **I** and **II** have been assigned to O_2^- radical anions adsorbed at slightly different angles onto Co^{3+} surface ions. Species **III** has been assigned to a O_2^- similar to species **I** and **II** but further stabilized by interaction with a neighboring Mg^{2+} cation, as attested by the perturbation of the ν_{OH} band assigned to an hydrogen bond between oxygen and a surface hydroxyl group (Figure 8).

The above results suggest that two kinds of cobalt Co_{5c}^{2+} cations are in fact present on the surface. The first one was irreversibly oxidized into Co^{3+} upon interaction with dioxygen and after transfer of O_2^- to the MgO matrix to form species **IV** (Figure 8). The second one, by contrast, is able to reversibly bind oxygen, giving species **III** or **I** and **II**, depending on the adsorption temperature. The latter cobalt species can be considered as an heterogeneous oxygen carrier since it is able to reversibly adsorb oxygen.

When nitric oxide on CoO–MgO, is adsorbed at room temperature, two new EPR signals are readily formed [25]. Signal **I'**, very similar to the signal obtained after NO adsorption on pure MgO under same conditions, is assigned to NO_2^{2-}. Signal **II'** becomes predominant at increasing adsorption time (Table III). By analogy with similar studies performed on NiO–MgO[24], the reaction of NO with the main surface species Co_{5c}^{2+} is expected to lead to a transfer of the NO π^* electron to the partially filled A_1 orbital of Co^{2+}. This electron transfer leading to a diamagnetic Co^+NO^+ species cannot explain the formation of species **II'**. The features of signal **II'** are more likely consistent with a zerovalent cobalt (d^9) formed by two possible reactions [3b]:

$$Co^{2+} O^{2-} + 2NO \rightleftharpoons Co^0(NO_2^-)(NO^+) \qquad (I)$$

$$Co^{2+} + 2NO \rightleftharpoons Co^0(NO^+)_2 \qquad (II)$$

In reaction I, the matrix assists the reduction of Co^{2+} to Co^+, which is then coordinated by NO to form the zero-valent cobalt complex, and in reaction II, a dinitrosyl complex is formed directly. EPR cannot discriminate between these two hypotheses.

By comparison with the NiO–MgO system [25], the authors have deduced that these reactions only involved very reactive Co_{lc}^{2+} ions at low coordinated (lc) positions of the crystal, such as corners, edges, steps, with $l \leq 4$ (Figure 7). This interpretation agrees with the g values of the Co^0 signal which indicate an orthorhombic symmetry and with the fact that 2 NO are admitted in the coordination sphere.

Hence, EPR spectroscopy has permitted a distinction to be made between different types of Co^{2+} on the surface of CoO–MgO solid solution. O_2 acts as a selective probe for the Co_{5c}^{2+} ions located on the (100) faces and NO adsorption characterizes Co_{lc}^{2+} with $l \leq 4$, located on corners, edges, steps (Figure 7).

6. Migration of Supported Ions into the Bulk of Oxide Supports

At high temperature, i.e., during activation treatment or catalytic reaction, the transition metal ions can migrate from the surface into the bulk of the oxide support, so they are no longer accessible to reactants or molecular probes.

Provided that the transition metal ions are paramagnetic, it is possible to use

EPR to distinguish surface from bulk ions, for the coordination symmetry of the latter is imposed by the host lattice. This case will be illustrated by two examples: Mo-doped TiO_2 (Section 6.1) and Mo-doped SnO_2 (Section 6.2). In order to facilitate the determination of the EPR parameters, most of the spectra have been recorded in both X and Q bands. The lines belonging to the hyperfine structures may therefore be separated from those belonging to the g tensors. Enriched ^{95}Mo ($I = 5/2$) has also been used to ease the determination of the hyperfine coupling constants. The analysis of the EPR parameters, made by comparison with those of known species or by modeling, provides information on the location of Mo.

It is also interesting to know the conditions for which the active phase is lost by migration into the support, and it will be shown below that migration can be monitored by EPR, and the Tammann temperature may even be determined (Sections 6.3, 6.4).

6.1. Mo-DOPED TiO_2

Mo-doped TiO_2 was obtained from a mixture of $MoCl_5$ ($8 \cdot 10^{-3}$ Mo atom per Ti) and $TiCl_4$ burned in an oxhydric flame at different temperatures [26], or by coprecipitation followed by calcination (0.6% Mo) [27]. Upon burning or calcination at 400 °C, Mo-doped anatase is generated, and exhibits an axial EPR signal at 77 K with $g_\| = 1.917$, $g_\perp = 1.828$, $A_\| = 83$ G, $A_\perp$ unresolved, characteristic of Mo^{5+} ions in a substitutional site with six O^{2-} neighbors in D_{2d} symmetry. Upon burning or calcination at 700 °C, Mo-doped rutile is obtained, and exhibits an orthorhombic EPR signal described by $g_{xx} = 1.912$, $g_{yy} = 1.810$, $g_{zz} = 1.787$, $A_{xx} = 73$ G, $A_{yy} = 30$ G, $A_{zz} = 36$ G, characteristic of Mo^{5+} ions in a substitutional site with six O^{2-} neighbors in D_{2h} symmetry (Figure 9). The evolution of the $g_\perp$ line can be used to monitor the anatase–rutile transformation upon increasing the calcination temperature. These attributions have been made by comparison of the EPR parameters with those of Mo^{5+} in rutile single crystal.

This analysis was complemented by ESCA and DRS studies which have provided information concerning the Mo^{5+} and Mo^{6+} distribution in the bulk and on the surface [27]. It was shown that there is a strong Mo surface enrichment in Mo-doped rutile.

It is also possible to obtain Mo ions in the bulk of rutile by impregnation followed by calcination at 800 °C, indicating that Mo is able to migrate from the surface to the bulk by diffusion [28].

Mo-doped rutile was reduced under hydrogen at 280 °C and then submitted to O_2 in order to eliminate the Mo^{5+} ions located on the surface by dipolar interaction. The EPR spectrum exhibits a new orthorhombic Mo^{5+} signal at 77 K: $g_{xx} = 1.897$, $g_{yy} = 1.920$, $g_{zz} = 1.857$, $A_{xx} = 37$ G, $A_{yy} = 57$ G, $A_{zz} = 68$ G, attributed to interstitial Mo^{5+}. According to the authors, they are supposed to arise from the reduction of Mo^{6+} ions located in interstitial sites while Mo^{5+} ions in substitutional sites would have been reduced into EPR silent Mo^{4+} [29]. This attribution has been confirmed by theoretical calculations using the crystal field approach for an nd^1 ion in rutile lattice [30, 31].

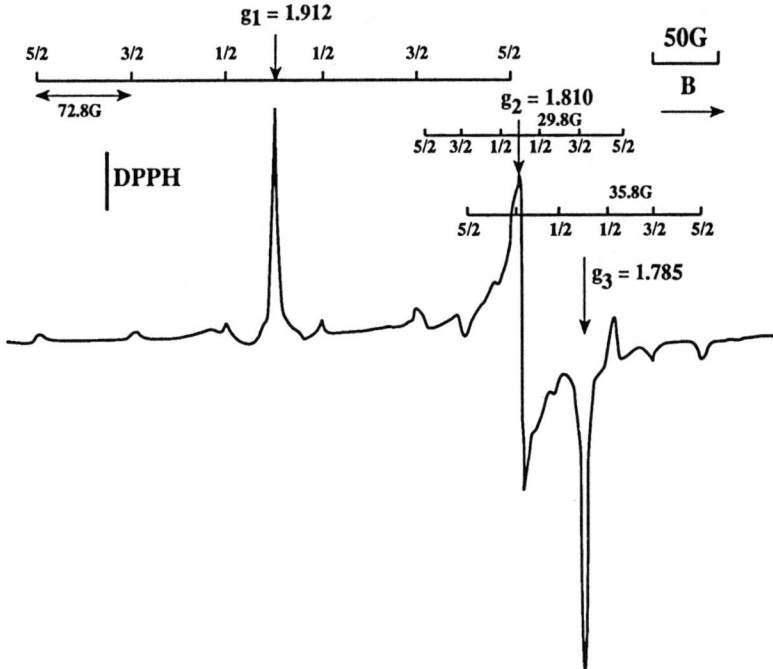

Fig. 9. EPR spectrum (X band, 77 K) of Mo^{5+} in Mo-doped TiO$_2$ (rutile) (from Ref. 27).

6.2. Mo-DOPED SnO$_2$

In the case of Mo-doped SnO$_2$ (10^{-2} Mo atom per Sn), the spectra are more complex since 16% of Sn (7.7% of ^{117}Sn and 8.7% of ^{119}Sn) possesses a nuclear spin $I = 1/2$. However, more information concerning the Mo^{5+} location has been obtained. Samples were obtained by burning a mixture of MoCl$_5$ in SnCl$_4$ or by impregnation of SnO$_2$ with molybdenum heptamolybdate followed in both cases by calcination at 800 °C [31, 32]. Enriched ^{95}Mo (97%) and Q band spectra were also used to improve the hyperfine structure of the Mo^{5+} signal (Figure 10a). Part of the molybdenum introduced is found to be stabilized in the pentavalent state at substitutional positions, based on experimental and theoretical analysis. The EPR spectra of Mo^{5+} are characterized by the following parameters: $g_{xx} = 1.891$, $g_{yy} = 1.835$, $g_{zz} = 1.923$, $A_{xx} = 24\,G$, $A_{yy} = 30\,G$, $A_{zz} = 70\,G$. In addition, two sets of superhyperfine structures are due to two tins located along the y axis of the unit cell ($a_{xx} = 281\,G$, $a_{yy} = 319\,G$, $a_{zz} = 279\,G$) and to four tins lying in the xy plane ($b_{xx} = 58\,G$, $b_{yy} = 50\,G$, $b_{zz} = 47\,G$) (Figure 10b).

6.3. IMPREGNATED V/TiO$_2$

This migration phenomenon was clearly illustrated in the case of the V/TiO$_2$ system ($I = 7/2$, 100%) [33]. The V/TiO$_2$ sample (1 wt.-% V) prepared by impregnation of rutile with ammonium metavanadate and drying at 80 °C, exhibits an

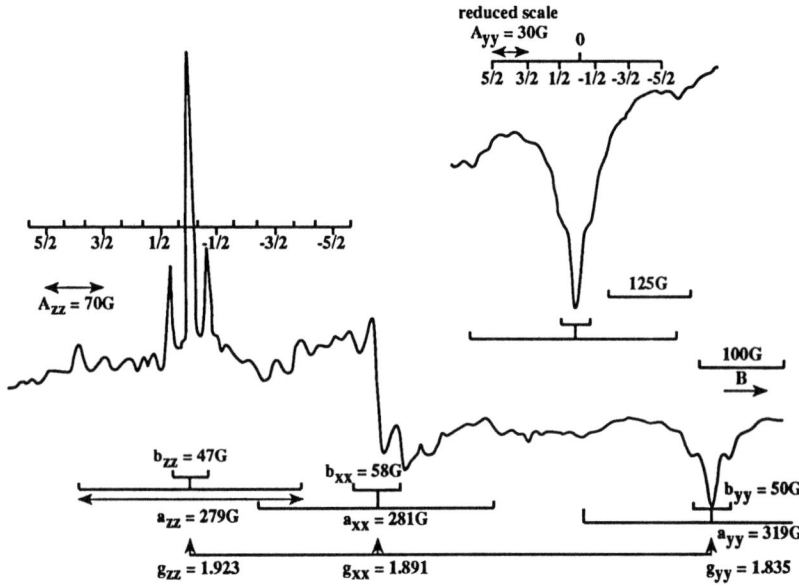
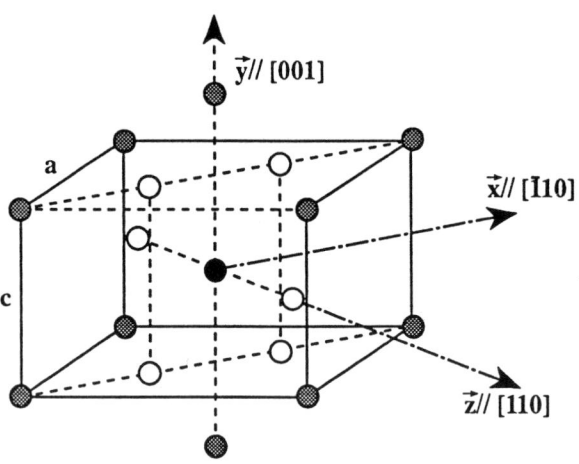

Fig. 10. (a) EPR spectrum (Q band, 77 K) of Mo^{5+} in Mo-doped SnO_2 (rutile). The upper part corresponds to the region around the g_{yy} component. (b) Unit cell of tetragonal SnO_2 with the addition of neighboring cations along the c axis. The solid circle is the Mo^{5+} central ion, the open circles are oxygen ions and the shaded circles are the Sn^{4+} ions (from Ref. 31).

EPR spectrum composed of two signals (Figure 11a). Simulation programs were used to separate the two signals (Figures 11b,c). The first one with an axial symmetry: $g_\parallel = 1.937$, $g_\perp = 1.968$, $A_\parallel = 175$ G, $A_\perp = 94$ G, is characteristic of surface VO^{2+} vanadyl ion and the second one, orthorhombic: $g_{xx} = 1.913$, $g_{yy} = 1.913$, $g_{zz} = 1.927$, $A_{xx} = 31$ G, $A_{yy} = 45$ G, $A_{zz} = 156$ G, is characteristic of V^{4+} ions located at substitutional TiO_2 sites. When the sample is calcined in air, the surface VO^{2+} signal disappears at about 400 °C because of the oxidation of surface vanadyl ions, and the bulk V^{4+} signal becomes more visible. By quantitative analysis of the spectrum, less than 0.1% of the vanadium is found in the bulk below 600°C. Above 700°C, the quantity of bulk vanadium notably increases and at 800°C, reaches 22% of the overall V (Figure 11d). Since sintering effects are negligible on the basis of specific surface area measurements, the increase of the bulk V^{4+} signal has been attributed to bulk diffusion from surface vanadium ions, which are supposed to be in the reduced state V^{4+}, to diffuse in the bulk. The temperature at which this diffusion becomes important (600 ± 15 °C) indicates the mobility onset of cations and cationic vacancies of TiO_2 and defines the Tammann temperature of this oxide (between $0.3\,T_m(K)$ to $0.5 T_m(K)$ where $T_m(K)$ is the melting temperature of the oxide (2109 K for TiO_2)).

6.4. IMPREGNATED Mn/MgO

Mn^{2+} supported on MgO, prepared by impregnation of basic magnesium carbonate with $Mn(NO_3)_2$, was also observed to migrate into the bulk upon evacuation at increasing temperature [34]. Two different Mn^{2+} EPR signals are observed ($3d^5$, $S = 5/2$, $I = 5/2$, 100% natural abundance): a broad unresolved spectrum due to dipolar broadening arising from Mn^{2+} ion clusters located at or near the surface, and a sharp resolved spectrum with six hyperfine lines corresponding to substitutional Mn^{2+} ions in the MgO lattice ($g = 2.0064$ and $A = 87$ G). Vacuum treatment in the range 500–1000 °C induces a decrease in the intensity of the EPR signal attributed to the cluster and an increase of the one due to substitutional Mn^{2+} ions. The changes in EPR spectrum become sharp at 1000 °C, indicating the onset of effective ionic diffusion processes. Meanwhile, the zero-field splitting term, D, resulting from axial distortion, decreases uniformly due to the progressive diffusion of the Mn^{2+} substitutional ions deeper into the bulk where they experience more cubic symmetry than near the surface. When treated in air instead of vacuum, all the Mn^{2+} ions have already diffused into the bulk at 1000 °C. According to the authors, this diffusion process is favored by sintering and crystallite reconstruction which both are enhanced when occurring in air because of the presence of water. Because sintering and crystallite reconstruction take place simultaneously with ion diffusion, the correlation between ion diffusion and Tammann temperature of MgO which was expected at about 1000 °C (melting point of MgO: 2800 °C), is not straightforward.

7. Ion Migration in Zeolitic Matrices

Zeolites are crystalline aluminosilicates possessing an open lattice framework. The crystal structure of Faujasite type zeolites such as X and Y zeolites consists of a

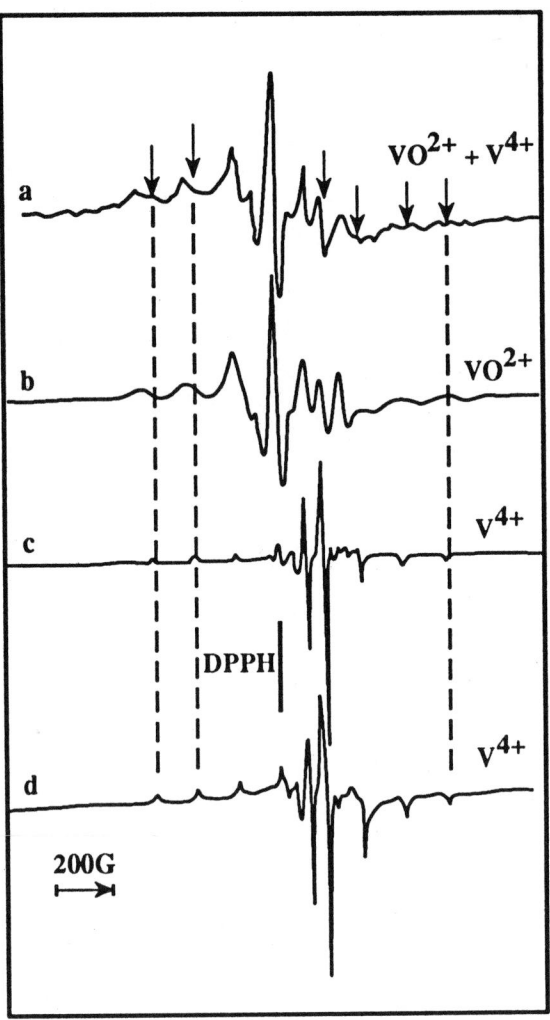

Fig. 11. EPR spectra (*X* band, 77 K) of V/TiO$_2$ impregnated sample: (a) spectrum of VO^{2+} and V^{4+} observed after drying at 80 °C; (b) simulation of the VO^{2+} signal; (c) simulation of the V^{4+} signal; (d) V^{4+} signal obtained after annealing at 850 °C (from Ref. 33).

tetrahedral arrangement of sodalite units (three-dimensional array of SiO$_4$ and AlO$_4$ tetrahedra in the form of a truncated octahedron) which are connected by hexagonal prisms. This arrangement induces the presence of large cavities, called supercages (Figure 12). The extraframework cations (which balance the negative charges created by the replacement of Si^{4+} by Al^{3+}) may be mainly located in the following positions:

– SI lies in the center of the hexagonal prism, and is surrounded by an octahedron of oxygens;

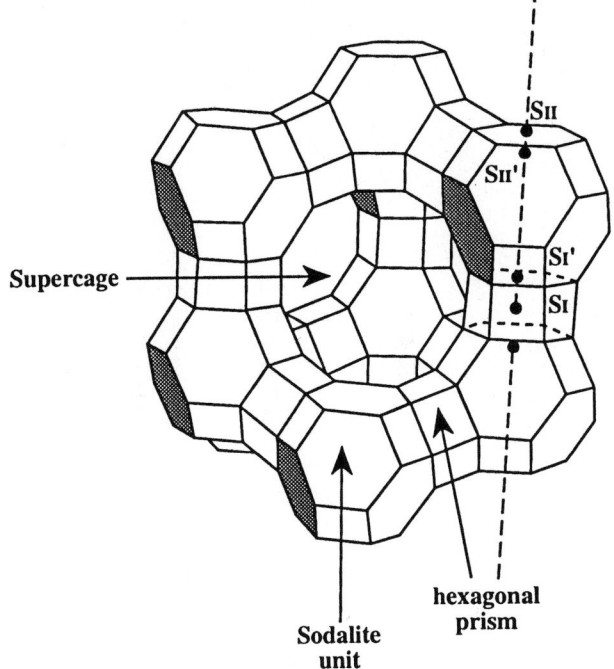

Fig. 12. Various cavities of a Faujasite zeolite and possible positions for extra framework cations.

- SI' lies in the sodalite cage at the center of the hexagonal window of the prism, and is coordinated to three framework oxygens atoms;
- SII lies in the supercage, at the center of the hexagonal window of the sodalite, and is coordinated to three framework oxygens atoms;
- SII' in the sodalite cage, is the mirror image of SII through an hexagonal window SI, SI' and SII' are called 'hidden sites' as they are inaccessible to most of the molecules.

The EPR and DRS studies on the location of transition metal ions in zeolites have been reviewed by Kazanskii and coworkers [35] and more recently by Schoonheydt [36].

7.1. DISTRIBUTION OF EXCHANGED CATIONS AMONG THE ZEOLITIC SITES

When the exchanged extraframework cations are paramagnetic, as in the case of Cu^{2+}, EPR can be used to study their distribution among the different zeolitic sites.

As mentioned above (see Section 1.2.b), the Cu^{2+} cations are mainly located in the supercage (SII) after the exchange of NaY zeolites. During thermal treatment at increasing temperature, the copper ions migrate from the supercage towards hidden sites where they coordinate to framework oxygens as they loose extraframework ligands such as water or ammines [37]. They move to SII' then

TABLE IV
Correspondence between the EPR signal and the location of Cu^{2+} ions in zeolites (from Ref. 37).

EPR signal of Cu^{2+} $I = 3/2$	$g_\| = 2.378$ $A_\| = 139 \cdot 10^{-4} \text{ cm}^{-1}$ $g_\perp = 2.060$ $A_\perp = 17 - 10^{-4} \text{ cm}^{-1}$	$g_\| = 2.347$ $A_\| = 157 \cdot 10^{-4} \text{ cm}^{-1}$ $g_\perp = 2.061$ $A_\perp = 19 \cdot 10^{-4} \text{ cm}^{-1}$
Location	SII'	SI'
Treatment temperature	100 °C	400 °C

TABLE V
Migration of Ni^I ions in the presence of CO in zeolites (from Ref. 40).

	A	B	C
EPR signal	$g_\| = 2.680$ $g_\perp = 2.096$	$g_\| = 2.160$ $g_\perp = 2.069$	$g_1 = 2.206, A_1 = 27.5 \text{ G}$ $g_2 = 2.071, A_2 = 33 \text{ G}$ $g_3 = 2.056, A_3 = 33 \text{ G}$
Assignment	$Ni^I(H_2)_x$	$Ni^I(H_2)_x$	$Ni^I(^{13}CO)_2$; $^{13}C\ I = 1/2$
Position	SI'	SII'	SII

to SI' sites, between 70 and 400 °C, and finally are mainly located in SI sites at temperatures higher than 400 °C (Figure 12). Table IV gives the correspondence between the site and the EPR signal.

The same distribution of Ni^{II} cations among the zeolitic sites was observed using DRS and X-Ray analysis [38].

7.2. MIGRATION OF EXCHANGED CATIONS IN PRESENCE OF POLAR COMPOUNDS

Ion migration from the hidden sites to the supercages can be also induced by adsorption of strongly polar molecules, as illustrated in the following example. The migration is evidenced with the use of CO as a probe.

NiCaY may be thermally reduced in the presence of hydrogen in the temperature range 120–200 °C. Only one EPR Ni^I signal, **A**, is observed, $g_\| = 2.680, g_\perp = 2.096$ [39]. This signal has been assigned to a $(O)_3Ni^I(H_2)_x$ complex ($x \leq 2$, O stands for silica surface oxygen atoms) located in SI'. In order to coordinate probe molecules or to act as catalysts, these Ni^I ions have to migrate from the SI' sites to the supercages. Indeed molecules such as CO, trialkylphosphines, pyridine are too bulky to enter the sodalite cage, even at high temperature.

Upon adsorption of CO (20 torr) at 77 K, signal **A** is partly replaced by signal **B** [40] (Table V). Upon warming at room temperature, signal **C** appears at the expense of **B**. The use of ^{13}CO shows that there is no superhyperfine structure involved in species **B**, whereas species **C** presents a superhyperfine structure corresponding to two CO ligands in the nickel coordination sphere. The ions originally in SI'(**A**), migrate to SII'(**B**) at 77 K to be closer to CO located in the supercage, but they do not interact directly with CO. Finally, the Ni^I ions migrate at room temperature to SII (**C**) where they can coordinate two CO ligands. CO was therefore able to induce the migration of the Ni^I ions from the hidden sites to the supercages.

8. Interaction between Paramagnetic Ions

As described above, EPR is suitable for the study of well-dispersed paramagnetic ions, but EPR can also be used to study the formation of ion pairs which may play a decisive role in catalytic reaction (see Section 8.2 below). However, when the concentration of spins increases, the EPR line broadens and may even disappear because of the interaction of clustering paramagnetic ions. This has been reported for example, for silica or alumina supported Cr [1, 41]. The changes in shape and intensity of the spectrum form a good criterion for studying the aggregation of the paramagnetic metallic ions and provide information on the state of dispersion of the supported catalysts. This, however, has the following counterpart: strongly interacting paramagnetic ions may be present but EPR silent. The following example illustrates the use of spin decoupling to regenerate and observe isolated ions.

8.1. SPIN DECOUPLING

In $Mo/\gamma-Al_2O_3$ or Mo/SiO_2 catalysts prepared by impregnation, the Mo^{5+} ions interacting through an oxygen bridge are EPR silent. Such aggregates can be quantified by EPR after removal of the oxygen bridge, by a suitable compound. The following ligand substitution reactions occur when the reduced sample is exposed to HCl or HBr (HX) at room temperature [42, 43]:

$$Mo^{5+}O + 2\,HX \rightarrow Mo^{5+}X_2 + H_2O$$

and

$$Mo^{5+}-O-Mo^{5+} + 2\,HX \rightarrow 2\,Mo^{5+}X + H_2O$$

The g and hyperfine tensors of the Mo^{5+} signal are modified in agreement with the substitution of oxide by halide ions in the coordination sphere of Mo^{5+} and the concentration of EPR visible Mo^{5+} therefore increases. The larger the enhancement upon reaction with halides, the lower the Mo^{5+} dispersion.

8.2. ION PAIRS

In CuY zeolites, as the Cu concentration increases, Cu^{2+} pairs are formed upon outgassing at 500 °C [44]. Two types of pairs, referred to as identical and nonlinear pairs, respectively, are observed (Table VI).

The identical pair involves the coupling by exchange interaction of two ions with same axial symmetry, identical spin Hamiltonian parameters and the same

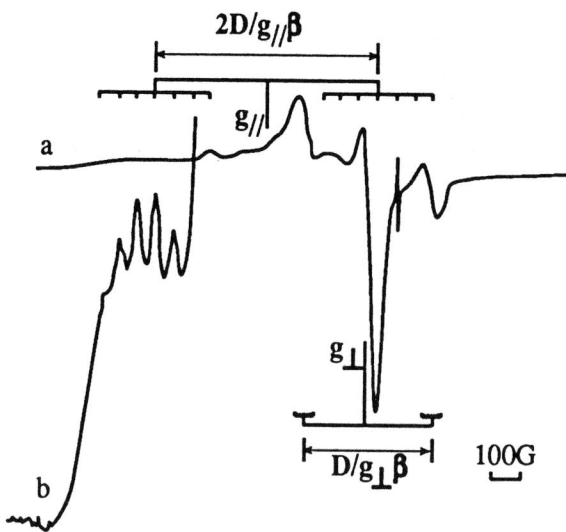

Fig. 13. EPR spectra (X band, 77 K) of $\Delta m_s = \pm 1$ transitions of identical Cu^{2+} pairs in a CuCaY zeolite evacuated at 500 °C. Only the positions of the hf lines of identical pairs are marked. Trace b is magnified 80 times compared to trace a (from Ref. 44).

principal axis. The corresponding fine structure ($S = 1$) is presented in Figure 13 and can be interpreted as indicated in Figure 14. The EPR spectrum exhibits two parts corresponding to the $\Delta m_s = +1$ and $\Delta m_s = \pm 2$ (half-field line) transitions. Assuming that the isotropic exchange interaction is small, a distance $d = 4.2$ Å between the Cu^{2+} ions can be evaluated from the experimental value $D = 0.0476$ cm^{-1} of the zero-field splitting. This is in agreement with the distance measured by X-ray diffraction between two cations located in two adjacent SI' positions on both sides of the same hexagonal prism (Figure 15). A less probable alternative for this identical pair would be two ions located in SI' and SII' sites in the same sodalite cage (Figure 15).

Conesa et al. [45] observed the same EPR signal in CuCeY zeolite. However, upon O_2 adsorption, the signal broadens, suggesting that at least one of the Cu^{2+} ions of the pair is located in the supercage, in a position accessible to oxygen (this latter cannot enter the sodalite cage at room temperature [46]). The EPR signal was therefore reassigned to two Cu^{2+} ions located in SII and SI' respectively and having different principal axes (Figure 15).

The spectrum of the nonlinear pairs consists of two parts: one isotropic main line ($\Delta m_s = \pm 1$) and one much weaker half-field line (Table VI). Modeling of an exchange coupled pair, in order to get an isotropic g value, allows one to predict that the angle between the symmetry axes of the ions is 110°. It is more likely that Cu^{2+} ions are at a distance of 3.9 Å. The exchange interaction is assumed to occur through the oxygen which is near the center of the sodalite unit.

Cu pairs have also been reported on other supports, such as silica [12]. But especially well-resolved EPR spectra of Cu^{2+} pairs were recently reported on thorium oxide [47] (Figure 16). The high dispersion of the Cu^{2+} pairs in the cubic

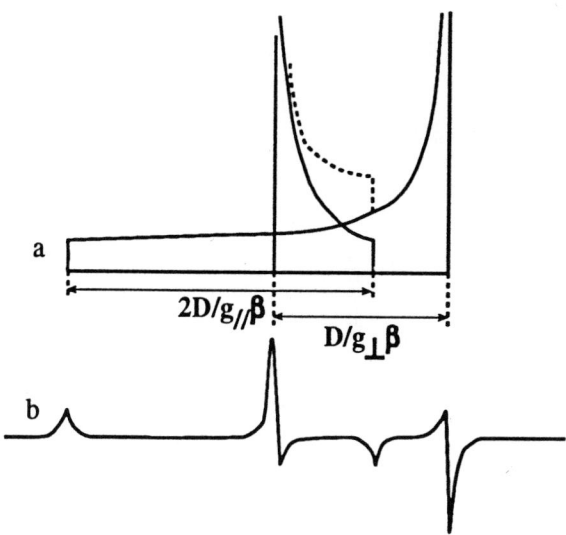

Fig. 14. Theoretical spectrum of Cu^{2+} identical pairs in a powder sample: (a) absorption spectrum, (b) first derivative spectrum. For simplicity no hf structure is included (from Ref. 44).

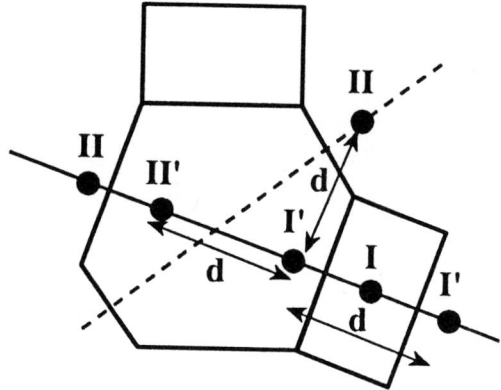

Fig. 15. Idealized sodalite unit and hexagonal prism. The solid circles indicate the possible positions of Cu^{2+} and therefore the possible assignments for the identical Cu^{2+} pair: II′–I′, II–I′ or I–I′ (from Ref. 45).

TABLE VI
EPR parameters of copper centers in Y-type zeolites (from Ref. 44).

Cu^{2+} center	Sample	g value			Zero-field splitting (10^{-4} cm^{-1})		Hyperfine splitting (10^{-4} cm^{-1})	
		$\Delta ms = \pm 1$		$\Delta ms = \pm 2$	D	E	$A_\parallel$	$A_\perp$
isolated Cu^{2+}	CuCaY evacuated at 200°C	2.320	2.066				176	24
	CuNaY evacuated at 200°C	2.367	2.074				133	24
identical Cu^{2+} pair	CuCaY evacuated at 500°C	2.345	2.066	4.13 ($\alpha = 90°$) 4.428 ($\alpha = 30°$)	476	0	150	unresolved[a]
nonlinear Cu^{2+} pair	CuNaY 54% exchanged or less	2.170		4.19				
	CuY 74% exchanged	2.170		4.19				

[a]Half field value $A = 134 \; 10^{-4}$ cm^{-1} ($\alpha = 30°$)

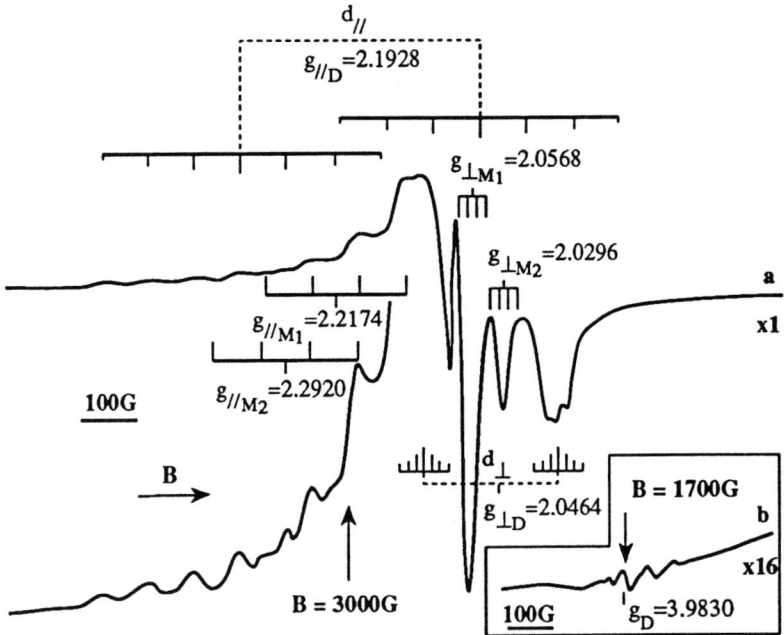

Fig. 16. EPR signal (X band, 293 K) observed from the Cu/Th = 1.5 sample: (a) signal for $\Delta ms = \pm 1$ (allowed transition); (b) signal for $\Delta ms = \pm 2$ (forbidden transition) (from Ref. 47).

lattice of high symmetry of thoria leads to highly resolved spectra. For a Cu/Th ratio of 0.25, the number of pairs is a maximum. The pair is made of two nonequivalent Cu^{2+} ions [48]. One of them is localized in a substitutional surface site (it can be thermally reduced by hydrogen) whereas the other one occupies a substitutional site in the bulk (it cannot be thermally reduced). From the EPR parameters, a separation distance of 5.1 Å is evaluated which is close to the unit cell dimension of the thoria lattice (5.6 Å).

The system is a catalyst for the selective dehydrogenation of isopranol into acetone. From comparison of EPR and DRS spectra before and after the reaction, it is suggested that the simultaneous presence of Cu^{2+}, Cu^+ and Cu^0 in adequate ratio (i.e. $(Cu^0 + Cu^+)/Cu^{2+} = 2$) is responsible for the catalytic activity [49].

The pairs seem to be involved in the selective hydrogenation of isoprene into methylbutene since the activity and selectivity correlate well with the amounts of pairs deduced from the EPR signals intensities [47].

9. Catalytic Reactivity

The purpose of this section is to describe studies in which EPR has contributed to elucidating either the nature of active sites or of intermediates in a catalytic reaction.

9.1. OLEFIN OLIGOMERIZATION

9.1.a. Characterization of the Site Before and After the Reaction

In the case of paramagnetic active sites, the comparison of the EPR spectrum before and after reaction is a good way to check if the catalyst is regenerated or poisoned.

The poisoning of Ni^I-based olefin oligomerization catalysts could be related to the formation of stable complexes of Ni^I with oligomers which further prevent the coordination and dimerization of the reactant. These complexes were revealed by EPR [50].

9.1.b. Quasi-in-situ Studies

As has already been mentioned, in order to obtain better resolved spectra it is often necessary to record them at low temperature (typically at liquid nitrogen temperature =77 K). As the catalytic reaction often occurs at higher temperature, it is generally impossible to carry out *in-situ* EPR studies. However, for reactions occurring at room temperature or lower, the EPR measurements can be performed almost in the catalytic conditions and will be designated as quasi-*in-situ* studies.

This was applied to the study of olefin dimerization catalyzed by silica-supported Ni^I complexes [51] (heterogeneous analogues of the homogeneous catalysts involved in industrial processes such as DIMERSOL [52] or SHOP [53] process). Two reaction pathways are generally proposed for olefin dimerization using nickel complexes:

- In the degenerated polymerization mechanism (**I**), two olefin molecules are successively inserted in the Ni-H then in the Ni-C bond of the intermediate [54]. Only one vacancy in the coordination sphere of the Ni center is required prior to the insertion of the reactant.
- The other reaction pathway, the so-called concerted coupling (**II**) involves the simultaneous coordination of two olefin molecules to the catalytic site and thus the formation of a metallacyclic intermediate [55]. Two vacancies are therefore required for the dimerization to occur. In both mechanisms, β-H elimination regenerates the catalytic center after desorption of the dimer.

9.1.b.1. *Complexes formed with reactants.* A knowledge of the structure of these complexes can provide useful insight into the reaction pathway, as illustrated below. The use of labelled reactants may allow one to count the number of ligands and to determine the type of bonding.

In the case of ethylene dimerization over nickel exchanged silica, supported Ni^I-ethylene complexes have been observed [17]. The absence of shf structure in the presence of $^{13}C_2H_4$ suggests that ethylene is bonded side-on to Ni^I.

In order to understand how many ethylene molecules are bonded to Ni^I for the dimerization to take place, CO poisoning experiments have been performed. Starting from complex **2** (see Section 5.1.a), complexes **5** or **6** were obtained depending on the ethylene pressure (see Scheme below).

EPR OF OXIDE SUPPORTED TRANSITION METAL IONS

[Reaction scheme showing complex **2** (Ni^I with 2 CO and 2 O ligands) reacting with ethylene at 250 torr to form complex **5** (Ni^I with COa, COe, C$_2$H$_4$, and 2 O ligands) — no oligomerization; and at 600 torr to form complex **6** (Ni^I with COa, 2 C$_2$H$_4$, and 2 O ligands) — oligomerization]

complex **5**:
$g_1 = 2.208$, $a_1(COa) = 55$ G, $a_1(COe) = 25$ G
$g_2 = 2.154$, $a_2(COa) = 55$ G, $a_2(COe) = 20$ G
$g_3 = 2.013$, $a_3(COa) = 46$ G, $a_3(COe) = 14$ G

complex **6**:
$g_1 = 2.290$, $a_1(COa) = 45$ G
$g_2 = 2.190$, $a_2(COa) = 45$ G
$g_3 = 2.007$, $a_3(COa) = 45$ G

From the relative order of the g tensor components ($g_1 \approx g_2 \approx g_\perp > g_3 \approx g_\parallel \approx g_e$) a trigonal bipyramidal or distorted octahedral symmetry (Table I) is expected for complexes **5** and **6**, whereas the number of CO ligands is deduced from the superhyperfine tensors of the EPR spectra obtained using ^{13}CO. The shape of the d_{z^2} orbital leads to assign the larger ^{13}C superhyperfine constant to the axial position and the smaller to the equatorial position. As Ni^I complexes admit at most five ligands in the coordination sphere, so structures **5** and **6** may be suggested. Chromatographic analysis shows that no dimerization occurs on complex **5**, whereas it does occur on complex **6**.

It therefore appears that two coordination vacancies are necessary for the dimerization to take place on silica-supported Ni^I complexes. This is in favor of the concerted coupling mechanism **II**.

9.1.b.2. Identification of intermediates [56]. In the following example, quasi-*in-situ* EPR was used to evidence the formation of a nickelacyclopentane intermediate and was thus determinant in identifying mechanism **II**.

The study was performed with the best catalyst for ethylene dimerization, that is Ni^I(PEt$_3$)$_2$(O)$_2$ (**D**) (where O stands for silica surface oxide ions). The electron-donating effect of the triethylphosphine ligands is responsible for the catalyst's high stability and selectivity for α-olefins.

Adsorption of 600 torr of ethylene onto **D** at 77 K leaves the EPR spectrum unchanged (Figure 17a): $g_\perp = 2.083$, $a_{\perp 1} = 85$ G, $a_{\perp 2} = 62$ G, $g_\parallel = 2.32$, $a_\parallel$ unresolved, with the shf structure resulting from the interaction of two inequivalent PEt$_3$ ligands (^{31}P, $I = 1/2$) with the unpaired electron. When the sample is warmed

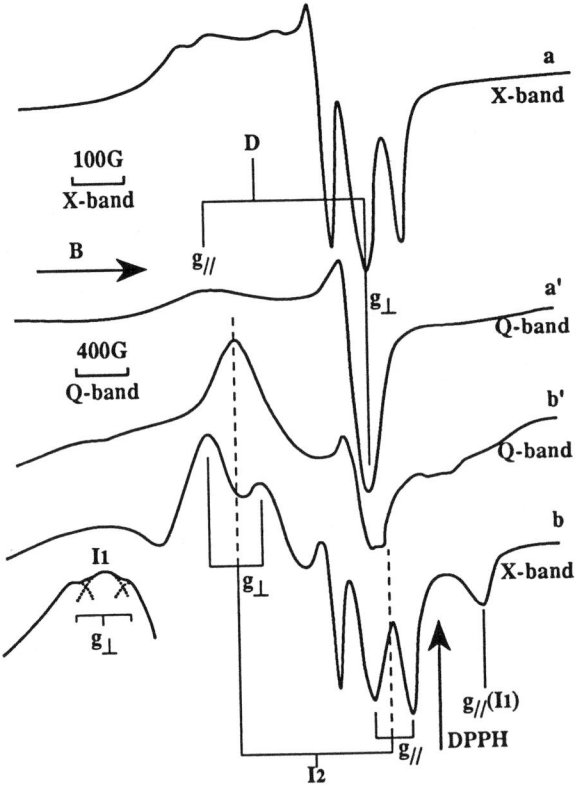

Fig. 17. EPR spectra (77 K) of $(O)_2Ni^I(PEt_3)_2/SiO_2$: (a) X band; (a') Q band; (b) X band after reaction with 600 torr of C_2H_4 1 min at 293 K; (b') Q band same treatment as (b) (from Ref. 56).

up to 25°C, the signal of complex **D** immediately decreases whereas two new signals I_1 and I_2 appear simultaneously (Figure 17b). Their intensities increase, then remain constant during several hours and finally decrease to zero at the end of the reaction. The g tensors of axial signals I_1 and I_2 are given below:

I_1: $g_\perp = 2.532$, $a_\perp = 100$ G I_2: $g_\perp = 2.256$, $a_\perp = 100$ G
$g_\parallel = 1.954$, no shf structure $g_\parallel = 2.066$, $a_\parallel = 70$ G

The ^{31}P shf structure of I_1, although poorly resolved in the **X** band, suggests a triplet for the perpendicular component. In the **Q** band, the perpendicular component is broadened but remains centered on $g_\perp = 2.532$, whereas no superhyperfine structure is noticeable on the parallel component. These results suggest an equivalent interaction of the unpaired electron with the two ^{31}P nuclei in the perpendicular direction.

The ^{31}P shf structure of I_2 is well resolved in the X band: each g tensor component is split into two lines suggesting a coupling of the single electron with only one ^{31}P.

For comparison, butadiene (600 torr) was adsorbed onto species **D**. A signal **J**

($g_\parallel$ = 1.954, $g_\perp$ = 2.532, $a_\perp$ unresolved, no shf structure on the parallel component) very similar to I_1 was observed and did not change with time, but no oligomerization reaction occurred. The relative order of the g components of signals I_1 and J is in agreement with a five-coordinate complex of Ni^I of D_{3h} symmetry and a d_{z^2} ground state. In the literature, butadiene is known to form π-complexes with metallic cations with the butadiene ligand in the meridional position for five-coordinate species [57]. By analogy, the following structure is proposed for complex J:

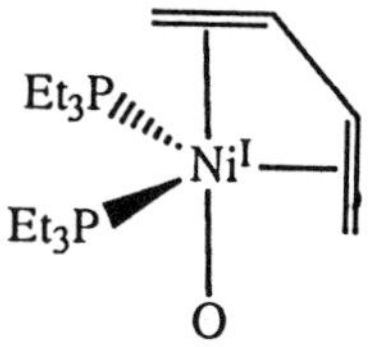

$$\text{Et}_3P\cdots\text{Ni}^I\cdots\text{(butadiene)} \quad \text{J}$$
$$\text{Et}_3P, O$$

By analogy, structure **a** may be assigned to signal I_1. However, the following equilibrium has been evidenced in homogeneous catalysis studies:

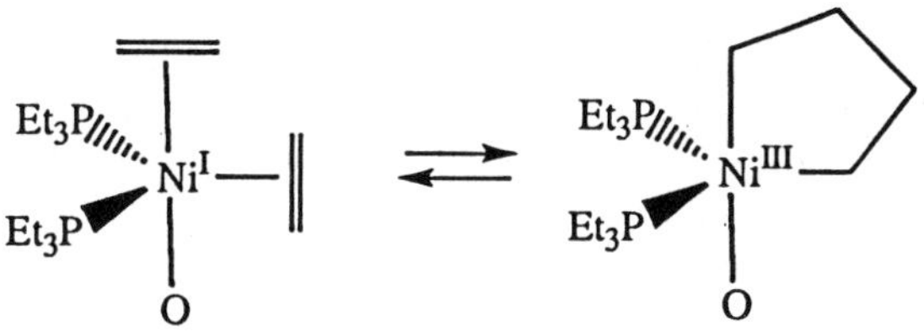

$$\text{a} \quad \rightleftharpoons \quad \text{b}$$

From the EPR parameters, it is not possible to choose between structures **a** and **b** as Ni^I and Ni^{III} can give very similar signals. Nevertheless, I_1 is relatively stable and it has been shown that in reaction involving concerted coupling mechanism, the diolefin complex is in equilibrium with the metallacyclopentane, the limiting step being the decomposition of the metallacyclic compound. This leads to assign I_1 to the metallacycle **b**, for which an EPR coupling with two phosphines ligands is expected in the perpendicular direction. By analogy, I_2 is assigned to the isomer in which the metallacycle is in the equatorial plane:

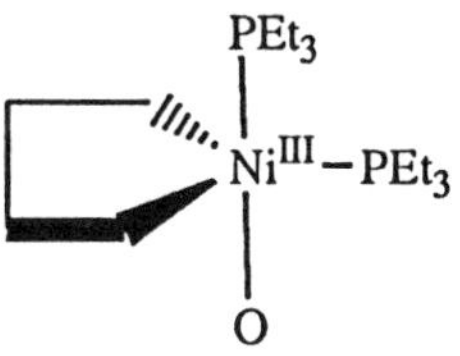

$$\text{(metallacycle)}\cdots\text{Ni}^{III}-\text{PEt}_3$$
$$\text{PEt}_3, O$$

Similar intermediates have been observed for propylene dimerization on Ni^I(PEt$_3$)(O)$_2$ [50]. Propylene is more bulky than ethylene, the number of phosphine ligands bonded to Ni^I is therefore limited to one in order to allow the simultaneous coordination of two propylene molecules to Ni^I. The concerted coupling mechanism has been confirmed by the distribution of reaction products which can be explained by a steric effect of the phosphine ligand on the stability of the intermediate metallacycle. The bulkiness of the trialkylphosphine ligand favors the formation of branched dimers whereas it is expected to favor the formation of linear dimers in the case of the degenerated polymerization mechanism [54a].

9.2. METHANOL OXIDATION ON Mo/SiO$_2$ CATALYSTS

As has been mentioned above (see Section 5.1.b), Mo has been shown to be much better dispersed on grafted than on impregnated samples and in stronger interaction with silica [10, 11, 14]. The catalytic behavior of these two types of catalysts has been compared in methanol oxidation reaction [58] which has been found to be structure sensitive [59, 60]. The reaction was also found to be dispersion sensitive [58]. Indeed, the main product for the grafted catalysts was methyl formate whereas the main one for impregnated catalysts was formaldehyde as for polycrystalline MoO$_3$. The study of the formation of methyl formate on grafted catalysts using different reaction mixtures, and kinetic calculations led to propose the following reaction mechanism [58]: methanol adsorbed on Mo^{6+} sites is first transformed into formaldehyde according to the same mechanism of oxidative dehydrogenation as that occurring on MoO$_3$ [61–63]. Because of the high Mo dispersion, instead of desorbing, as in the case of low dispersed catalysts or MoO$_3$ where Mo sites interact each others, CH$_2$O spills over onto silica. There it further reacts with methoxy groups and formed methyl formate via a hemiacetal intermediate. Mo sites, reduced during formaldehyde formation, are reoxidized by O$_2$ [58].

An EPR study was performed aiming at establishing the elementary steps of this mechanism [21]. The catalytic process was modeled beyond the rate determining step, which is supposed to be the hydrogen abstraction from methyl of CH$_3$OH. The chosen cycle models the reaction between CH$_3$OH and adsorbed O$^-$ ions, known to easily abstract hydrogen. These ions are generated by UV excitation of molybdenyl Mo^{6+}=O^{2-} bonds at low temperature.

When the oxidized grafted Mo/SiO$_2$ samples, onto which methanol was first adsorbed, were UV irradiated at 77 K, two signals were detected by EPR, one due to Mo^{5+} ions ($g_\perp = 1.946$, $g_\parallel = 1.90$) and a sharp triplet attributed to the hydroxymethyl radical ˙CH$_2$OH (Figure 18) ($g_{av} = 2.003$, the hyperfine parameters, $A_{xx} = 23$, $A_{yy} = 29$ and $A_{zz} = 13$ G, have been obtained by simulation). The following reaction scheme involving a ligand-to-ligand hydrogen transfer in the last step, was proposed:

$$Mo^{6+}=O^{2-} \xrightarrow{CH_3OH}_{300\ K} \overset{\overset{\displaystyle CH_3OH}{|}}{Mo^{6+}=O^{2-}} \xrightarrow{h\nu}_{77\ K} [\overset{\overset{\displaystyle CH_3OH}{|}}{Mo^{5+}-O^-}]^*$$

$$\xrightarrow{h\nu}_{77\ K} {}^\cdot CH_2OH_{ads.} + Mo^{5+}-OH^- \tag{III}$$

The last step is in agreement with the observed increase of both the Mo^{5+} and ˙CH$_2$OH EPR signals upon irradiation.

Above 140 K, the ˙CH$_2$OH radical becomes unstable. Its decay can be explained by a redox process with Mo^{6+} ions whose reduction is attested by the increase of the Mo^{5+} signal intensity and the formation of adsorbed formaldehyde:

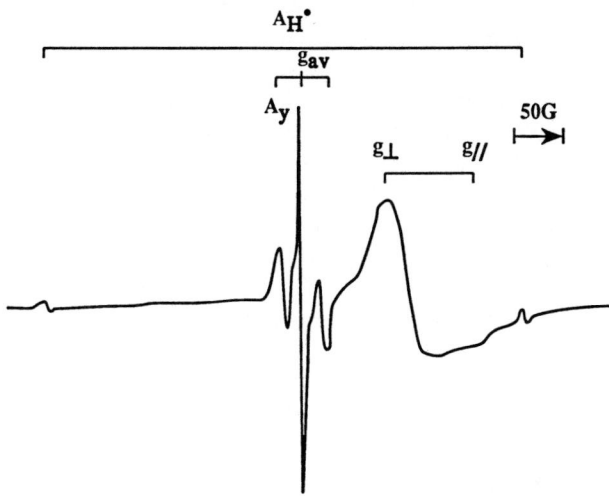

Fig. 18. EPR spectrum (X band, 77 K) of $\cdot CH_2OH$ radicals obtained after UV irradiation at 77 K of Mo/SiO$_2$ sample in the presence of CH$_3$OH (from Ref. 21).

$$Mo^{6+}=O^{2-} + \ ^{\cdot}CH_2OH \xrightarrow{T > 140 \text{ K}} CH_2O-Mo^{5+}-OH^- \qquad (IV)$$

Other experiments involving adsorption of methanol onto [Mo^{6+}–O$^-$] obtained by N$_2$O adsorption onto reduced Mo/SiO$_2$ catalysts, followed by its thermal decomposition:

$$Mo^{5+} + N_2O \xrightarrow{25\,°C} [Mo^{5+}-N_2O] \xrightarrow{100°C} [Mo^{6+}-O^-] + N_2 \qquad (V)$$

have confirmed processes according to Equations III and IV.

This EPR study is consistent with the proposed reaction mechanism. Indeed, because of the high Mo dispersion of the grafted Mo/SiO$_2$ catalysts, the decay of the $^{\cdot}CH_2OH$ intermediates by the redox process (Equation IV) implies that they have to migrate onto silica in order to find other Mo=O ions. Side reactions with support surface groups, such as ≡SiOCH$_3$ arising from the dissociative adsorption of CH$_3$OH onto silica, can occur during migration, giving rise to the formation of methyl formate. For bulk MoO$_3$ or impregnated Mo/SiO$_2$ catalysts whose Mo adsorption sites are in close interaction, a neighboring Mo can be easily found by $^{\cdot}CH_2OH$ to deactivate according to Equation IV, and CH$_2$O is the main product of methanol oxidation.

9.3. OXIDATIVE DEHYDROGENATION OF ETHANE ON Mo/SiO$_2$ CATALYSTS

The oxidative dehydrogenation of ethane over molybdenum supported on silica has been investigated using nitrous oxide as the oxidant at temperatures of 280–350 °C [64]. The main products over impregnated Mo/SiO$_2$ catalysts were C$_2$H$_4$, CO$_2$ and H$_2$O. The gaseous products at 280 °C after 100 min of reaction mainly comprised ethylene (about 80%).

On the basis of EPR and ESCA analyses, it appears that Mo is weakly reduced during catalytic reaction and that the active site is a coordinatively unsaturated Mo^{5+} ions which is present as a small fraction of the total Mo (probably less than 0.3%). In a separate experiment, the O^- ion ($g_\perp = 2.019$, $g_\parallel = 2.002$) obtained by N_2O adsorption onto unsaturated Mo^{5+} ions then decomposition (Equation V), is shown to react rapidly with ethane at 25 °C and form an intermediate species which fully desorbs by 300 °C and gives ethylene in amounts comparable to the original O^- concentration. In addition, the maximum concentration of O^- obtained by EPR by decomposition of N_2O, was of the same order of magnitude as the amount of C_2H_4 formed initially in the reactions of $N_2O-C_2H_6$ mixtures at 280 °C. It was proposed that O^- can abstract hydrogen from ethane on Mo/SiO_2 catalysts (as in the previous example, see Equation III) and act as the active form of oxygen in the dehydrogenation reaction:

$$C_2H_6 + Mo^{6+}-O^- \rightarrow \cdot C_2H_5-Mo^{6+}-OH^-$$

No paramagnetic ethyl radicals were observed, probably because of their rapid oxidation by Mo^{6+} ion to form the ethyl cation:

$$\cdot C_2H_5-Mo^{6+}-OH^- \rightarrow C_2H_5^+-Mo^{5+}-OH^-$$

A proton from $C_2H_5^+$ is probably given to an oxide ion to form another OH^-:

$$C_2H_5^+-Mo^{5+}-OH^- + O^{2-}_{surf.} \longrightarrow C_2H_4-Mo^{5+}-OH^- + OH^-_{surf.}$$

The final step of the mechanism involves the desorption of the products. The reaction of O^- with C_2H_6 was shown to be fast since no paramagnetic intermediates were detected; hence, it was proposed that the slow step of the reaction was the desorption of ethylene:

$$C_2H_4-Mo^{5+}-OH^- \longrightarrow C_2H_4(gas) + Mo^{5+}-OH^-$$

The active site can be regenerated by the removal of OH^- ions:

$$Mo^{5+}-OH^- + OH^-_{surf.} \rightarrow Mo^{5+} + H_2O + O^{2-}_{surf.}$$

10. Conclusions

The present review has shown that EPR spectroscopy can be a powerful tool to identify, at a molecular level, the different steps of catalyst preparation, and of catalytic reactions.

  (i) Deposition of paramagnetic transition metal ions onto a support may be monitored, and the coordination sphere of the metallic center can be characterized by EPR.

  (ii) The catalyst may be characterized after activation (thermal oxidation or reduction):

- the distribution among the different sites in zeolites can be determined;
- the dispersion of the active phase may be appreciated;
- the unsaturation degree of the active site may be evaluated using probe molecules such as water or ^{13}C enriched carbon monoxide.

  (iii) The catalytic mechanisms may be investigated by studying the elementary steps of the catalytic reaction, as illustrated for methanol oxidation over Mo/SiO_2

catalysts whose EPR results have extended the reaction mechanism proposed on the basis of kinetic data. In addition, reaction intermediates may be isolated in *quasi-in-situ* conditions as in the case of olefin oligomerization catalyzed by Ni/SiO_2 systems.

Due to the high sensitivity of EPR spectroscopy, very low concentrations can be studied. However, if the EPR spectrum does not exhibit any signal, it does not mean that there is no paramagnetic species in the sample. Indeed, short relaxation times can make the signal invisible, and dipolar interactions or surface site heterogeneities can broaden the signal beyond detection. On the other hand, only paramagnetic species can be probed, particularly those which are isolated and contain one unpaired electron. One has therefore to keep in mind that EPR often gives an incomplete picture of the main processes occurring on catalytic surfaces, so the use of other analysis techniques is generally required to complement the EPR assignments.

References

1. P. E. O'Reilly: *Adv. Catal.* **12**, 31 (1960).
2. (a) R. F. Howe: *Colloids and Surfaces A* **72**, 353 (1993).
   (b) M. Che and L. Bonneviot: *Z. Phys. Chem.*, *Neue Folge* **112**, 113 (1987).
   (c) M. Che and Y. Ben Taarit: *Adv. Colloid. Interf. Sci.* **23**, 179 (1985).
   (d) J. Védrine: in *Characterization of Heterogeneous Catalysts*, F. Delannay, Ed., Marcel Dekker Inc, New York (1984).
   (e) Z. Sojka: *Proceedings of the 11th Specialized Ampère Symposium on Magnetic Resonance in Catalysis*, Menton, France (1993).
   (f) J. H. Lunsford: in *Catalysis Sciences and Technology*, J. R. Anderson and M. Boudart, Eds., Springer Verlag, Berlin, Vol. 8 (1987).
   (g) M. Che and E. Giamello: in *Catalyst Characterisation: Physical Techniques for Solid Materials*, B. Imelik and J. C. Védrine, Eds., Plenum Press, New York (1994).
3. (a) J. E. Wertz and J. R. Bolton: in *Electron Spin Resonance*, McGraw-Hill (1972).
   (b) M. Che and E. Giamello: in *Studies in Surface Science and Catalysis*, J. L. G. Fierro, Ed., Elsevier, Amsterdam, 1989, **57**, B265 (1990).
   (c) J. H. Lunsford: *Adv. Catal.* **22**, 265 (1972).
4. K. Dyrek and M. Che: submitted to *Magnetic Resonance Review*.
5. (a) H. G. Hecht and T. S. Johnston: *J. Chem. Phys.* **46**, 23 (1967).
   (b) B. A. Goodman and I. B. Raynor: *Adv. Inorg. Chem. Radiochem.* **13**, 135 (1970).
6. M. Che, B. Canosa, and A. R. Gonzalez-Elipe: *J. Phys. Chem.* **90**, 618 (1986).
7. M. Che, J. C. McAteer, and A. J. Tench: *J. Chem. Soc., Faraday Trans. I* **4**, 2378 (1978).
8. (a) K. Dyrek, A. Madej, E. Mazur, A. Rokosz, and M. Rusiecka: *Wiss. Z. Friedrich-Schiller Univ. Jena, Naturwiss. R.* **37**, 781 (1988).
   (b) K. Dyrek, A. Rokosz, and A. Madej: *Appl. Magn. Res.* **6** 309 (1994).
9. T. T. Chang; *Magn. Res. Rev.* **9**, 65 (1984).
10. M. Che, C. Louis, and J. M. Tatibouët: *Polyhedron* **5**, 123 (1986).
11. C. Louis, and M. Che: *J. Catal.* **135**, 156 (1992).
12. M. Amara, M. Bettahar, L. Gengembre, and D. Olivier: *Appl. Catal.* **35**, 153 (1987).
13. P. Peigneur, J. H. Lunsford, W. De Wilde, and R. A. Schoonheydt: *J. Phys. Chem.* **81**, 1179 (1977).
14. C. Louis, M. Che, and M. Anpo: *J. Catal.* **141**, 453 (1993).
15. C. Louis, M. Che, and M. Anpo: *Res. Chem. Interm.* **15**, 81 (1991).
16. M. Che and A. J. Tench: *Adv. Catal.* **32**, 1 (1983).
17. L. Bonneviot, D. Olivier, and M. Che: *J. Mol. Catal.* **21**, 415 (1983).
18. M. Kermarec, D. Delafosse, and M. Che: *J. Chem. Soc., Chem. Commun.* 411 (1983).
19. C. Louis and M. Che: *J. Phys. Chem.* **91**, 2876 (1987).
20. M. Che, M. Fournier, and J. P. Launay: *J. Chem. Phys.* **71**, 1954 (1979).

21. Z. Sojka and M. Che: *J. Phys. Chem.* (submitted).
22. E. Giamello, Z. Sojka, M. Che, and A. Zecchina: *J. Phys. Chem.* **90**, 6084 (1986).
23. Z. Sojka, E. Giamello, M. Che, A. Zecchina, and K. Dyrek: *J. Phys. Chem.* **92**, 1541 (1988).
24. E. Giamello, E. Garrone, P. Ugliengo, M. Che, and A. J. Tench: *J. Chem. Soc., Faraday Trans. I* **85**, 3987 (1989).
25. E. Giamello, E. Garrone, E. Guglielminotti, and A. Zecchina: *J. Mol. Catal.* **24**, 59 (1984).
26. M. Che, G. Fichelle, and P. Mériaudeau: *Chem. Phys. Letters* **17**, 66 (1972).
27. J. C. Védrine, H. Praliaud, P. Mériaudeau, and M. Che: *Surf. Sci.* **80**, 101 (1979).
28. P. Mériaudeau, B. Clerjaud, and M. Che: *J. Phys. Chem.* **87**, 3872 (1983).
29. P. Mériaudeau: *Chem. Phys. Letters* **72**, 551 (1980).
30. T. Shimizu: *J. Phys. Soc. Jpn.* **23**, 848 (1967).
31. P. de Montgolfier, P. Mériaudeau, Y. Boudeville, and M. Che: *Phys. Rev.* **14B**, 1788 (1976).
32. P. Mériaudeau, M. Che, and A. J. Tench: *Chem. Phys. Letters* **31**, 547 (1975).
33. A. Davidson and M. Che: *J. Phys. Chem.* **96**, 9909 (1992).
34. E. G. Derouane, J. Thelen, and V. Indovina: *Bull. Soc. Chim. Belg.* **82**, 657 (1973).
35. (a) V. D. Atanasova, V. A. Shvets, and V. B. Kazanskii: *Russ. Chem. Rev.* **50**, 209 (1981).
    (b) I. D. Mikheikin, G. M. Zhidomirov, and V. B. Kazanskii: *Russ. Chem. Rev.* **41**, 468 (1972).
36. R. A. Schoonheydt: *Catal. Rev.-Sci. Eng.* **35**, 129 (1993).
37. I. R. Leith and H. F. Leach: *Proc. Roy. Soc. Lond.* **A330**, 247 (1972).
38. M. Briend-Faure, J. Jeanjean, G. Spector, D. Delafosse, and F. Bozon-Verduraz: *J. Chim. Phys.* **79**, 489 (1982).
39. R. A. Schoonheydt and D. Roodhooft: *J. Phys. Chem.* **90**, 6319 (1986).
40. M. Kermarec, D. Olivier, M. Richard, M. Che, and F. Bozon-Verduraz: *J. Phys. Chem.* **86**, 2818 (1982).
41. D. E. O'Reilly, F. D. Santiago, and R. G. Squires: *J. Phys. Chem.* **73**, 3172 (1969).
42. S. Abdo, A. Kazusaka, and R. F. Howe: *J. Phys. Chem.* **85**, 1380 (1981).
43. S. R. Seyedmonir and R. F. Howe: *J. Chem. Soc. Faraday Trans. I* **80**, 87 (1984).
44. C. Chao and J. H. Lunsford: *J. Chem. Phys.* **57**, 2890 (1972).
45. J. C. Conesa and J. Soria: *J. Phys. Chem.* **82**, 1575 (1978).
46. O. R. Flentge, J. H. Lunsford, P. A. Jacobs, and J. B. Uytterhoeven: *J. Phys. Chem.* **79**, 354 (1975).
47. R. Bechara, A. D'huysser, C. F. Aissi, M. Guelton, J. P. Bonnelle, and A. Abou-Kais: *Chem. Mater.* **2**, 522 (1990).
48. (a) A. Abou-Kais, R. Bechara, D. Ghoussoub, C. F. Aissi, M. Guelton, and J. P. Bonnelle: *J. Chem. Soc, Faraday Trans.* **87**, 631 (1991).
    (b) A. Abou-Kais, R. Bechara, C. F. Aissi, and M. Guelton: *Chem. Mater.* **3**, 557 (1991).
49. A. Abou-Kais, R. Bechara, C. F. Aissi, and J. P. Bonnelle: *J. Chem. Soc., Faraday Trans.* **89**, 2545 (1993).
50. C. Lepetit, M. Kermarec, and D. Olivier: *J. Mol. Catal.* **51**, 95 (1989).
51. C. Lepetit, M. Kermarec, and D. Olivier: *J. Mol. Catal.* **51**, 73 (1989).
52. Y. Chauvin, J. F. Gaillard, D. V. Quang, and J. W. Andrews: *Chem. Ind.* 375 (1974).
53. P. W. Glockner, W. Keim, and R. F. Mason: *US Patent*, 3 647 914 (1971) to Shell Oil Company.
54. (a) B. Bogdanovic: in *Advances in Organometallic Chemistry*, Academic Press, New York, **17**, 105 (1979).
    (b) P. W. Jolly and G. Wilke: in *The Organic Chemistry of Nickel*, Academic Press, New York, Vol. 2, p. 21 (1975).
55. S. J. McLain and R. R. Schrock: *J. Am. Chem. Soc.* **100**, 1474 (1978).
56. F. X. Cai, C. Lepetit, M. Kermarec, and D. Olivier: *J. Mol. Catal.* **43**, 93 (1987).
57. A. Saltzer: *Chemia* **38**, 421 (1984).
58. C. Louis, J. M. Tatibouët, and M. Che: *J. Catal.* **109**, 354 (1988).
59. J. M. Tatibouët and J. E. Germain: *J. Catal.* **72**, 375 (1981).
60. J. M. Tatibouët, J. E. Germain, and J. C. Volta: *J. Catal.* **82**, 240 (1983).
61. N. Pernicone: in P. C. H. Mitchell (Ed.), *Proc. 1st Int. Conf. Chemistry and Uses of Molybdenum*, Reading, 1973, CLIMAX, p. 155 (1974).
62. R. Miranda, J. S. Chung, and C. O. Bennett: *Proc. 8th Int. Congr. Catal.*, Berlin, *1984*, Verlag Chemie, Weinheim, Vol. 3, p. 347 (1984).
63. J. N. Allison and W. A. Goddard III: *J. Catal.* **92**, 127 (1985).
64. M. B. Ward, M. J. Lin, and J. H. Lunsford: *J. Catal.* **50**, 306 (1977).

Study of Catalytic Site Structure and Diffusion of Radicals in Porous Heterogenous Systems with ESR, ENDOR and ESE

R. I. SAMOILOVA, A. D. MILOV, and YU. D. TSVETKOV
Insitute of Chemical Kinetics and Combustion, Novosibirsk 630090, Russia

(Received: 15 December 1993; accepted: 5 July 1994)

Abstract. The generation of paramagnetic products by adsorption of quinones on activated catalysts has been used for the diagnostic of Lewis acid sites. It has been shown that the application of ENDOR, ESE, and 2 mm-band ESR are extremely effective methods for studying the nature of observed radical species and their environment. Two applications of the ESE method for studying the diffusion of spin probes in porous media are considered. The measurement of effective diffusion coefficients of radical probes in specimens of various heterogeneous systems is described. It has been found that effective diffusion coefficients depend strongly on the mean value of silica gel pore sizes and the mobility of the probe inside the pore.

Key words. ESR, ENDOR, ESE, Lewis acid sites, paramagnetic complexes, zeolites, diffusion processes.

1. Studies of Catalytic Sites

1.1. INTRODUCTION

When quinones (particularly anthraquinone) are employed as probe molecules to study active sites of a surface, the commonly observed ESR spectra exhibit only a single, featureless line. Nevertheless, the unresolved quinone spectrum can contain important information concerning the environment of the radical. In that case ENDOR, ESE and 2 mm-band ESR are more suitable methods. The most commonly used is Electron Nuclear Double Resonance (ENDOR) spectroscopy [1, 2]. The ENDOR spectrum is the spectrum of nuclear transitions in a paramagnetic center which is removed from the state of ESR saturation by an additional irradiation of the radiofrequency magnetic field. The parameters of the hyperfine and quadrupole interactions with nuclei (of the order of a few MHz and less) can often be determined from the ENDOR spectra. Another method which is a conventional one now is pulsed ESR spectroscopy (ESE method) [3]. Compared to ESR, which is a frequency domain spectroscopy, ESE is a time domain spectroscopy in which the times of the spin system evolution (T_1, T_2) are recorded directly. In the ESE experiment the unresolved structure of the line resulting from weak hyperfine interactions with surrounding nuclei manifests itself as electron spin echo envelope modulation (ESEEM) [4]. This spectroscopic method has significantly expanded the experimental possibilities of the investigation of weak hyperfine interactions due to the environment of magnetic nuclei around a paramagnetic center.

Finally, 2 mm-band ESR has greatly increased the resolution of ESR for such spectral characteristics as the g value. Deviations of the g value characterize a spin–orbital interaction, which is one of the most important structural characteristics. For free radicals these deviations are, unfortunately, small and are usually of the order of $g = (1-10) \cdot 10^{-4}$, which leads to a line shift in the X-band of only 0.1–1 mT. In the 2 mm-band this shift is already 1.5–15 mT [5, 6]. In this review we consider the results of some investigations, refering to quinone paramagnetic products on a series of surfaces and catalysts using the above methods.

1.2. DIAGNOSTICS OF LEWIS ACID SITES ON ZEOLITES

Depending on properties of catalysts and variations of possible locations, coordinations and oxidation states of Al^{n+} ions in zeolites, the redox reactions of adsorbed quinones can produce the different kinds of radical species. Since the radical reactions of quinones have been studied in detail, it is possible to compare the parameters of their spectra in zeolites with earlier observations [7, 8]. It is known that the catalytic activity of zeolites (H-form) is due to the presence of Lewis and Brønsted acid sites. Considering the electron transfer reaction which occurs on the activated surface, we are more interested in those which involve the formation of stable paramagnetic species as a primary product in the reaction between active site and the adsorbed molecule. From this point of view quinones are promising and effective indicators. The protonated and diprotonated radical forms of quinone seem to be related to the proton transfer from Brønsted acid site to the adsorbed quinone molecule. Besides, the electron transfer reaction between a Lewis acid site and a quinone molecule produces a charge-transfer complex. The purpose of our investigation is to separate these two situations.

Experimental. The zeolite activation and adsorption procedures were carried out as described in [9, 10]. We used Y-type zeolites in Na- and H-forms, Na-mordenite and H-ZSM-5 with different Si/Al ratios, and reagents such as 1,4-benzoquinone, chloranil, fluoranil and o-chloranil. To identify the radical species resulting from quinone adsorption on zeolites, we compared their spectra with those of the synthesized radical cations, anions and complexes of the same quinones in frozen solutions. The method used for generating the radicals has been described in [11, 12]. The ESR and ENDOR spectra were recorded with a *Bruker ESP-300* spectrometer; the ESEEM measurements were performed on a home-made spectrometer described elsewhere [13].

1.2.1. ENDOR and ESEEM Measurements

As can be seen from Table I, the g factors of quinone radicals formed on HY zeolite are close to those observed in the reactions of quinones with $AlCl_3$ in nonpolar solvent and are substantially less than those of radical anions in frozen solutions. There are no structural hydroxyl groups in NaY zeolite. Thus, the activation of this zeolite fails to produce Brønsted and Lewis sites. That is why the radicals formed on the surface of NaY zeolite coincide in their spectral parameters with quinone radical anions formed upon the adsorption of quinones on oxides [14]. Similar spectra have been observed during the adsorption of quinones

TABLE I
The values of the g factor of paramagnetic particles, formed by the reaction of quinones in various media

Type R	Method	T (K)	$g_{iso} \pm 0.001$
R$^+$	Benzoquinone in CF$_3$SO$_3$H	77	2.0037
		Room	2.0032
	Benzoquinone with AlCl$_3$ in toluene-d_8	77	2.0036
	Hydroquinone with AlCl$_3$ in CH$_3$NO$_2$	Room	2.0035
	Fluoranil in CF$_3$SO$_3$H	77	2.0050
		Room	2.0036
	Fluoranil with AlCl$_3$ in toluene-d_8	77	2.0036
	Chloranil in CF$_3$SO$_2$H	77	2.0046
	Chloranil with AlCl$_3$ in toluene-d_8	77	2.0048
R$^-$	Benzoquinone with DTNa[a] in alkaline C$_2$D$_5$OD	77	2.0044
	Benzoquinone with DTNa[a] in C$_2$H$_5$OH	77	2.0047
	Fluoranil with KJ in acetone-d_8	77	2.0059
		Room	2.0051
	Tetrafluorohydroquinone in alkaline C$_2$D$_5$OD	77	2.0059
	Chloranil with KJ in acetone-d_8	77	2.0061
	Chloranil with DTNa in alkaline C$_2$D$_5$OD	77	2.0061
	Reduction chloranil with Na in DME[b]	Room	2.0058
	Benzoquinone on HY zeolite	Room	2.0031
	Fluoranil on HY zeolite	Room	2.0033
	Chloranil on HY zeolite	Room	2.0041
	Fluoranil on NaY zeolite	Room	2.0056
	Different quinones on Al$_2$O$_3$	Room	2.0057

[a] DTNa = dithionite Na.
[b] DME = dimethoxyethane.

on the surface of oxides activated in vacuum at 1270 °C. Two mechanisms can generally be considered [14]: one is electron transfer from the oxygen vacancies formed under thermal treatment in vacuum; the other is the hydroxylation of quinone in the alkaline hydrated film, as observed for alkaline solutions.

In the ENDOR spectra (Figure 1) narrow lines at the proton (ν_H = 14.5 MHz) and fluorine (ν_F = 13.8 MHz) Larmor frequencies are caused by a weak hfi with distant proton and fluorine nuclei. The wide lines in the frequency regions around 10 and 18 MHz in spectrum 2 are caused by the ring proton of benzoquinone radicals, and in spectrum 3 by the nuclei of fluoranil radicals. An additional line in the spectra is observed at a frequency of about 6 MHz (Figure 1). This line is independent of the type of substituents (H, F, Cl) on the quinone carbon atoms. In the last case, the position of this line in the spectrum varies within 5.6–6.8 MHz. The same situation has been observed when paramagnetic products were generated in the reactions of o-chloranil with different activated zeolites in the H-form. o-Quinones should be more suitable since they form chelate paramagnetic complexes with metals of different groups in various solvents [15, 16]. Such complexation occurs with some transfer of spin density from the ligand to the Al^{3+} ion. A low value of spin density on the Al^{3+} indicates that the complex is mainly ionic in nature, namely that the radical and the metal are bound by electrostatic interaction. Accordingly, the paramagnetic complex generated in solutions of o-quinone

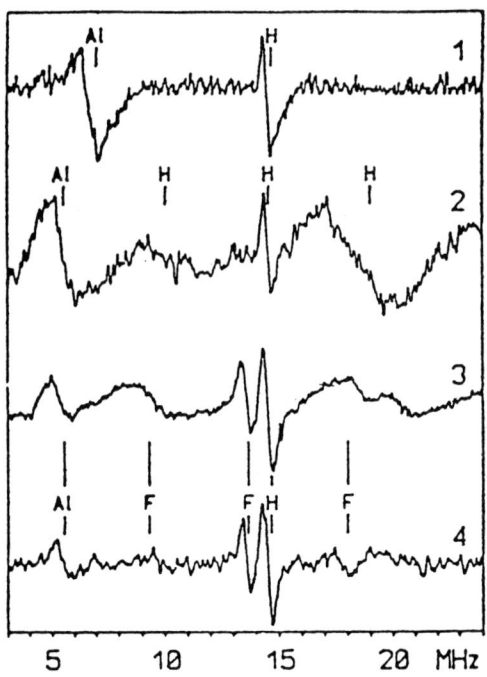

Fig. 1. ENDOR spectra of radicals arising on the adsorption of 1 – chloranil, 2 – p-benzoquinone, 3 – fluoranil from solution on HY zeolite (298 K); 4 – in the reaction of fluoranil with $AlCl_3$ in toluene (77 K).

and $AlCl_3$ can serve as a good model for the formation of similar complexes with Lewis acid sites on catalysts.

A typical ESR spectrum obtained from the adsorption of o-chloranil vapors on activated zeolite is shown in Figure 2c. For comparison, the ESR spectra of o-chloranil with $AlCl_3$ are recorded in a toluene-d_8 solution at room temperature and at 140 K (Figure 2a,b). The room temperature spectrum consists of six lines with a spacing of 0.25 mT due to the coupling to the ^{27}Al nucleus ($I = 5/2$), which is in good agreement with the values reported previously [12]. At 140 K the fine structure is lost and the spectrum exhibits a singlet line with a width of 1.3–1.4 mT, similar to that observed in zeolites. ENDOR spectra of zeolite HY after o-chloranil adsorption along with the ENDOR spectrum of o-chloranil/$AlCl_3$ solution are shown in Figure 3. These are similar to the spectra of adsorbed p-quinone molecules in HY zeolite. At this stage we assign the peak at 6.5–7 MHz to the high frequency line of a hyperfine split doublet, centered about the ^{27}Al Larmor frequency. This implies a ^{27}Al hyperfine splitting of 7.2 MHz (0.25 mT), which is in good agreement with the splittings observed in the ESR spectrum at room temperature. The magnitude of the hfi constant on ^{27}Al estimated from this line position is 6–7 MHz for o-chloranil, 5 MHz for p-benzoquinone, 3.8 MHz for fluoranil, and 6 MHz for chloranil paramagnetic complexes on HY.

The observation of ^{27}Al hfi is not unique for o-chloranil adsorption on activated

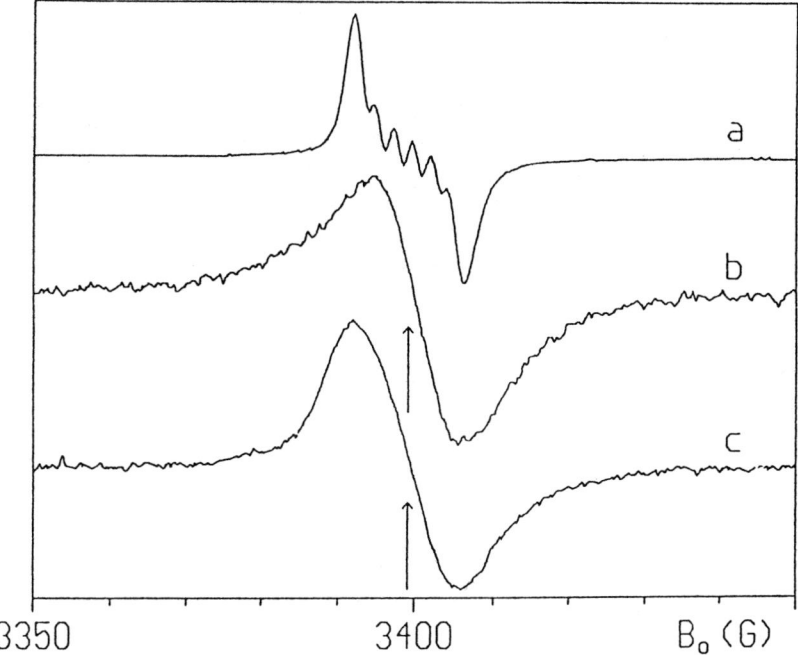

Fig. 2. ESR spectra of o-chloranil complex with $AlCl_3$ in toluene-d_8 recorded at room temperature (a) and 140 K (b). ESR spectrum formed after o-chloranil adsorption from vapors on HY zeolite at room temperature (c). The arrows depicted on spectra (b) and (c) indicate the field positions corresponding to the observation of the ENDOR spectra.

HY zeolite and has also been observed in Al_2O_3, H-mordenite and H-ZSM-5. The ENDOR results are given in Table II. The concentration of the paramagnetic centers, as obtained from the integration of the ESR spectrum, for each of the catalysts studied, is also listed in Table II. Note the correlation between the number of paramagnetic centers and the total ^{27}Al content of the catalysts.

To confirm the above assignment we have performed ESEEM experiments, which show a better sensitivity to low frequencies than the ENDOR method. Figure 4 shows the three-pulse FT-ESEEM spectra measured from the sample where o-chloranyl was adsorbed on HY from hexafluorobenzene solution at two different τ values, 0.2 and 0.3 μs. Both of the spectra show proton and fluorine lines, as observed in the ENDOR spectra, and broad lines at 6.5–7 MHz. The spectrum measured at $\tau = 0.3\ \mu$s contains an additional line at ~1 MHz, whereas the spectrum recorded at $\tau = 0.2\ \mu$s shows a peak at the ^{27}Al Larmor frequency. The absence of this line in the $\tau = 0.3\ \mu$s spectrum is due to the well-known suppression effect [17]. This is also the reason for the absence of the 1 MHz line in the spectrum recorded at $\tau = 0.2\ \mu$s. The approximately symmetric positions of the peaks at 1 and 6.5 MHz around the ^{27}Al Larmor frequency and their τ dependence, confirm their assignment to an ^{27}Al nucleus ENDOR transitions given by:
$\nu_{\alpha(\beta)} = |\nu_{Al} \pm a/2|$, where α and β correspond to different electron spin states with

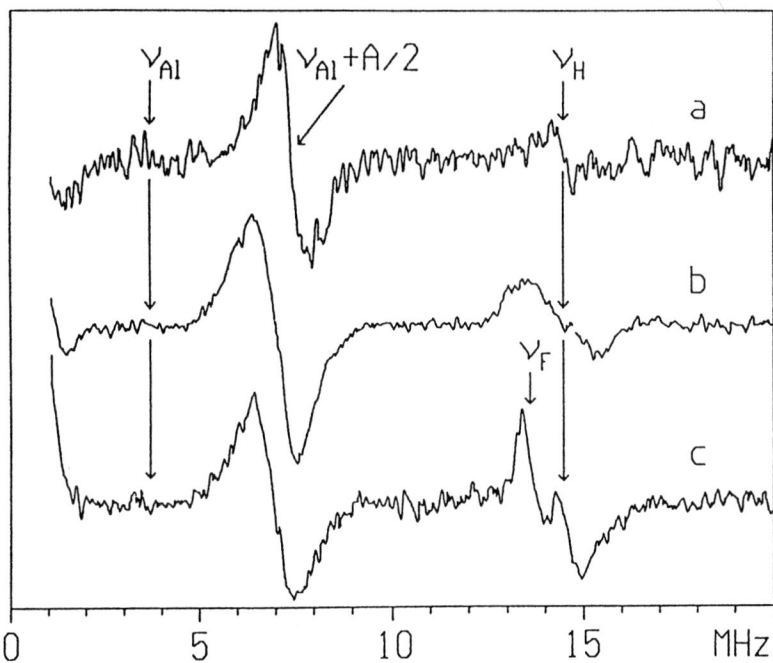

Fig. 3. ENDOR spectra: of o-chloranil complex with AlCl$_3$ in toluene-d_8 recorded at 140 K (a), of HY zeolite after o-chloranil (b) and hexafluorobenzene (c) adsorption from vapors, recorded at 200 K.

TABLE II
Catalysts characteristics, ^{27}Al isotropic hyperfine constants, and the concentration of o-chloranil paramagnetic complexes for the different catalysts investigated

Catalysts	SiO$_2$/Al$_2$O$_3$	Window size (Å)	ENDOR$^a_{Al}$ MHz (g) (±0.1 g)	Complex concentration $* 10^{18}$ spin/g (±50%)
γ-Al$_2$O$_3$	–	–	5.6 (2.0)	20
ZSM-5(1)	50			0.8
ZSM-5(2)	35	5.5	6.0 (2.1)	1.2
ZSM-5(3)	19.5			2.5
H-Mordenite	1.2–1.45	6.7 × 7.0 2.9 × 5.7	6.2 (2.2)	4.6
HY	2–2.5	8	6.2 (2.2)	9
AlCl$_3$ in toluene-d_8	–	–	7.0 (2.5)	–

an isotropic hyperfine constant $a \simeq 5.5$–6 MHz. Since $I_{Al} = 5/2$, five ENDOR frequencies should be observed for each manifold. However, when the quadrupole coupling constant of the ^{27}Al nuclei is small, all lines overlap and a doublet is observed [18]. The peak at the ^{27}Al Larmor frequency is attributed to ^{27}Al nuclei

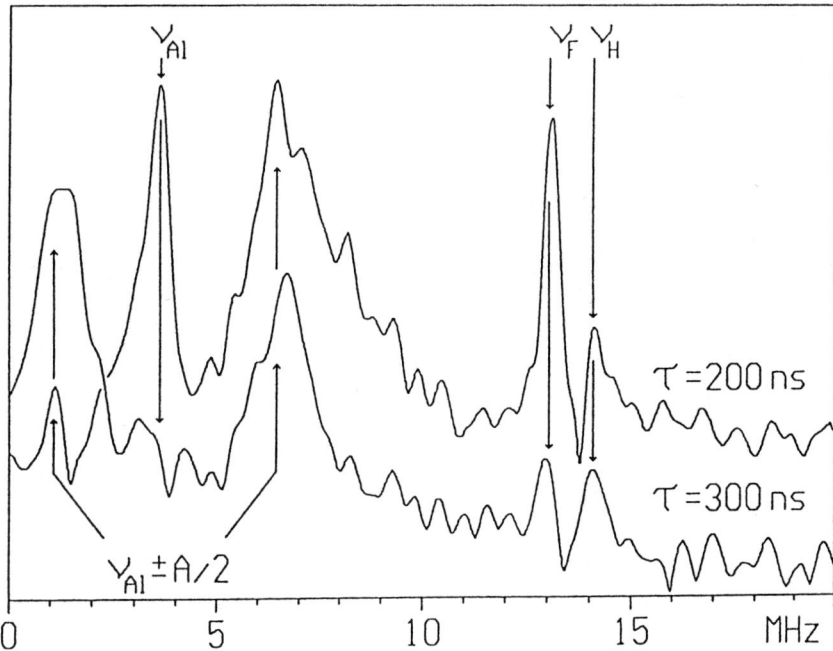

Fig. 4. Stimulated magnitude FT-ESEEM spectra of HY zeolite with o-chloranil adsorbed from hexafluorobenzene with $\tau = 0.2$ μs and 0.3 μs, recorded at 4.2 K.

in the vicinity of the unpaired electron ($r < 8$ Å) [19], which are not directly involved in the complex formation but are weakly coupled by a dipolar interaction. This is confirmed by the disappearance of the peak at $\tau = 0.3$ μs $\approx 1/\nu_{Al}$ [17].

Although we have established the existence of complexes between framework Al^{3+} and radicals generated from quinone adsorption, the nature of the Al^{3+} sites and the mechanism for the radical formation are still not clear. The existence of two types of Al sites in decationated zeolites is well established [20]. One type is the tetrahedrally coordinated framework Al and the second one is the extraframework Al, produced by partial dealumination of the zeolite during the processes of decationation and dehydroxylation. The latter appears as octahedral Al [21]. A third possible type of Al sites is the so-called three-coordinated Al or unsaturated Al sites, generated as a consequence of the dehydration of decationated zeolites at high temperature, as proposed by Vytterhoeven et al. [22]. The existence of these sites was highly controversial, and it has recently been concluded that they do not exist as a stable form of Al^{3+}. Accordingly, one must assign the Al^{3+} participating in the paramagnetic complex to extraframework Al^{3+} formed by thermal treatment. Another question that should be addressed is whether the Al^{3+} Lewis sites are located in the complexation, on the external or internal surface of zeolites. In the case of H-ZSM-5, the channels are too small to accommodate the o-chloranil and interact with Lewis acid sites on the external surface. HY and H-Mordenite, however, have a larger pore opening, which probably allows the penetration of *o*-chloranil. This is supported by the increase of the concentration of the paramagnetic centers following a slight temperature increase (after adsorption) in HY and H-Mordenite as opposed to ZSM-5.

1.2.2. 2 mm-*Band ESR Spectroscopy of Paramagnetic Quinone Complexes on Zeolites*

In magnetic fields of 5 T, the *g* factor resolution in the 2 mm-based ESR spectra is substantially higher than in X-band ESR. Thus it is possible to determine a small anisotropy of a radical *g* tensor in disordered samples and to measure its principal values.

Figure 5 (a,b,c) depicts the 2 mm-band and X-band ESR spectra of the radical anions of *p*-benzoquinones in frozen glassy ethanol. The spectrum shape is typical of disordered particles with a rhombic anisotropy of the *g* tensor. The three lines in the spectra of *p*-benzoquinone and chloranil radical anions correspond to the canonical orientations at which the external magnetic field is parallel to each of the *g* tensor principal axes. The principal *g* tensor values g_{ii} ($i = x, y, z$) may be directly determined from the component position. The principal axes for *p*-benzoquinone radical anions are chosen so that the minimum principal g_{zz} value corresponds to the direction perpendicular to the ring plane, and the direction for the maximum g_{xx} corresponds to that of the C=O bond.

In the 2 mm-band ESR spectrum of fluoranil radical anion the *z* orientation demonstrates an additional multiplet structure consisting of five lines, with intensities are close to 1:4:6:4:1, and the splitting between which is 1.65 mT. Such a structure is due to the *hfi* between the unpaired electron and four equivalent fluoranil nuclei. The X-band ESR spectra of *p*-benzoquinone and chloranil are in the form of asymmetric singlets. For fluoranil, the X-band spectrum is more complex owing to the hyperfine structure of fluoranil nuclei. The principal values for the *g* tensors of *p*-benzoquinone radical anions determined from the 2 mm-band ESR spectra are listed in Table III and are consistent with the previous ones [23]. Figure 5d depicts the 2 mm-band ESR spectra of diprotonated *p*-benzoquinone radical in frozen CF_3SO_3H solution at 1650 K. As for the radical anions, in the spectrum of radical cations a rhombic anisotropy of the *g*-tensor is resolved. Its principal values are also presented in Table III. Figure 6 presents the 2 mm-band and X-band ESR spectra of complexes of quinones with aluminium chloride fixed by freezing solutions down to 77 K and spectra of paramagnetic complexes formed upon adsorption of *p*-quinones on the activated HY zeolite surface. The *g* factor anisotropy of these complexes is axial, in contrast to the rhombic anisotropy of radical anion and cation. The parameters typical of these spectra are given in Table III. As the change of Δg^{hal} for chloranil anion in solution and in complexes is small, it might be concluded that the complexes on zeolite are bound to the matrix via oxygen, as radical anions in the alcohol solvent. This is sure to weaken the influence of halide contribution to the *g* tensor.

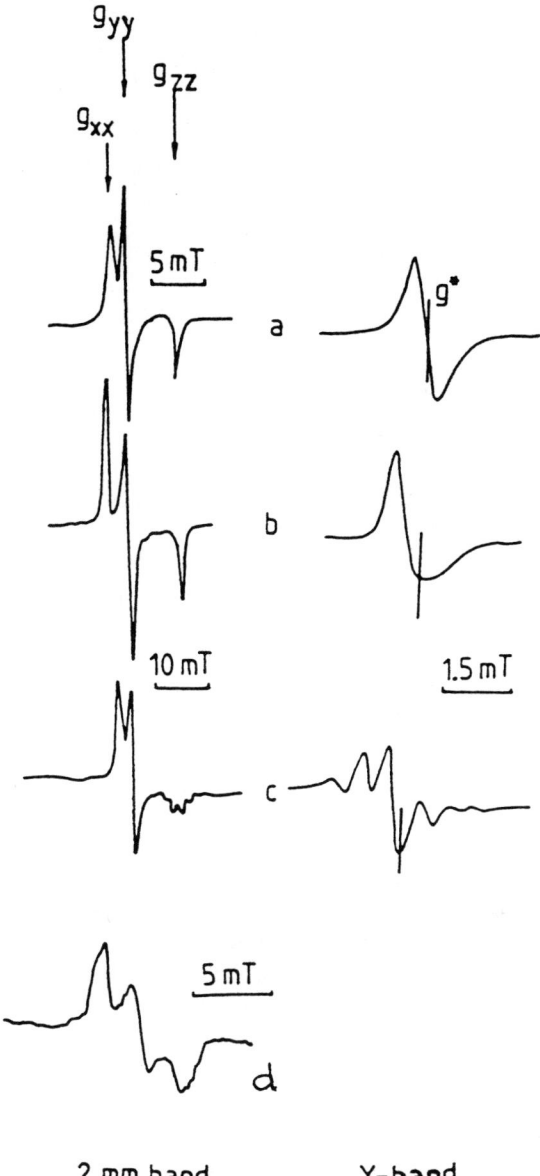

Fig. 5. 2 mm-band and X-band ESR spectra of p-benzoquinone (a), chloranil (b), fluoranil (c) radical anions in frozen ethanol solutions, and diprotonated p-benzoquinone radical cations (d) in frozen CF_3SO_3H. The vertical line on the X-band spectra corresponds to $g = 2.0045$.

TABLE III
Magnetic resonance parameters of p-quinone radicals

	p-Benzoquinone	Fluoranil	Chloranil
Anion radicals			
X-band, $\langle g \rangle$	2.0047	2.0059	2.0061
2-mm band	2.85		
$(g_{yy} - g_{zz}) \cdot 10^3$		3.98	4.39
$(g_{xx} - g_{yy}) \cdot 10^3$	1.13	1.20	2.18
Q-band[a]	–		
$(g_{yy} - g_{zz}) \cdot 10^3$	3.0	–	4.39
$(g_{xx} - g_{yy}) \cdot 10^3$	1.2	–	1.77
g	2.0047		2.00564
Cation radicals			
X-band, $\langle g \rangle$	2.0037	2.0050	2.0046
2-mm band			
$(g_{yy} - g_{zz}) \cdot 10^3$	1.35	–	–
$(g_{xx} - g_{yy}) \cdot 10^3$	1.18	–	–
Complexes with AlCl₃			
X-band, $\langle g \rangle$	2.0036	2.0036	2.0048
2-mm band			
$(g_\perp - g_\parallel) \cdot 10^3$	1.89	1.57	–
p-Benzoquinones on HY			
X-band, $\langle g \rangle$	2.0031	2.0033	2.0041
2-mm band			
$(g_\perp - g_\parallel) \cdot 10^3$	1.69	1.65–1.75	3.85

2. ESE Diffusion Measurements in Porous Media

The results of the investigations reported in this section concern molecular translation measurements. When diffusion processes are under study in porous media, such as silica gels, activated coals, zeolites, etc., it is interesting to obtain information on both the mechanism of diffusion (capillary, quasihomogeneous, etc.) and the values of effective diffusion coefficients and their dependence on the structural parameters of the porous medium. In [25, 26] a proposal was made to perform studies of radical diffusion in solutions based on the dependence of CW ESR spectrum shapes on the radical concentration. However, these concentration changes in ESR lineshapes are considerably masked by a large inhomogeneous spectrum width. The use of the concentration dependence of paramagnetic relaxation times is more efficient and reliable for these purposes. The ESE method is convenient for studying relaxation processes that exclude the influence of inhomogeneous broadening of ESR lines.

In this review we consider two types of the application of this technique for studying the diffusion of spin probes in porous media. The first approach [27] is based on the two-pulse ESE technique applied to study the dependence of the primary echo signal amplitude on the time of probe diffusion into a porous granule. In this case the ESE signal amplitude at a fixed time τ between the mw pulses exciting the echo responds to the change in spin–spin interactions upon diffusion of paramagnetic centers. The second one is the ESE tomography method [28],

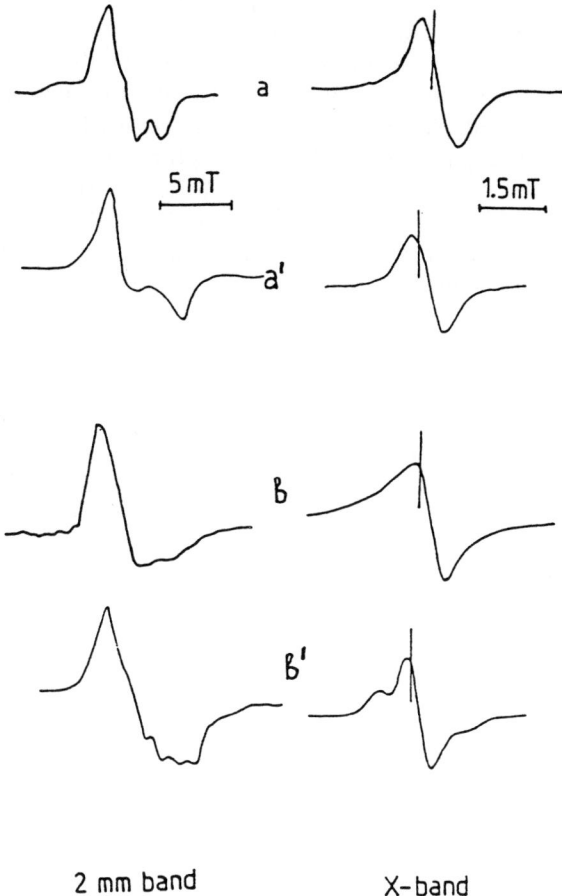

Fig. 6. 2 mm-band and X-band ESR spectra of the complexes of p-benzoquinone (a), fluoranil (b) with AlCl₃ in frozen solutions and spectra after adsorption of p-benzoquinone (a') and fluoranil (b') on HY zeolites.

based on the recording of echo signals for a sample situated in an inhomogeneous external magnetic field. This makes it possible to determine the concentration profile of radical probes in the specimen under study at various diffusion times and to estimate the effective diffusion coefficients for various heterogeneous systems. We describe the procedure of measurements in considerable detail because the method is still new to many.

2.1. DIFFUSION CONTROLLED CHANGE OF ESE SIGNAL INTENSITY (ESE-DCI)

The paramagnetic relaxation times in both liquid and frozen glassy solutions depend on radical concentration. This dependence is defined by the mechanism of relaxation related to the interaction of radicals, i.e. magnetic dipole–dipole, and exchange interactions. In solids, the dipole–dipole interaction is the main concentration-dependent relaxation mechanism.

In two-pulse ESE, the signal appears after the action on the spin system of two mw pulses separated by a time interval τ. The intensity or amplitude of the ESE signal is proportional to radical concentration and decreases due to time-dependent relaxation processes. The change of concentration profile due to diffusion will lead to a dependence of the echo signal amplitude on the diffusion time t if the time interval between mw pulses is fixed at $\tau = \tau_0$. The form of this dependence will be defined by both a specific system geometry and the parameters of radical diffusion mobility. When the system geometry is known, the dependences can actually give information about diffusion processes. The dV_i amplitude of echo signal induced by radicals situated in the ith volume element, dv_i, has the form:

$$dV_i = Ac_i \cdot \exp[-f(\tau) - F(c, \tau)] \cdot dv_i, \tag{1}$$

where A is an apparatus constant; $c_i = dN_i/dv_i$ is the radical concentration in the volume element dv_i; $\exp(-F(c, \tau))$ is the concentration-dependent echo signal decay; and $\exp(-f(\tau))$ is the echo signal decay due to other relaxation mechanisms. The amplitude of the echo signal from the whole sample can be calculated by integrating (1):

$$V = A \cdot \exp[-f(\tau)] \cdot \int_0^\infty c \cdot \exp[-F(c, \tau)] \cdot (dv/dc) \, dc. \tag{2}$$

The value $c \cdot dv/dc$ in (2) is determined by a specific concentration profile of radicals in the system under study. The time dependence of the concentration profile will lead to the dependence of the echo signal amplitude at the given τ value on diffusion time t. Consider now the two cases of the time dependence of ESE signal amplitude, i.e. radical diffusion from solution into a solvent-filled capillary and into spherical granules. Upon diffusion into a capillary, one end of which is sealed and the other open, the radical concentration at the open end $c = c_0$. Providing the radical concentration in the capillary is zero at $t = 0$, the concentration profile is of the form [29]:

$$c(x, t) = c_0[1 + s(x, t)], \tag{3}$$

$$s(x, t) = \frac{4}{\pi} \cdot \sum_{n=0}^{\infty} \left(\frac{(-1)^{n+1}}{2n+1}\right) \cdot \exp\left[-\frac{\pi^2(2n+1)^2 Dt}{4L^2}\right] \cdot \cos\left(\frac{\pi(2n+1)x}{2}\right),$$

where $x = l/L$, and l is the distance between the closed capillary end and the given point, L is the capillary length, and D is the radical diffusion coefficient.

To consider the radical diffusion into a spherical porous granule we shall use the quasihomogeneous diffusion model [30, 31]. In this case, the diffusion equation is of the usual form but with an effective diffusion coefficient D_1 that differs from the radical diffusion coefficient D for the solvent [29]. The concentration profile, averaged over the microstructure inside the granule, will acquire the form:

$$c(z, t) = c_0[1 + s_1(z, t)], \tag{4}$$

$$s_1(z,t) = \frac{2}{\pi z} \cdot \sum_{n=1}^{\infty} \left(\frac{(-1)^n}{n}\right) \cdot \exp\left[-\frac{\pi^2 n^2 D_1 t}{R^2}\right] \cdot \sin(\pi n z),$$

when $z = r/R$, and r is the distance from the granule center to the given point, R is the granule radius, D_1 is the effective coefficient of radical diffusion inside the granule. The concentration of radicals c in solution filling the pores of granules is ϵ times less than the radical concentration of pure solution, where ϵ is the granule porosity, or the specific volume of pores. Taking this into account and substituting (3) and (4) into (2) we derive the following expressions for ESE signal amplitude. For the spherical granule

$$V(t) = V(\infty) \cdot \int_0^1 [1 + s_1(z,t)] \cdot \exp[-F(c,\tau)] \cdot z^2 \, dz, \tag{5}$$

where

$$V(\infty) = A \cdot \frac{4\pi R^3}{3} \cdot \epsilon c_0 \cdot \exp[-f(\tau) - F(c_0, \tau)].$$

For diffusion into capillary

$$V(t) = V(\infty) \cdot \int_0^1 [1 + s(x,t)] \cdot \exp[-F(c,\tau)] \cdot dx, \tag{6}$$

$$V(\infty) = A v_0 c_0 \cdot \exp[-f(\tau) - F(c_0, \tau)],$$

where v_0 is the capillary volume.

The concentration dependence of magnetic relaxation rate in glassy frozen (77 K) solutions of the probe of the stable radical of 2,2,6,6-tetramethyl-4-oxypiperidine-1-oxyl (TEMPON) in methanol has been studied earlier [32]. Two mechanisms determine the phase relaxation in this system, namely the instantaneous diffusion the contribution of which is proportional to $\exp(-\alpha c \tau)$ and the spectral one that is proportional to $\exp(-\gamma c^2 \tau^2)$. In the general case the function $F(c,t)$ in (5, 6) can be represented as:

$$F(c,t) = \alpha c \tau + \gamma c^2 \tau^2. \tag{7}$$

Special relaxation measurements made in glassy frozen solutions of TEMPON in methyltetrahydrofuran (MTHF) yield the following values for numerical coefficients: $\alpha = 1.15 \cdot 10^{-13}$ cm^3 s^{-1}, $\gamma = 4 \cdot 10^{-27}$ cm^6 s^{-2}.

Thus, since the form of $F(c,t)$ function is known (7), relation (5) and (6) can be used for numerical calculations of the dependence of V on either $D_1 t/R^2$ or Dt/L^2 parameters, respectively, for granule or capillary, for the fixed τ value. These curves are expected to be different for granule and capillary, and a comparison with the experimental curve will allow conclusions to be drawn on the diffusion mechanism in a porous medium – whether it is mostly capillary or quasihomogeneous. In addition, the effective diffusion coefficients can be determined.

2.2. ESE-TOMOGRAPHY (ESE-T)

CW ESR-tomography is the determination of a macroscopic distribution of paramagnetic particles in space by analysing the shapes of their spectra in the inhomogeneous external magnetic field. The resolution ability of the method in that case is determined by the relation of ESR line width to the inhomogeneity of external magnetic field. Since in CW ESR the lines can be broadened to 1–10 mT due to anisotropy of the g factor and hfi, the gradients of the external magnetic field must be large enough, which is limited by technical potentialities. The theory of ESE–T and preliminary experimental data are given in references [33, 34].

The possibility of applying the ESE technique to determine the macroscopic distribution of spins throughout the sample is based on the artificial creation of additional phase relaxation in spin systems due to short (of time less than τ) switch-on of the inhomogeneous (within-sample) magnetic field. Under the action of the pulse of inhomogeneous magnetic field, the frequency of spin precession changes by a value which depends on the position of these spins in the sample. As a result, when the ESE signal appears there is no complete compensation of spin dephasing. This can lead to both a decrease of ESE signal amplitude compared to its value in the absence of the inhomogeneous magnetic field pulse, and a change of the full ESE signal phase. By analysing the behavior of the ESE signal, depending on the parameters of the inhomogeneous magnetic field pulse, one can obtain information about the regularities of the macroscopic spatial distribution of spins throughout the sample.

CW ESR tomography has successfully been used in a number of studies [35, 36]. It is evident, however, that ESE–T has definite advantages, particularly owing to its high resolution ability. Compared to the CW ESR technique, in which the resolution is determined by the relation of inhomogeneous linewidth ΔH^* to the gradient of external magnetic field H_z, in the ESE–T the homogeneous broadening $\Delta H = 1/(\gamma T_2)$ is relevant rather than the inhomogeneous width, which increases the ESE–T resolution by 2–3 order of magnitude. For example, for hydrocarbon radicals, with $\Delta H^* \simeq 0.1 - 1$ mT and $T_2 = 10^{-6}$ s, if the gradient is 10 mT/cm, the ESR–T has a resolution of 10^{-1}–10^{-2} cm, and the ESE–T resolution increases up to $5 \cdot 10^{-4}$ cm. Note also that pulse gradients of a magnetic field are technically easier to create than the one which is stationary for the time necessary to record a CW ESR spectrum.

In ESE–T experiments the magnetic field gradient is directed only along the z axis, coinciding with the direction of the external magnetic field H_0, and according to [34], we experimentally determine the $V(q)$ value:

$$V(q) = V_1(q) - i \cdot V_2(q) \tag{8}$$

where $V_1(q)$ and $V_2(q)$ are the intensities of echo signal components that are phase-shifted by $\pi/2$. This value, $V(q)$, is related to the function of spatial distribution of paramagnetic centers $n(z)$ and parameters a and q that characterize the magnetic field gradient. The gradient is created by Helmholtz coils (connected in opposite directions) through which current $\mathbf{J}$ is passed during time period t. The q value is defined as the charge upon pulse $q = \int_0^{\Delta t} \mathbf{J}(t) \cdot dt$, and for constant current in the pulse (rectangular pulse) $q = \mathbf{J}\,\Delta t$. Parameter a is the empirical constant that connects the magnitude of magnetic induction along z with the current in the pulse. This parameter can be calculated. However, it is more convenient to determine a from a calibration experiment for known q and spatial distribution function of spins along the z direction $n(z)$ of a sample. In this case, one should take into account the peculiarities of coil geometry that create H_z gradient, sample geometry, resonator design, etc. Thus, we experimentally determine $V(q)$, a, q values that are connected via the relation

$$\frac{V(q)}{V(0)} = \int_{-\infty}^{+\infty} n(z)\cdot \exp(-i\cdot aqz)\cdot dz, \tag{9}$$

where $V(0)$ is the ESE signal amplitude in experiments without magnetic field gradient. Inverse Fourier transformation of the experimentally determined function (9) gives the spatial distribution function $n(z)$:

$$n(z) = \frac{a}{2\pi}\cdot \int_{-\infty}^{+\infty}\left(\frac{V(q)}{V(0)}\right)\cdot \exp[i\cdot aqz]\cdot dq. \tag{10}$$

For diffusion in porous spherical granules we can once again use Equation (4) describing the changes of $n(z)$ upon quasihomogeneous diffusion:

$$n(z) = 2\pi \cdot \int_z^R c(r)\cdot r\,dr, \tag{11}$$

$$n(z,t) = 2\pi R^2 c_0 \cdot \left(\frac{1-z^2}{2} + s_2(z,t)\right),$$

$$s_2(z,t) = \frac{2}{\pi^2}\cdot \sum_{n=1}^{\infty}\left(\frac{(-1)^n}{n^2}\right)\cdot \exp\left[-\frac{\pi^2 n^2 D_1 t}{R^2}\right]\cdot [\cos(\pi n z) - (-1)^n], \tag{12}$$

where $z = r/R$, and R is the granule radius, c_0 is the concentration of radical probes at the granule boundary, and D_1 is the effective diffusion coefficient. The diffusion coefficient D_1 can be determined from a comparison of the theoretical dependence for $n(z,t)$ for the different diffusion times to experimental tomograms.

2.3. DETERMINATION OF THE DIFFUSION COEFFICIENTS OF RADICAL PROBES IN POROUS MEDIUM

2.3.1. *ESE–DCI Technique*

ESE measurements were carried out using an X-band ESE spectrometer [32].

TABLE IV

Structural parameters of silica gels and the values of effective diffusion coefficients ($D_1 \cdot 10^5 \, \text{cm}^2 \, \text{s}^{-1}$) of TEMPON radicals in methanol and MTHF filling the pores of different silica gels, determined by ESE–DCI and ESE–T methods

Silica gels	ESE–T (methanol)	ESE–T (MTHF)	ESE–DCI (MTHF)	ϵ/S (nm) $*$ (R) (cm)
KSK-2	2.0 ± 0.5	1.0 ± 0.35	0.85 ± 0.17	3.7 (0.18)
KsK-2,5	–	–	0.57 ± 0.12	2.5 (0.15–0.2)
KSS-3	0.62 ± 0.2	0.18 ± 0.1	–	1.46 (0.2)
KSM-5	0.23 ± 0.05	0.03 ± 0.02	0.03 ± 0.006	0.84 (0.15)
KSM-6S	0.034 ± 0.02	0.006 ± 0.004	0.0014 ± 0.0003	0.47 (0.11–0.15)

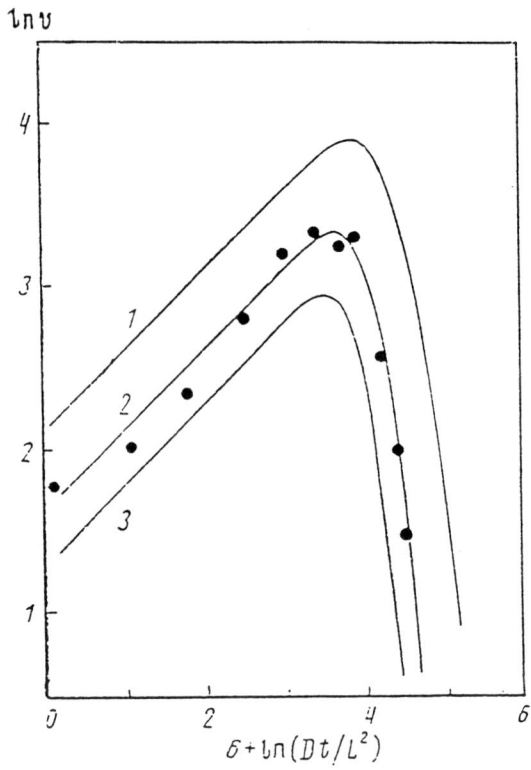

Fig. 7. The calculated (curves) and measured (dots) ESE–DCI results from TEMPON diffusion into glass capillaries permeated with MTHF. Calculations at $c_0\tau$ (cm^{-3} s): (1) $-4 \cdot 10^{13}$, (2) $-6 \cdot 10^{13}$, (3) $-8 \cdot 10^{13}$. Experimental data at $c_0\tau = 6 \cdot 10^{13}$ cm^{-3} s.

Details of the experiments are described in [27, 28]. Spherical granules of silica gels, standard materials for studying adsorbtion, chromatography, and catalysis [37], were used as porous specimens. The data on their structural parameters are listed in Table IV. As model system we studied specially prepared, sealed (at one end) glass capillaries with a channel radius of 10^{-2} cm and length $L = 0.5$ cm.

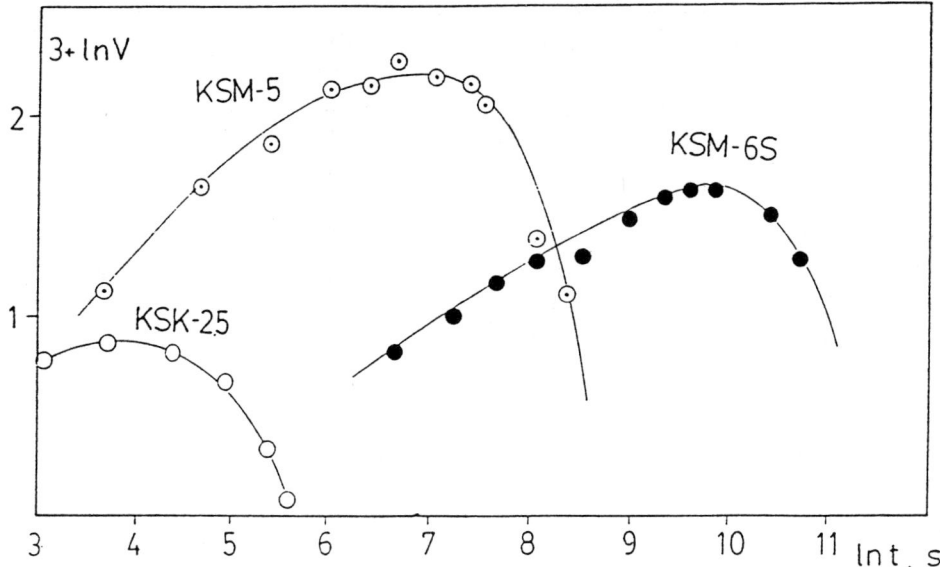

Fig. 8. The calculated and measured ESE–DCI results for TEMPON diffusion into porous spherical granules of different silica gels permeated with MTHF. Calculations and experiment at $c_0\tau = 6 \cdot 10^{13}$ cm^{-3} s.

Relations (5), (6) and (7) were used to calculate the dependences of V on $D_1 t/R^2$ and Dt/L^2, for granules and capillaries, respectively. It is convenient to represent the calculated dependences in $\ln(V)$ vs $\ln(D_1 t/R^2)$ or $\ln(Dt/L^2)$ coordinates, and to compare them with experimental data, given in $\ln(V)$ vs $\ln(t)$ coordinates. When experimental and calculated dependences coincided, the values of either D_1/R^2 or D/L^2 were determined from the value of the shift along the time axis. Figure 7 shows the calculated and measured time dependences of ESE signal amplitude for TEMPON diffusion into MTHF filling glass capillaries. The experimental data are satisfactorily described by the calculated dependence. The coefficient of TEMPON diffusion into MTHF determined from the shift of coordinate axes is $1.8 \cdot 10^{-5}$ cm^2 s^{-1}. Figure 8 depicts the experimental data on the dependence of $\ln(V)$ on $\ln(t)$ for porous spherical granules of silica gel samples. The experimental data in these coordinates are observed to coincide with the calculated dependence for quasihomogeneous diffusion into spherical granules with the effective diffusion coefficient D_1. The values of diffusion coefficients obtained from the position of maxima on the theoretical and experimental dependences are given in Table IV for all systems studied.

2.3.2. ESE–T Method

The tomograms given in Figure 9a and the calculation, involving relations (9–12) make it possible to determine D_1 at different times from the begining of radical probe diffusion into the spherical granule. However, the experimental process can

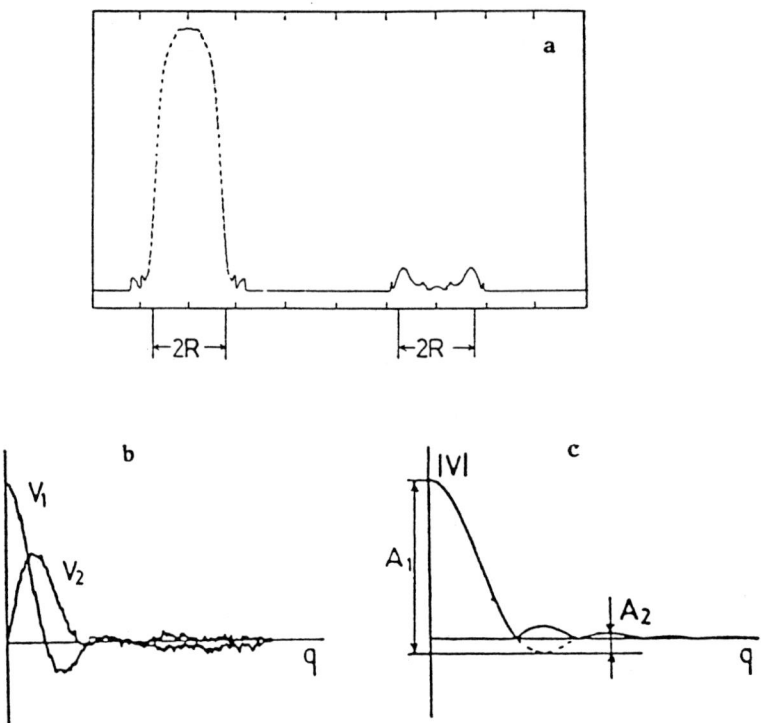

Fig. 9. Radical distribution function along the z axis in silica gel KSM-6S ball granule, permeated with ethanol (a). Right-hand picture: corresponding *radial* distribution calculated with inverse Abel transformation. Diameter of the sample is 2.6 mm, diffusion time 18 min. The dependence of echo signal amplitude on the value of the charge passing through anti-Helmholtz coils per pulse: (b) the experimental components of signals V_1 and V_2 from KSM-5 silica gel with $\ln(D_1 t/R^2) = -1.5$, (c) echo signal module calculated from Equations (9)–(12) with $\ln(D_1 t/R^2) = -1.9$. A dashed line corresponds to the value $(-|V|)$.

be simplified by not using the space distribution function $n(z, t)$, but by direct computations of the experimental functions V_1 and V_2 that describe the dependence of the ESE signal amplitude on parameter q (the field gradient) (Figure 9b). According to (9–12), these are oscillating, damped curves. Figure 9c shows the modulus of ESE signal amplitude $|V| = (V_1^2 + V_2^2)^{1/2}$. The diffusion coefficient D_1 can be determined using the ratio A_1/A_2 between the amplitudes of the first and second maxima rather than the absolute value $|V|$. The result of the calculation of this ratio as the function of parameter $\ln(D_1 t/R^2)$ is given in Figure 10. This calibrating diagram was used to determine D_1 from the experimentally determined A_1/A_2 ratio. The D_1 values obtained by the ESE–T method for different silica gels are also listed in Table IV.

On the basis of the experimental values of diffusion coefficients of radical probes in porous media we have obtained information about the influence of structural parameters of these materials on the effective value of the translational diffusion coefficient D_1. We shall consider silica gel granules as the solid material, with

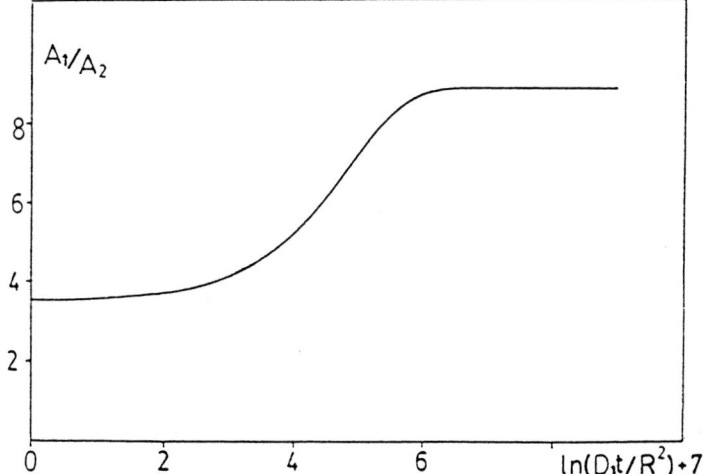

Fig. 10. The calculated dependence of parameter A_1/A_2 on the value of $\ln(D_1 t/R^2)$.

volume pierced by microcapillaries and whose diffusion properties depend on one parameter only, i.e. the mean value of the inlet window diameter of capillary d. For this model, the relation of specific volume of capillaries (pores) ϵ to their specific surface corresponds to the inlet window diameter $\bar{d} = \epsilon/S$ (see Table IV). The latter is the average parameter that characterizes each type of silica gel using independent adsorbtion measurements of ϵ and S. As follows from these data, when the pore diameter decreases, the diffusion coefficient D_1 of TEMPON probe decreases for MTHF faster than for methanol. It can be concluded that translational motion of the radical probe depends not only on the steric parameters of the porous medium but also on the mobility of the probe in the solvent that fills the pores. For the maximum pore diameter in silica gels of the KSK-2 type, the D_1 values are close to the corresponding values of D in pure solvents. A sharp decrease of D_1/D for silica gels KSM-5 and KSM-6S is likely due to the fact that the steric hindrances in the case of small pores can be accompanied by the decrease in the mobility owing to the effective interaction of solvent with pore surface. This is observed for rotational and progressive mobility of nitroxyl radicals in silica gels [38, 39].

3. Conclusions

Some of the applications of ENDOR, ESE and 2 mm-band ESR spectroscopic methods to the investigation of catalytic systems have been considered. Adsorption of quinones on activated H-forms of zeolites yields the paramagnetic complexes with Al^{3+} Lewis acid sites. Combinations of the different spectroscopic methods are extremely effective for their characterization.

Two methods based on the ESE technique have been developed and used to measure the diffusion coefficients of radical probes in silica gels permeated with organic solvents (methanol and MTHF). It was shown that application of the quasihomogeneous diffusion model for porous materials can be used for the description of the probe transport. It was found that effective diffusion coefficients strongly depend on mean value of silica gels pore sizes and on the probe mobility inside the pores.

Acknowledgments

We thank A. V. Astashkin, V. I. Gulin, S. A. Dikanov, D. Goldfarb and A. A. Sukhoroslov for their valuable research contributions to this work, and A. M. Tyryshkin for taking time to read the manuscript and give his comments.

References

1. L. Kevan and L. D. Kispert: *Electron Spin Double Resonance Spectroscopy*, John Wiley & Sons, New York (1976).
2. H. Kurrech, B. Kirste, and W. Lubitz: *Electron Nuclear Double Resonance Spectroscopy of Radicals in Solutions*, VCH Publishers, New York (1988).
3. *Time Domain Electron Spin Resonance*, L. Kevan and R. N. Schwartz (eds), Wiley Interscience, New York (1979).
4. S. A. Dikanov and Yu. D. Tsvetkov: *Electron Spin Echo Envelope Modulation Spectroscopy*, CRC Press, Boca Raton (1992).
5. O. Ya. Grinberg, A. A. Dubinski, and Ya. S. Lebedev: *Russian Chem. Rev.* **52**, 850 (1983).
6. Yu. D. Tsvetkov: in N. D. Yordanov (ed.), *Electron Magnetic Resonance of Disordered Systems*, World Scientific, Singapore, p. 15 (1989).
7. *CRC Handbook of ESR Spectra from Quinones and Quinols*, J. A. Pedersen (ed.), CRC Press, Boca Raton (1985).
8. Landolt-Bornstein: *Numerical Data and Functional Relationships in Science and Technology*, Vol. 9, Springer-Verlag, New York (1980).
9. A. V. Astashkin and R. I. Samoilova: *Zeolites* **11**, 282 (1991).
10. R. I. Samoilova, A. V. Astashkin, S. A. Dikanov, D. Goldfarb, and E. V. Lunina: *Colloids and Surfaces, Part A*, **72**, 29 (1993).
11. G. A. Abakumov, E. S. Klimov, and G. A. Razuvaev: *Izv. Akad. Nauk USSR, Ser. Khim.* **8**, 1827 (1975); *ibid.* **5**, 1199 (1972).
12. C. C. Felix, and R. C. Sealy: *J. Am. Chem. Soc.* **104**, 1555 (1982).
13. D. Goldfarb, J.-M. Fauth, Y. Tor, and A. Shanzer: *J. Am. Chem. Soc.* **113**, 1941 (1991).
14. Yu. D. Pimenov, V. E. Kholmogorov, and A. N. Terenin: *Dokl. Akad. Nauk. USSR* **163**(4), 935 (1965); V. Ya. Lodin, V. E. Kholmogorov, and A. N. Terenin: *ibid.* **160**(6), 1347 (1965).
15. S. L. Kessel, R. M. Emberson, P. G. Debrunner, and D. N. Hendrickson: *Inorg. Chem.* **19**, 1170 (1980).
16. M. E. Cass, N. R. Gordon, and C. G. Pierpont: *Inorg. Chem.* **25**, 3962 (1986).
17. W. B. Mims: *Phys. Rev. B* **6**, 3543 (1972).
18. K. Matar and D. Goldfarb: *J. Chem. Phys.* **96**, 6464 (1992).
19. K. Matar and D. Goldfarb: in N. D. Yordanov (ed.), *Electron Magnetic Resonance of Disordered Systems*, World Scientific, Singapore, p. 237 (1991).
20. N. P. Rodes and R. Rudham: *J. Chem. Soc. Faraday Trans.* **89**(14), 2551 (1993).
21. G. Engelhard and D. Michel: *High Resolution Solid-State NMR of Silicates and Zeolites*, Ch. V, Wiley Interscience, New York (1987).
22. J. B. Vytterhoeven, L. G. Christner, and W. K. Hall: *J. Phys. Chem.* **69**, 2117 (1965).
23. B. S. Prabhananda, C. C. Felix, J. S. Hyde, and A. Walvekar: *J. Chem. Phys.* **83**, 6121 (1985).
24. K. B. Yoon: *Chem. Rev.* **93**, 336 (1993).
25. P. Devaux and H. M. McConnell: *J. Am. Chem. Soc.* **94**, 4475 (1972).
26. S. A. Dzuba, V. I. Popov, V. M. Moralev, and Yu. D. Tsvetkov: *J. Phys. Chem.* (*Russian*) **11**, 2188 (1987).

27. R. I. Samoilova, A. D. Milov, and Yu. D. Tsvetkov: *J. Chem. Phys.* (*Russian*) **9**, 541 (1990)
28. A. A. Sukhoroslov, R. I. Samoilova, and A. D. Milov: *Appl. Magn. Res.* **2**, 577 (1991).
29. N. S. Koshliakov, E. B. Gliner, and M. M. Smirnov: *Basic Differential Equations of Mathematical Physics*, Fismatisdat, Moscow (1962).
30. O. A. Malinovskaya, V. S. Beskov, and M. G. Slinko: *The Modelling of Catalytic Processes on Porous Granules*, Nauka, Novosibirsk (1975).
31. L. M. Heifits and A. V. Neimark: *Polyphase Processes in Porous Media*, Khimia, Moscow (1982).
32. K. M. Salikhov, A. G. Semenov and Yu. D. Tsvetkov: *Electron Spin Echo and its Applications*, Nauka, Novosibirsk (1976).
33. A. D. Milov and A. Yu. Pusep: *USSR Patent No.* 1105794 (1984).
34. A. D. Milov, A. Yu. Pusep, S. A. Dzuba, and Yu. D. Tsvetkov: *Chem. Phys. Lett.* **119**, 421 (1985).
35. F. Demsar, T. Walczak, P. D. Morse, G. Bacic, Z. Zolnai, and H. M. Swartz: *J. Magn. Res.* **76**, 224 (1988).
36. V. Ewert, T. Herling, and W. Schneider: *Exper. Techn. Phys.* **36**, 289 (1988).
37. N. V. Keltsev: *The Basis of Adsorption Technique*, Khimia, Moscow (1984).
38. H. L. Weissberg: *J. Appl. Phys.* **34**, 2636 (1972).
39. I. I. Ioffe, V. A. Rechetov, and A. M. Dobrovolskii: *Heterogeneous Catalysis*, Khimia, Leningrad (1985).

Theoretical Studies of Core Ionization, Excitation and De-excitation of Adsorbates

MASAHIDE OHNO
Institute de Chimie B6, Sart Tilman, B4000 Liege 1, Belgium

(Received: 15 December 1993; accepted: 5 July 1994)

Abstract. The change of the local electronic structure of the adsorption site is manifested in the XPS, XAS and DES spectra of a molecule adsorbed on a metal surface. Based on recent molecular orbital many-body calculations of core hole spectra of single metal molecules such as NiCO, a systematic interpretation of the core ionization, excitation and de-excitation processes of adsorbates is given.

Key words. Core ionization, excitation, de-excitation adsorbates, many-body calculations

1. Introduction

Since the first X-ray photoelectron spectroscopy (XPS) observation of extra core hole (O $1s$) shakeup features of adsorbed molecules, namely CO and O_2 on Ru(001), which are not seen in the spectra of the free molecules [1], several studies have been performed on the increased satellite intensity from coordinated molecules, both adsorbed on surfaces and in transition metal (TM) carbonyl complexes [2–7]. A very strong satellite, 'giant satellite', was observed as a broad feature at about 5 to 6 eV above the lowest energy peak. The first theoretical model which could explain the existence of giant satellites in adsorbate XPS spectra was mainly developed by Schönhammer and Gunnarsson (SG) [8–11]. The model is an extension of that by Kotani and Toyozawa, who introduced an Anderson model picture for screening in the core hole spectra of La and Ce metals [12]. The basic idea of the Anderson impurity model is to describe the effects of the interaction of a correlated localized state with a noncorrelated band. In this picture a local $4f$ level is unoccupied but becomes 'pulled down' below the Fermi level by the core hole potential so that there is a finite probability of the $4f$ level becoming occupied in the final state. In the static limit of the SG model, extramolecular screening of an adsorbate core hole is caused by CT (charge transfer) from the substrate metal band to a previously unoccupied adsorbate valence level (e.g. 2π in CO or N_2) which is pulled below the Fermi level by the Coulomb potential of the core hole. The total energy of the final state will thus be lowered if an electron is transferred from the Fermi level to the screening level. The probability of transferring an electron to the screening level becomes larger as the coupling of the level to the substrate is increased, hence transferring spectral weight from the 'unscreened' peak having an unoccupied screened level to the 'well screened' peak with an occupied screened level. For the XPS spectra of CO/Cu, SG parametrized [13] the coupling between the adsorbate and substrate in terms of metal sp and d DOS. The screening orbital is the same and the position of the peaks are related to the energy-level positions in the metal from which the electron is transfered. They concluded that the shape of the valence DOS of the metal substrate can dramatically influence the form of the XPS spectrum of the adsorbate.

Linear model molecules, such as NiCO and NiN$_2$, have been used as model systems to study the core hole XPS spectra of CO(N$_2$) on a Ni metal surface [14–22]. Bagus and Hermann (BH) performed an *ab initio* Hartree–Fock (HF)ΔSCF calculations of NiCO and NiN$_2$ to determine the main line satellite line energy separations and spectral intensity ratios [14]. The role of the 2π orbital in the CT screening by a molecular orbital (MO) approach is as follows. The 2π orbital splits into a bonding and an antibonding orbital with respect to the metal by mixing with metal valence electrons. In the neutral ground state the (occupied) bonding orbital will mostly have metal character, while the presence of a core hole in the adsorbate will enhance the CO character to this orbital, thus the CT to fully screen the core hole in the molecule. BH interpreted the main line as a 'screened state' (shakedown) and the satellite as an 'unscreened state' (Koopmans state, KT state) because of a fairly large CT in the main line state [14]. However, Saddei *et al.* [18] analyzed the BH results and pointed out that the lowest energy state is the KT state and the satellite line is due to the metal $4s$ excitation. BH used an incorrect ground state to boost the main line intensity.

An instance of considerable disagreement in the interpretation of the spectral features is that of N$_2$ adsorbed on Ni. Fuggle and Menzel interpreted the splitting at lower B.E. as resulting from two different N atoms which are present, if the N$_2$ is adsorbed 'head on' coordinated to the metal atom [23]. Umbach employed the SG method and concluded that it is mainly the substrate valence DOS which determines the core hole spectrum profile [24]. He interpreted the lowest energy state as the screened state and the splitting as due to the CT screening from different substrate bands (band structure effect). However, the GVBCI (generalized valence bond configuration interaction) and CNDO (complete neglect of differential overlap) Green's function calculations show that the splitting is attributed to the inequivalent N atoms [18, 20]. Furthermore, the interpretation by Umbach was criticized by Freund *et al.* [20]. They pointed out that collections of the N $1s$ core hole spectra of N$_2$ adsorbed on different kinds of TM surfaces are very similar, although the DOS differ considerably from each other. The assignment of peak splitting by the inequivalent N atoms was also supported by Egelhoff [25] who provided independent experimental evidence by determining the angular dependence of the intensities of the split lines. Moreover, a recent high resolution photoelectron diffraction spectroscopy study of N$_2$/Ni(100) separated the N $1s$ XPS spectrum into the inner N(N$_a$) and outer N(N$_b$) atom ionized spectra and gave decisive evidence [26]. Recently the SG model used for CO/Cu was criticized by Lovric *et al.*, who claimed that the SG model used the parameter values which describe an unphysically strong adsorbate/substrate interaction [27]. These are certainly not in favour of the SG model, which has enjoyed a vast popularity during the last decade.

Despite substantially different interpretations provided by different approaches, until recently there does not seem to have been any further experimental and theoretical attempt to clarify the discrepancies. Recently the XPS core spectra of

CO/Ni(100) and N_2/Ni(100) systems were remeasured at high resolution [26, 28]. The spectra show a number of newly resolved satellite lines which had not previously been theoretically predicted. A large part of the spectral intensity is found to go to the shakeup/off satellite lines, extending over more than 75 eV, showing that inclusion of the most prominent satellite peak is not sufficient for the spectral intensity determination. Previous molecular approaches failed to give a proper description of these spectral features. This does not imply that the small cluster model of the adsorption site is inadequate. On the contrary, even single metal atom clusters have been extensively and successfully employed to model 'on top' chemisorption of small molecules and reproduce well the qualitative aspects of both the local electronic structure of the adsorption site and the behavior of the ground state properties [29, 30]. Moreover, the core hole spectra of closed shell molecular carbonyls show a close resemblance to the core hole spectra of the adsorbate [28, 31, 32]. So there appears to be significant evidence of a strongly localized nature for the excitations in the adsorbates and the adequacy of a small molecular cluster approach for a description of their basic features. Inadequate reproduction of the experimental features, or even different interpretations afforded by older cluster approaches, may well be attributed to limitations of the scheme employed, notably insufficient treatment of many-body effects in both ground and ionized states, rather than to an intrinsic fault of the finite cluster model. Of course, size effects may be important, and may be systematically examined by enlarging the cluster employed, although this is computationally expensive.

For the incorporation of many-body effects, it is an advantage of the many-body techniques such as the Green's function method and CI method beyond the simple SCF MO approach, e.g. the ΔSCF method, that one can deal simultaneously with local metal, local ligand and metal-ligand CT excitations. Notable attempts to include many-body effects have been the case of the GVB-CI approach which is, however, better suited to the description of the ground state properties [19, 20], and a Green's function method calculation (the diagrammatic two particle-hole Tamm–Dancoff approximation) based on a semiempirical CNDO Hamiltonian [18] which is hindered by inaccurate matrix elements and eigenvalues.

Recently, extensive systematic *ab initio* MO Green's function calculations of the valence photoemission spectra of NiCO, NiN_2, PdCO, PtCO, $NiNH_3$, $NiPF_3$, TM carbonyls and TM nitrosyls were performed, obtaining good agreement with experiment [33]. Moreover, extensive and systematic *ab initio* MO CI calculations of the XPS core spectra of NiCO, NiN_2, PdCO, Pd_2CO, PdN_2 and $NiNH_3$ were also performed [34–37]. The calculations provide a fairly good description of the spectral features, including the newly resolved satellite features. The CI calculations were performed at the level of the $3h2p$ (three hole-two particle) CI scheme. This scheme includes the metal–ligand CI excitations as additional excitations to the $2h1p$ configurations, e.g. the many-body effects such as the $2h2p$ double excitations in the presence of a core hole, the relaxation of two holes and one particle and the screening of the $1h1h$ and $1h1p$ interactions in the $2h1p$ configurations.

In the present article, by referring to our publications for the details of the method, numerical results and detailed discussions on the spectral features of the

XPS core spectra of various adsorbate/substrate systems [34–37], we focus on the cooperative core-hole screening mechanism which provides a systematic description of strong adsorbate XPS core spectral feature changes due to variations of the adsorbate/substrate systems. The mechansim is based on the s–$d\sigma$ hybridization and promotion mechanism of local metal configurations which also plays an important role in the metal-ligand (ML) bonding.

Recently, angle resolved near edge X-ray absorption fine structure (NEXAFS) spectroscopy or X-ray absorption spectroscopy (XAS) has been used to study molecular geometry changes upon chemisorption and the electronic structure of the adsorbed molecules through the transition from the adsorbate core to the low lying unoccupied orbitals [38, 39]. XAS is used to probe the unoccupied states in the presence of a core hole. In the case of chemisorption the empty states into which the core electron is excited are modified by the hybridization between the adsorbate and the substrate in the final state. In the present work we discuss relaxation in the resonantly excited state based on the recent CI calculations of the excitation energies and the oscillator strengths [40].

We then proceed to a discussion on the de-excitation (core hole decay) of the resonantly core excited state, namely the spectral features of de-excitation electron spectroscopy (DES) (or resonant photoemission, resonant Auger emission, autoionization) spectra of adsorbates. The resonantly excited state is neutral and may decay by autoionization, namely participant decay and spectator decay. In the former, the initially excited electron participates in the decay, leading to a single hole configuration, identical to the main line state observed by photoemission spectroscopy (PES), while in the latter the excited electron remains in the final state and results in the $2h1p$ state, corresponding to the satellites obtained in the PES spectrum. For chemisorbates, the strong coupling with the substrate allows for CT on the timescale of the core excitation (ionization) process (10^{-17} s), leading to a complete screening of the core hole before its decay starts. So the DES spectrum becomes almost identical to the normal Auger spectrum [41–45]. However, for weakly coupled physisorbed systems, the CT is much slower and comparable to the core hole lifetime (10^{-15} s). The CT may go to the substrate (charge delocalization) if the lowest core hole state is ionic, and from the substrate if the lowest state is neutral. Then the CT (charge delocalization) will only be manifested in the autoionization and Auger spectrum respectively [46–48]. In the present article we focus on the competition between the CT and the core hole decay, and its influence on the autoionization spectrum features [49].

2. Core Hole Screening

2.A. BINDING MECHANISM

The binding of the adsorbate/substrate systems such as CO/Ni and N_2/Ni is usually divided into a σ part and a π part [50]. The σ bonding is represented by a donation of electrons from the adsorbate to empty orbitals of σ character on the TM atom. The π bonding is produced by the back-donation from the filled metal $d\pi$ orbital into the empty π^* orbital of the adsorbate. These two types of bonding are optimised at different bond distances, resulting in a negatively charged metal atom. At short distances the σ interaction becomes less attractive, until finally the repulsive part takes over due to the contact between the lone pair and the d shell. The π bonding becomes more efficient and the metal atom begins to become positively charged. The π bonding leads to stronger bond than the σ interaction so that in systems like CO/Ni, where both types of interaction are possible, π bonding will dominate. Near-degeneracy effects play an important role in the description of the bonding mechanism and also that of the core hole screening [30, 34–37, 51, 52].

The most important near degeneracy effect arises as a consequence of an efficient hybridization between the $4s$ orbital and the singly occupied $3d$ electron of the same symmetry which are singlet coupled. This is possible for the $^1\Sigma^+$ state which is the ground state of NiCO (NiN$_2$, PdCO, PdN$_2$) and is dominated by a closed shell configuration. In this configuration the hybridization is between $d\sigma$ and s, the orbitals being essentially $\sigma = s - d\sigma$ hybridized perpendicular to the Ni—C bond and $\sigma^* = s + d\sigma$ along the bond. The σ hybrid has an increased density in the plane perpendicular to the Ni—C bond and a decreased density in the Ni—C bond region. By increasing the occupation of the σ hybrid, the σ repulsion between the ligand and metal, the magnitude of which is directly related to the amount of the valence $s\sigma(sp\sigma)$ character on the adsorption site metal atom, can be reduced ($s - d\sigma$ hybridization [51, 52]). In addition, the $3d\sigma$ orbital is much more compact than the $4s$ and the metal occupation d^{n+2} can mix in with $d^{n+1}s^1$; this mixing of the spatially compact d^{n+2} occupation greatly reduces the repulsion ($s - d\sigma$ promotion [51, 52]). In NiCO, where Ni is strongly perturbed by a ligand, the hybrid pointing away from the ligand (σ) tends to be doubly occupied, with the other hybrid empty (σ^*). The dominating σ^2 configuration effectively shuffles away the σ electrons from the bonding region, thereby allowing for the buildup of a strong π overlap. The smaller σ repulsion in the $^1\Sigma^+$ state is due to the hybridization described above, and makes it possible for the molecule to come close to the Ni atom. In NiCO, the $3d$ orbitals are strongly involved in the bonding to ligands, which does not mean, however, that the $3d$ population changes significantly. In NiCO the mixture of s and $d\sigma$ in the hybrid orbitals puts more weight on d in the strongly occupied σ orbitals giving total $3d$ and $4s$ population of 9.2 and 0.5, respectively [51]. This agrees well with the $3d$ population of bulk Ni metal (9.4). According to a recent cluster calculation of the XPS and XAS spectra of Ni metal, a ground state with weights of 14.3% d^8, 48.1% d^9 and 37.6% of d^{10} gives good agreement with experiment ($d = 9.23$) [53]. The loss of a $4s$ electron is mostly a CT effect and, to a lesser extent, is due to the mixing with the d^{10} state, which should also have increased the d population.

2.B. CORE HOLE SCREENING

In Tables I and II we summarize the experimental and theoretical XPS core hole spectra of CO/Ni(100) (NiCO), N$_2$/Ni(100) (NiN$_2$) and PdCO. In Table III we summarize the Mulliken population analysis of the ground and ionized states of NiN$_2$ and PdCO. For the interpretation of the core hole spectra of adsorbates, we turn to a cluster model since it embraces most of the relevant physics and is helpful in discussing the qualitative features of the *ab initio* MO calculation results

TABLE I

Theoretical core hole spectra of NiCO and NiN$_2$ by the 2h–2p/3h2p CI method and experimental spectra of CO and N$_2$ adsorbed on a Ni metal surface

Level	Exp		Theory (3h2p CI)			
	E (eV)	I	E (eV)	I	I_{rel}	Configuration
1σ	0.	0.36	0.	0.370	100.	$1s^{-1}$
O$_{1s}$	5.5		1.59		1.2	$\pi - \pi^*$
	8.5		5.83	0.103	28.0	$\sigma - \sigma^*$
	15.0		8.94	0.090	24.3	$\pi - \pi^*$
	26.0	0.64	9.71		7.3	$\pi - \pi^*$
	36.0		10.23		3.9	$\pi - \pi^*$
	45.0		10.87		0.1	$\sigma - \sigma^*$
	55.0		11.45		4.4	Same
			12.63		0.3	$\delta - \delta^*$
			12.71		0.8	$\delta - \delta^*$
			12.86		0.3	$\pi - \pi^*$
2σ	0.	0.29	0.	0.392	100.	$1s^{-1}$
C$_{1s}$	2.1		2.03		4.8	$\pi - \pi^*$
	5.5		7.15		1.1	$\sigma - \sigma^*$
	9.5	0.71	7.34	0.147	37.6	$\sigma - \sigma^*$
	33.0		8.69		1.3	$\pi - \pi^*$
			9.32		0.3	$1\pi - \pi^* + \pi - \pi^*$
			10.23		14.5	$\pi - \pi^*$
			10.45		12.7	Same
			11.51		1.5	$1\pi - \pi^*$
			11.98		0.1	$\pi - \pi^*$
			12.29		1.2	$1\pi - \pi^* + \pi - \pi^*$
			12.95		0.6	$\sigma - \sigma^*$
Inner N$_{1s}$	0.	0.16	0.	0.1557	42.95	$1s^{-1} + \pi - \pi^* + \pi^2 - \pi^{*2}$
	5.3		1.88		0.6	$\pi - \pi^* + \pi^2 - \pi^{*2} + 1s^{-1}$
			5.85		6.98	$\sigma - \sigma^*(+\sigma\pi - \sigma^*\pi^*)$
		0.84	5.93	0.3625	100.	$\sigma - \sigma^* + \pi - \pi^*(+\sigma\pi - \sigma^*\pi^*)$
			9.15		0.14	
			10.09		0.68	$\pi - \pi^*$
			11.75		0.30	$\sigma - \sigma^* + \pi - \pi^*$
	15.0		12.92		0.33	$\sigma - \sigma^*$ + double exc.
			13.28		0.07	Same
			13.30		0.04	$\sigma - \sigma^*$
						$\sigma - \sigma^*$
Outer N$_{1s}$	0.	0.11	0.	0.1354	36.74	$1s^{-1} + \pi - \pi^* + \pi^2 - \pi^{*2}$
	2.1		3.08		0.89	$\pi - \pi^* + \pi^2 - \pi^{*2} + 1s^{-1}$
	5.8		6.92	0.3685	100.	$\sigma - \sigma^* + \sigma\pi - \sigma^*\pi^*$
	8.5	0.89	9.80		3.88	$\sigma - \sigma^* + \pi - \pi^*$
			11.63		1.16	$\sigma - \sigma^* + \pi - \pi^*$
			12.99		0.52	$\sigma - \sigma^* + 1\pi - \pi^*$
	15.0		13.37		0.53	$\pi - \pi^* + \pi^2 - \pi^{*2}$
			13.88		0.21	$\sigma - \sigma^* + 5\sigma - \sigma^*$
			14.46		0.18	$\pi - \pi^*$

CORE IONIZATION, EXCITATION AND DE-EXCITATION

TABLE II
Theoretical corehole spectra of PdCO calculated by the $2h2p/3h2p$ CI method

Level	Theory ($3h2p$ CI) at 3.65 au			
	E (eV)	I	I_{rel}	Configuration
1σ	0.0	0.5023	100.0	$1s^{-1}$
O_{1s}	4.31		0.2	$\pi - \pi^*$
	8.03	0.0533	10.6	$\sigma - \sigma^*$
	8.63	0.0684	13.6	$\pi - \pi^* (\sigma - \sigma^*)$
	11.30		0.5	$\sigma - (n)\sigma^*$
	12.68		0.2	Same
	12.78		0.5	Same
	13.48		1.0	$\pi^2 - \pi^{*2}$
2σ	0.0	0.5469	100.0	$1s^{-1}$
C_{1s}	4.81		1.1	$\pi - \pi^*$
	9.08	0.1274	23.3	$\pi - \pi^*$
	11.54		1.3	$1\pi - \pi^*$
	12.32		0.2	$\pi^2 - \pi^{2*} + 1\pi^2 - \pi^{2*}$
	14.63		0.2	$\sigma - \sigma^*$

TABLE III
Mulliken population analysis of the ground, core ionized and core excited states of NiN$_2$ and PdCO

State (NiN$_2$)	$3d\sigma$	$3d\pi$	$3d\gamma$	Total $3d$	$4s$
Ground state SCF	0.7556	3.9494	4.0	8.705	1.496
Ground state $2h2p$ CI	0.8812	3.9128	4.0	8.794	1.346
N_a^{-1} main	1.9072	3.4491	4.0	9.3562	0.183
π satellite[a]	1.9145	2.8175	4.0	8.732	0.171
σ satellite	1.0636	3.416	4.0	8.4791	0.993
N_b^{-1} main	1.9223	3.2492	4.0	9.1715	0.155
π satellite[a]	1.9208	2.7716	4.0	8.6924	0.149
σ satellite	1.0579	3.2738	4.0	8.3317	0.955
Ground state $2h2p$ CI[b]	1.1511	3.8784	4.0	9.0295	0.880
$N_a^{-1} 1s^{-1}$ (SCF)	1.9054	3.4376	4.0	9.343	0.170
$1s^{-1} \to \pi^*$ (SCF)	0.6738	3.9832	4.0	8.657	1.666
$1s^{-1} \to \pi^*$ ($2h2p$ CI)	0.7868	3.9602	4.0	8.747	1.484
$N_b^{-1} 1s^{-1}$ (SCF)	1.9236	3.2914	4.0	9.215	0.158
$1s^{-1} \to \pi^*$ (SCF)	1.1354	3.9156	4.0	9.051	1.000
$1s^{-1} \to \pi^*$ ($2h2p$ CI)	1.0888	3.9452	4.0	9.034	0.920

State (PdCO)	$4d\sigma$	$4d\pi$	$4d\gamma$	Total $4d$	$5s$
Ground state SCF	1.8502	3.8068	4.0	9.657	0.032
Ground state $2h2p$ CI	1.8272	3.7648	4.0	9.592	0.085
C $1s^{-1}$ main	1.963	3.184	4.0	9.147	0.059
π satellite[a]	1.9528	2.9072	4.0	8.860	0.084
π satellite[c]	1.7084	3.3906	4.0	9.099	0.265
O $1s^{-1}$ main	1.8872	3.1948	4.0	9.082	0.031
π satellite[a]	1.9898	2.8932	4.0	8.883	0.079
σ satellite	1.168	3.547	4.0	8.715	0.833
π satellite[c]	1.6902	3.4368	4.0	9.127	0.146

[a] 2 eV π satellite (NiN$_2$), 4.3(4.8) eV π satellite (PdCO).
[b] Ground state used for the calculation of the XAS excited states.
[c] 8.6(9.1) eV π satellite (PdCO).

and the physical origin of the many-body effects which enter into the problem of the core ionization in adsorbates. As the orgin of many-body effects we consider the local metal $s - d$ population changes in the metal–ligand systems, namely the $s - d\sigma$ hybridization and promotion mechanism ($s - d\sigma$ mechanism), by which the d population increases and the s population decreases [34–37, 54]. The ground state local metal configuration consists of $d^{10}(d\sigma^2 d\pi^4 d\delta^4 s^0)$, $d^9 s^1 (d\sigma^1 d\pi^4 d\delta^4 s^1)$ and $d^8 s^2 (d\pi^4 d\delta^4 s^2)$ configurations. With respect to d^{10}, the energies of $d^9 s^1$ and $d^8 s^2$ are Δ_σ and $2\Delta_\sigma + U$, respectively. Here $\Delta_\sigma = E(d^9 s^1) - E(d^{10})$ is the $s - d\sigma$ excitation energy ($\Delta_\sigma = -1.71$ and 1.07 eV for Ni and Pd atoms, respectively [55]) and U is the effective Coulomb $d - d$ interaction (about 5 eV). For the Ni–ligand systems (NiCO, NiN$_2$), the local metal configuration are low-lying, in the order of $d^9 s^1$, d^{10} and $d^8 s^2$, while for the Pd ligand (PdCO), they are in the order of d^{10}, $d^9 s^1$ and $d^8 s^2$. This explains why the weight of d^{10} increases for the Pd–ligands and decreases for the Ni–ligands, in other words, the larger s population for the Ni–ligands than for the Pd–ligands. The contribution of $d^8 s^2$ to the ground state is rather small but not negligible.

The consequence of the interaction between d^{10} and $d^9 s^1$ by the $s - d\sigma$ mechanism is the forming of bonding and antibonding, σ and σ^* orbitals, which both consists of admixtures of these two configurations. The π back-bonding in the ground state may result in $d^9 (d\sigma^2 d\pi^3 d\delta^4)L$ and/or $d^8 (d\sigma^1 d\pi^3 d\delta^4)s^1 L$ (here L denotes a π^* ligand electron). The d^{10} and $d^9 L$ configurations are separated by the πCT energy $\Delta_\pi = E(d^9 L) - E(d^{10})$ (about 5 eV for CO/Ni). The consequence of the interaction between $d^{10}(d^9 s^1)$ and $d^9 L$ ($d^8 s^1 L$) is the forming of bonding and antibonding π, π^* orbitals which both consists of admixtures of these two configurations.

Upon ionization of the ligand core, both π and σ relaxation may occur. The π relaxation leads to an overall $d\pi$ (metal) to π^* (ligand) CT. The $\mathbf{c}d^{10}$ ($\mathbf{c}d\sigma^2 d\pi^4 d\delta^4$) and $\mathbf{c}d^9 L$ ($\mathbf{c}d\sigma^2 d\pi^3 d\delta^4 L$) are separated by $\Delta_\pi - Q$. Here $\mathbf{c}$ denotes a core hole and Q is the effective core hole $- \pi^*$ electron Coulomb interaction. In the final state the interaction between $\mathbf{c}d^{10}$ and $\mathbf{c}d^9 L$ gives rise to bonding and antibonding combinations. The energy difference is decreased, which results in an increased hybridization in the final state. As the π bonding is mainly responsible for the metal-ligand (ML) bonding, the ML coupling strength increases in the core ionized state (more exactly, the π CT main line state) than in the neutral ground state. The *ab initio* MO CI calculations [34–37] show that the π relaxation leads to a significant $d\pi$ loss by the π CT, but to a modest relaxation of the bonding π orbitals. Although the relaxation following core hole formation leads to a full screening and a whole unit of electron transfer in the MO approach, this is accomplished by a modest change in the ligand participation in the bonding orbitals. Thus, according to the π CT model, the lowest energy state is expected to take the largest spectral intensity and is close to the Koopmans theorem (KT) state.

In the conventional π CT screening model employed so far, the role of σ relaxation ($s - d\sigma$ mechanism) in both ground state and core ionized state has not been realized and completely neglected. We consider the σ correlation. In this case we consider four different local metal configurations, $\mathbf{c}d^{10}$, $\mathbf{c}d^9 L$, $\mathbf{c}d^9 s^1$ and $\mathbf{c}d^8 (d\sigma^1 d\pi^3 d\delta^4)s^1 L$, the energies of which are $-E_c$, $-E_c + \Delta_\pi - Q$, $-E_c + \Delta_\sigma$

and $-E_c + \Delta_\pi + \Delta_\sigma + U - Q$, respectively. Here E_c is the core ionization energy. When the π CT is small, in the final state, $\mathbf{c}d^{10}$ and $\mathbf{c}d^9s^1$ will dominate. The energy difference ($s - d\sigma$ promotion energy) between $\mathbf{c}d^{10}$ and $\mathbf{c}d^9s^1$ is the same as that between d^{10} and d^9s^1. When a core hole is created, the core hole attraction does not lead to any change in the ordering of the configuration energy. Thus, unless the $s - d\sigma$ hybridization strength in the final state changes, all strength should go to the main line state (no satellite lines). By contrast, when the π CT is larger in the final state, $\mathbf{c}d^9(d\sigma^2 d\pi^3 d\delta^4)$L and $\mathbf{c}d^8(d\sigma^1 d\pi^3 d\delta^4)s^1$L dominate. The energy difference between $\mathbf{c}d^9$L and $\mathbf{c}d^8s^1$L is $\Delta_\sigma + U$, while that between the d^{10} and d^9s^1 in Δ_σ. For the Ni ligands Δ_σ is negative, while $\Delta_\sigma + U$ is positive. Thus the energy ordering of the final ionic state local metal configurations is opposite to that of the initial state one. For the Pd-ligands both Δ_σ and $\Delta_\sigma + U$ are positive and the energy ordering does not change. For the Ni ligands, when a core hole is created, the core hole attraction leads to a change in the configuration energy ordering. $\mathbf{c}d^9$L is pulled down below the $\mathbf{c}d^8s^1$L. For both ligands, the main line state (lowest energy state) corresponds to a final state which consists of the low lying $\mathbf{c}d^9$L, while the satellite state corresponds mainly to a final state with a mixture of $\mathbf{c}d^8s^1$L and $\mathbf{c}d^9$L. In other words to allow the core ionized state to be bound to a metal atom, it is important to include electronic relaxation processes on the metal atom. The change of the energy ordering, the electronic relaxation on the metal atom, results in a smaller overlap between the ground state and the core ionized state, leading to a large satellite intensity [54]. The rule 'no satellite–no correlations' is quite true here, too. Note that if there is not sufficient π CT, the energy ordering of the local metal configuration in the ionized state does not change from that in the ground state.

We now have to consider how in the case of Ni-ligands the low lying d^9 $(d\sigma^1 d\pi^4 d\delta^4)s^1$ turns into $\mathbf{c}d^9(d\sigma^2 d\pi^3 d\delta^4)$L. Upon ligand core ionization, d^{10} and d^9s^1 turn into $\mathbf{c}d^9$L and $\mathbf{c}d^8s^1$L, respectively, by losing one $d\pi$ electron. The $s - d\sigma$ mechanism renders the latter into $\mathbf{c}d^9L$. This can be seen as the σ shakedown process. The d^8s^2 turns into $\mathbf{c}d^7s^2$L by losing one $d\pi$ electron and becomes $\mathbf{c}d^8s^1$L. The latter eventually becomes $\mathbf{c}d^9$L. The s–$d\sigma$ mechanism is associated with the s population and the local metal configuration changes. The local metal σ relaxation depletes the s population by the s–$d\sigma$ mechanism. For a small d population system (Ni-ligands) where d^9s^1 (and d^8s^2) is more dominant than d^{10}, the s–$d\sigma$ mechanism compensates the loss of d electron to the ligand by the π CT. Thus the d population is increased, despite large π CT, and the metal configuration in the lowest energy state becomes close to d^9. One may also see this in a reverse way; without the s–$d\sigma$ mechanism, there is no π CT screening. So there is a cooperative mechanism between the σ local metal relaxation and the π relaxation. For the Ni-ligands, the s–$d\sigma$ mechanism induces a large relaxation in the σ (and σ^*) orbitals because the mechanism changes the energy ordering of $\mathbf{c}d^9$L and $\mathbf{c}d^8s^1$L. For NiN_2, as the s population is large and the d population is small, the s–$d\sigma$ mechanism is energetically less expensive and the change in the metal σ orbital upon core ionization becomes so significant that it gives almost equal overlap with σ and σ^* relaxed orbitals, although the relaxation of other orbitals (e.g. π orbitals) is found to be modest and comparable to the case of NiCO and PdCO. Because of this, the overlap between the Koopmans and ΔSCF relaxed $1h$ configuration is very small.

This is the main reason for a small intensity of the lowest energy state, which is still dominated by the relaxed $1h$ configuration. The s–d population of the lowest energy state is very close to that of the ΔSCF relaxed core hole state. For NiCO (CO/Ni) the lowest energy state is still close to the KT state and takes the largest spectral intensity. For both Pd and Ni ligands, the π CT from d^9s^1 will be more expensive than that from d^{10} by U. So the π relaxation is mainly associated with d^{10}. For the Pd ligands, in the ground state d^{10} is much more dominant than for the Ni ligands, so the π relaxation dominates, leading to cd^9L in the lowest energy state. As the π relaxation is modest, the lowest energy state takes the largest spectral intensity and close to the KT state. This is in accord with the notion that the local metal configuration energy ordering does not change in the presence of a ligand core hole [54].

The XPS core spectra of CO/H/Ni system (top site) show a decrease in the main line intensity in comparison to that of CO/Ni (top site) [56]. This is possibly due to an increase of the 5σ-metal repulsion, as a result of reduction in the CO 5σ to metal $(sp + d)$ σ donation (an increase in metal $sp\sigma + d\sigma$ character) [54]. This is caused by the indirect interaction between H and CO by which valence electron occupation is increased [57]. With an increase of the coordination number, the spectra of CO/H/Ni system show an increase in the main line intensity, reflecting the decrease of $sp\sigma$ occupation [54].

2.C. GIANT SATELLITE

For N_2/Ni, the 5 eV giant shakeup satellite takes the largest spectral intensity and becomes the main line. This state is a $2h1p$ state with respect to the lowest energy state, characterized by an additional σ to σ^* excitation; however, this is close to the KT state (unrelaxed $1h$ state) because the relaxation of π orbital is very modest. It is the state which, expanded by the ground state orbitals, contains the largest weight of the KT state, although the latter becomes so widely spread over several states that it does not dominate any single state. Oversimplifying, one can say that the KT configuration redistributes itself equally between the σ^2, $\sigma^1\sigma^{*1}$ and σ^{*2} configurations, although the latter becomes washed out by the CI in the higher energy tail of the system. Clearly no single state is dominated by the KT configuration. In this sense, there is no clearly defined 'main line' or KT state, in the spectrum. This may be considered as an example of complete breakdown of the one-electron picture, well known in the inner shells of atoms and in the inner valence region of molecules [58, 59], but very unusual in the core spectra of first row atoms. In fact the mechanism here is quite different, being 'assisted' by the metal atom, which provides the strong relaxation associated with the $3d$, $4s$ quasi-degeneracy. As it is closer to the KT state, it also appears legitimate to talk of this state as the 'unscreened' state. If the s–$d\sigma$ mechanism is hampered by an increase in the d population (a decrease of the s population), the 5 eV giant shakeup state should take the 'whole' spectral intensity (one may consider the 'main line' state as the σ shakedown state). This appears to be the case with the XPS spectrum of N_2 adsorbed on Ni–Ti alloy where only the 'unscreened' state is observed at 405 eV (5 eV giant satellite energy position) [60]. The $2p$ XPS spectra of Ni alloys show the sharp decrease in Ni 6 eV satellite intensity (which is assigned to $2p^{-1}3d^9$) due to the Ni $3d$-band filling [61]. In the case of CO/Ni, the 5 eV satellite is σ to σ^* shakeup satellite with respect to the KT-like main line state. The 5 eV satellite of CO/Ni and N_2/Ni has been interpreted by others as the CT shakeup state [14] or the KT state [19, 20, 24] (or for CO/Ni local metal $4s$ excitation according to the analysis of the results of Reference [14] by the authors of Reference [18]) or the π screening orbital to Rydberg-like level (3π) shakeup [26, 28]. For CO/Pd the σ shakeup satellite disappears because of much less d^8s^2 and d^9s^1 in the ground state in comparison to the Ni ligands. This is expected to be the case with CO/Pt, too. The composition of the final ionic state is quite similar among NiCO, NiN_2 and PdCO. Therefore the pronounced core hole spectral feature changes are due to the slight differences in the local metal s–d populations.

2.D. π SATELLITE

For the π CT shakeup, the excitation from d^{10} is favoured. Consequently, the spectral weight of the π CT shakeup satellite observed at 2.1 and 8.5 eV above the main line [26, 28] reflects the weight of d^{10} in the ground state. For Pd ligands the ground state of which is dominated by d^{10}, the π shakeup satellite dominates, while for N_2/Ni which has much smaller weight of d^{10}, the π shakeup satellite intensity decreases. The authors of References [26 and 28] also interpreted the 2.0 eV satellite as the π CT shakeup state, using the equivalent core approximation (ECA). For N_2/Ni Umbach [24] claimed this structure as being due to the substrate band effect. Freund and coworkers [20] interpreted this satellite as the lowest energy ionic state resulting from different spin coupling of the core and the valence electrons in the ionized state, namely the $2h1p$ CT shakedown state.

2.E. SUM RULE [33–36]

It is shown that for adsorbate and TM carbonyls the main line intensity of the valence and core hole spectra of a free molecule is distributed to the lowest energy state of a smaller intensity and several satellite states by the π CT and σ excitations. This explains the major spectral feature differences between the free molecule and the adsorbate (coordinated molecule).

2.F. SCREENING WITHOUT LOW LYING π ORBITALS [37]

For NH_3 adsorbed on Ni metal surface there is no low lying strong π acceptor orbital. The absence of the π CT screening mechanism forces a pure σ CT screening from Ni sp to NH_3 Rydberg and antibonding orbitals, generating the appearance of significant σ to σ^* satellites, albeit with a different origin. See Reference [37] for details.

3. $1s$ to π^* Resonant Excitation [40]

In Table IV we collect experimental ionization energies (referred to the vacuum level for the adsorbates) and $1s$ to π^* resonant excitation energies for free and

TABLE IV
Core ionization[a] and resonance excitation energies (in eV) for free and adsorbed molecules

System	Core ionization energy		$1s \to 2\pi^*$ excitation energy	
	C	O	C	O
CO	296.2[b]	542.6[b]	287.4[c]	534.2[c]
CO/Ni	292.1[d]	538.4[d]	287.5[e]	533.5[e]
ΔE	4.1	4.2	−0.1	0.7
	N_a (inner)	N_b (outer)	N_a (inner)	N_b (outer)
N_2	409.9[b]	409.9[b]	401.0[f]	401.0[f]
N_2/Ni	406.9[g]	405.6[g]	401.0[e]	400.4[e]
ΔE	3.0	4.3	0.0	0.6

[a] Referred to the vacuum level, workfunction of 6.2 eV added to the experimental energies.
[b] Reference [62].
[c] Reference [63].
[d] Reference [28].
[e] Reference [64].
[f] References [65, 66].
[g] Reference [26].

adsorbed CO and N_2 [26, 28, 43, 62–66]. In Table V we summarize the resonant excitation energies and oscillator strengths (f values) of NiCO and NiN$_2$ calculated by different approximations [40].

3.A. EXCITATION ENERGY

The agreement with experiment is good when the core relaxation is included. The resonant energy shifts between the free and adsorbed molecules are small. It is also shown that the chemical shifts are small and the localized extramolecular π CT relaxation is mainly responsible for the ionization and affinity energy shifts [26, 40]. For N_2/Ni, the $N_a - N_b$ excitation energy splitting is 0.6 eV, while the KT energy difference is only 0.1 eV. The splitting in the XPS ionization peak (1.3 eV) and that in the XPS resonant peak (0.6 eV) are due to the nonequivalent N atoms [26, 40]. For a comparison of the excitation energy between free and adsorbed molecules, it was recently argued that no correction was made for the fact that the open shell character of the core excited free molecule leads to exchange split singlet and triple states and that only the dipole allowed transition to the singlet state is observed in the XAS spectrum. When the hybridization broadening of the empty π^* orbital with the substrate band is comparable to the exchange splitting, the latter is quenched in the XAS spectrum of the adsorbate. The configuration average excitation energies of a free molecule was recommended for a comparison, rather than the singlet resonant excitation energy [67]. As supporting evidence one may consider the lack of any exchange splitting in the XPS spectra of NO on Ni(100) [68]. However, a theoretical study of the geometries and binding mechanism of NiNO by the CASSCF method shows that there is no open shell character on the NO and consequently there will be no multiplet splitting in the NO derived XPS spectra of chemisorbed NO [69]. So the lack of

CORE IONIZATION, EXCITATION AND DE-EXCITATION 73

TABLE V
Theoretical and experimental 1s-to-π^* excitation energies (in eV) and f-values for the CO/Ni and N$_2$/Ni systems

Excitation	Exp.[a]	Method A	Method B	Method C	Method D	2h2p CI
Energy						
C 1s → π^*	287.5	277.8	287.6	286.5	287.1	
O 1s → π^*	533.5	526.7	532.9	533.5	532.1	
N$_a$ 1s → π^*	401.0	391.2	400.3	401.4	401.1	400.8
N$_b$ 1s → π^*	400.4	389.2	400.4	400.4	400.0	399.8
f-Value						
C 1s → π^*		0.234	0.115	0.096	0.116	
O 1s → π^*		0.125	0.055	0.041	0.042	
N$_a$ 1s → π^*		0.171	0.148	0.026	0.057	0.1018
N$_b$ 1s → π^*		0.205	0.029	0.024	0.041	0.1067
N$_b$/N$_a$ ratio	1.42	1.20	0.20	0.92	0.72	1.05

[a] References [43] and [64]. Method A; the 1h1p/1h1p CI method with the ground state orbital approximation. Method B; the 2h2p/2h2p CI method with the ground state orbital approximation. Method C; the 1h1p/1h1p CI method with the core ionized state SCF relaxed orbital approximation. Method D; the 2h2p/2h2p CI method with the core ionized state SCF relaxed orbital approximation. 2h2p CI; the 2h2p/2h2p CI method with the core excited state SCF relaxed orbital approximation.

exchange splitting in the XPS spectra of NO/Ni does not indicate the quenching of the exchange splitting in adsorbates. Upon chemisorption one can expect that the mixing of adsorbate and substrate states delocalizes the free molecular orbitals. As a result the splitting decreases because the smaller the spatial overlap of 1s and π^* is, the smaller is the singlet–triplet splitting. The splitting is larger for the C 1s excitation than for the O 1s excitation [69].

The ML binding (chemisorption) energy difference between the core ionized (excited) state and the neutral ground state is given by the ionization (excitation) energy difference between free molecule and the adsorbate [40]. For CO/Ni the chemisorption energy increases in the order of the C 1s resonantly excited state, the neutral ground state, the O 1s resonantly excited state, the C 1s ionized state and the O 1s ionized state. For N$_2$/Ni, it is in the order of the N$_a$ resonantly excited state, the neutral ground state, the N$_b$ resonantly excited state, the N$_a$ core ionized state and the N$_b$ core ionized state. The larger overlap of the π^* orbital with the metal substrate results in a stronger ML interaction in the final state and a smaller excitation (ionization) energy. So the ML coupling in the N$_b$ core ionized (excited) state is stronger than that in the N$_a$ one [26, 40, 43]. This can also be explained by ECA [28, 64, 70–72], by which an atom with a core hole excited is considered equivalent to an atom of the next element in the periodic table so that Ni—N—N with a N$_a$ 1s hole is equivalent to Ni—O—N, while Ni—N—N with a N$_b$ 1s hole is equivalent to Ni—N—O. For NiN$_2$ with a N$_a$1s hole, since O has higher electronegativity than N, the amplitude on the N$_a$ is increased in the occupied 1π orbital and is decreased in the orthogonal 2π orbital. In contrast, for NiN$_2$ with a N$_b$1s hole, the N$_2$ 2π orbital is more heavily concentrated on the N$_a$ atom than for the ground state NiN$_2$ and a NiN$_2$ with a N$_a$ 1s hole. Since ML interaction will be maximized when the Ni($d\pi$)–N$_2$(2π) overlap is optimised, the π bonding is stronger for the N_b core excited (ionized) state than that for the N_a one. Consequently the relaxation energy shift for the N_b excitation (ionization) is larger than that for the N_a one due to the stronger π CT coupling. The same explanation is applicable to CO/Ni where the O excited (ionized) state is more strongly coupled than in the C excited (ionized) state. In the resonantly excited state, there is a little change in the occupied π and σ orbitals. The change is smaller in the N_a excited state than in the N_b one because of weaker ML coupling in the former state than in the latter [40].

The coupling in the N_a excited state is the same as that in the ground state so the change in the occupied orbitals should be small. The s–d population in the N_a excited state shows a small s population increase and a d population decrease in comparison to the ground state, possibly inducing the reversed s–$d\sigma$ mechanism, which slightly increases the σ repulsion and reduces the π back-donation due to repulsion between the localized $d\pi$ electrons on the metal atom and $2\pi^*$ electron (weaker coupling). On the other hand, the s–d population in the N_b resonantly excited state shows a substantial decrease in the s and an increase in the d population, still indicating the presence of the s–$d\sigma$ mechanism which increases slightly the $d\pi$ back-donation because the N_a site is still available for the $d\pi$ back-donation (stronger coupling)[40]. The increased electron–electron repulsion due to the presence of the resonantly excited π^* electron hinders the $d\pi$ back-donation. In other words, the adsorbate/substrate coupling in the resonantly excited state is much weaker than in the core ionized state. This is also the case with CO adsorption [40].

Egelhoff [73] used the Born–Harber cycles with the ECA to determine important thermochemical quantities such as the adsorption energy. He considered the N_2/Ni to show that NO is moderately strongly chemisorbed N-end down but physisorbed O-end down. In his estimate, he used the XPS N core ionized states as reference states for the NO adsorption on Ni metal surface. However, we consider that the $1s$-to-π^* resonantly excited states rather than XPS core ionized states are more appropriate for the ECA [40]. In fact the ECA of the core excited molecule is the neutral system with the core hole species replaced by the $Z+1$ atom, while in the case of core ionization the ECA equivalent is the $Z+1$ valence positive ion. Using the recent XPS and NEXAFS data and an N_2 adsorption energy of 11 kcal/mol [74], the use of the XPS states of N_2/Ni system leads to an adsorption energy of 47.9 kcal/mole for the NO N-end down and one of 17.9 kcal/mole for the NO O-end down, while use of the resonant excitation energies leads to 24.8 kcal/mole and 11 kcal/mole, respectively. The experimental adsorption energy for the NO N-end down is about 25 kcal/mole [75].

3.B. OSCILLATOR STRENGTH [40]

When the relaxation of the core hole and excited electron and the screening of the bare hole–particle Coulomb interaction are neglected (method A), the f-value reflects the empty π^* charge distribution on the $C(N_a)$ and $O(N_b)$ atoms in the ground state, besides the obvious change in atomic $1s$-to-π^* transition matrix elements. While in the ground state the empty π^* orbital is more localized at the C atom site than at the O site, the opposite holds for N_a and N_b from the calculated intensity. The ground state population and analysis gives an empty N $2p$ population of 2.09 on N_b and 1.86 on N_a, respectively [40]. There is almost no reduction in the f-value from the free molecule to the coordinated molecule (the C and O f-values for free CO obtained by method A is 0.235 and 0.129, respectively. The f-value for free N_2 obtained by method A is 0.38 and the sum of N_a and N_b f-values for NiN_2 obtained by method A is 0.378 [40]).

The hole and particle relaxation, the screening of bare hole–particle Coulomb interaction and the $1h1p$ shakeup excitations can be included at the first step, by the $1h1p$ excitations in the presence of the $1h1p$ primary excitation, using the ground state orbitals (method B). Inclusion of such many-body effects significantly improves the excitation energies and f-values to the low-lying empty orbitals (in the case of NiCO). This way of accounting for relaxation effects is related to the final state rule [76, 77]. It is well known that the presence of the core hole in the final state has an effect on the XAS line shape. According to the final state rule the spectrum is determined by the final state DOS, supplemented by an enhancement close to the threshold. Allowance of relaxation by method B or by employing relaxed orbitals (methods C and D) causes a large reduction in the f-value. The reduction is larger for NiCO than for free CO. Rühl and Hitchcock [78] determined the f-values for the $1s$-to-π^* excitation in TM carbonyls. There is quite satisfactory agreement with their estimate of the absolute f-value of $Ni(CO)_4$ (0.087 and 0.045 for C and O, respectively), as well as on the magnitude of reduction from free CO value. The reduction of the O f-value from free CO molecule to the TM carbonyls is smaller than that of the C one. This is consistent with a smaller variation of the O_{2p_z} contribution to the π^* orbital [78]. Thus the electron resonantly excited to π^* and Rydberg derived levels tends to be more localized (less reduction in the f-value) in the O $1s$ excitation than in the C $1s$ excitation because the π^* and higher Rydberg orbitals are less influenced by the coupling with the metal substrate for the O $1s$ excitation than for the C $1s$ excitation.

For both free N_2 and NiN_2, the relaxation causes also a large reduction. Moreover, in the case of NiN_2, it reverses the ratio between the N_a and N_b f-values around $1:0.8$ with the exception of method B, where the ratio is greatly diminished. By studying the DES spectral feature dependence on the photon excitation energy for the N_2/Ni system, a deconvolution of the $1s$-to-π^* XAS spectrum of the N_2/Ni system, which is a superposition of N_a and N_b contributions, has recently been proposed. The relative spectral intensity ratio of 1.4 ± 0.2 was obtained for the N_b/N_a [43, 79]. This disagrees with the prediction. The effect of relaxation is to redistribute the f-value, but the total intensity is conserved (sum rule) [76]. In the case of NiN_2, a use of the core ionized state SCF relaxed orbitals results in a large reduction in the f-values (inaccurate f-values) because a large intensity goes to the shakeup satellites. This has its origin in different core relaxation between the core ionized and core excited states. In contrast to the case of NiCO, the larger difference in the results obtained even at the $2h2p$ CI level (methods B and D) for NiN_2 points out the importance of an optimal choice of the orbital set if convergent results are to be obtained at practical excitation levels. Thus, to obtain an accurate estimate of the f-value and excitation energy for the $1s$-to-π^* excitation for NiN_2, the $1h1p$ core excited state SCF relaxed orbitals are employed [40]. At the ΔSCF level the N_a and N_b resonant excitation energies are almost identical because of near degeneracy of the KT energies in the ground state (about 0.1 eV difference). The same is true with method B, which is similar to the ΔSCF method. By the $2h2p$ CI scheme, the experimental excitation energies with the N_a–N_b excitation energy difference of 0.6 eV are well reproduced. The f-values obtained by the ΔSCF method and the $2h2p$ CI method are similar, except for a slight increase in the latter. The f-value ratio (N_b/N_a) is 1.08 and 1.05, respectively which is slightly reduced from the empty N_{2p} population ratios in the ground state (1.12 and 1.07, respectively), while the experimental value is 1.4 ± 0.2. The increase of N_{2p} population at each excitation atomic site from the ground state to the N_a and N_b core excited states obtained by the ΔSCF ($2h2p$ CI) method are 1.032(0.997) and 1.184(1.061), respectively, giving the N_b/N_a ratio of 1.15(1.06) which is close to the empty N_{2p} population ratio in the ground state. The f-value represents the empty N_{2p} population of the π^* orbital in the ground state (the sum rule). The use of the core ionized state SCF relaxed orbitals leads to a dramatic change not only in the f-values but also in their N_b/N_a ratios [40]. The N_{2p} population in the N_b core ionized state is slightly larger than that in the N_a core ionized state, in contrast to the reverse situation in the neutral ground state. So one can expect a smaller f-value for the N_b excitation by using the core ionized SCF relaxed orbitals. The π resonance for the adsorbate is considered to have a smaller intensity than that for the free molecule. The experimental f-value for free N_2 is 0.195, while for NiN_2 the sum of N_a and N_b f-values at the SCF and $2h2p$ CI levels are 0.194 and 0.209, respectively. There is almost no difference in f-value between free N_2 and NiN_2.

3.C. DISAPPEARANCE OF THE GIANT SATELLITE IN THE XAS SPECTRA OF ADSORBATES [40]

The resonantly excited π^* electron acts as a screening charge and hampers the extra molecular π CT relaxation. As a result the π shakeup satellites are suppressed in the XAS spectra. The excess of the π^* charge hinders the increase in the $d\pi$ back-donation, which occurs in the case of core ionized systems, because of the increased electron–electron repulsion. Consequently the s–$d\sigma$ mechanism is also hampered. The excess of the π^* charge could even give rise to the 'reversed' s–$d\sigma$ mechanism by which the s population increases and the d population decreases. This could give rise to significant σ-to-σ^* satellites. However, there is no efficient mechanism such as the 'reversed' π^*–$d\pi$ (ligand-to-metal) CT mechanism which could promote the 'reversed' s–$d\sigma$ mechanism. As the s–$d\sigma$ population of the core excited state does not change from that of the ground state, the σ local metal shakeup excitation is much more suppressed than in the case of the core ionization. The giant satellite is absent in the XAS spectrum because of the absence of the s–$d\sigma$ mechanism which is a consequence of the hindrance of the cooperative screening mechanism by the presence of the resonantly excited π^* electron. It is also demonstrated that the satellite feature in the XAS spectrum of adsorbate does indeed depend on the s–d population in the ground state [40]. The XAS spectra of NiCO obtained by using the core hole relaxed SCF orbitals, do not differ much from those calculated by the ground state SCF orbitals. However, the NiN_2 spectra show strong satellite features. This is because, in order to reach the resonantly excited state from the core ionized state, there is significant relaxation in the metal σ orbital, which regains a large s population at the expense of the d participation, giving a situation much closer to that found in the ground state. The smaller the s population is, the less expensive becomes the 'reversed' s–$d\sigma$ mechanism. From the viewpoint of the s–d population for NiCO (CO/Ni), the XPS main line state is the π CT state and reaching the XAS state, which is close to the ground state, does not much need the 'reversed' s–$d\sigma$ mechanism. However, for NiN$_2$ (N$_2$/Ni), the XPS lowest energy state where the s population is greatly depleted, is the σ 'shakedown' state. To reach the XAS state from the XPS state, the 'reversed' s–$d\sigma$ mechanism must be switched on. This explains why there are strong satellite lines and f-values become very much reduced [54]. Note that the relaxation energy of the resonantly excited state is well described by employing the core ionized state relaxed SCF orbitals.

In this section it is argued that the π CT and σ–σ^* satellite lines are smaller or absent in the XAS spectra, in contrast to their importance in the XPS spectra. One may explain this in a different manner [54]. In the XAS final state the interaction between **c**d^{10}L and **c**$d^9 s^1$L character gives rise to bonding and antibonding combinations. In the final state the s–$d\sigma$ promotion energy is the same as that of the ground state. Under the assumption that the hybridization is not modified in the final state, all intensity goes to the bonding combination of the final state (main line state). This explains the absence of the giant shakeup satellite in the XAS spectrum. For the π CT shakeup, in the XAS final state the interaction between **c**d^{10}L (**c**$d^9 s^1$L) and **c**d^9L^2 (**c**$d^8 s^1$L^2) gives rise to bonding and antibonding combinations. The energy difference between **c**d^{10}L (**c**$d^9 s^1$L) and **c**d^9L^2 (**c**$d^8 s^1$L^2) changes from that between the initial ground state d^{10} ($d^9 s^1$) and d^9L ($d^8 s^1$L) by $U_{LL} - Q$. Here U_{LL} is the π^*–π^* electron–electron Coulomb interaction. Possibly $U_{LL} \simeq Q$ so that the energy difference does not change much. Consequently, the π CT hybridization in the XAS final state does not differ much from that in the ground state. This explains why the ML coupling strength (which is mainly due to the π bonding) in the XAS final state is close to that in the ground state and also why the π shakeup satellite is very small. In general, the changes in the s–$d\sigma$, $d\pi$–π hybridization will give rise to σ and π satellite structures. In XAS process the change in hybridization is considerably smaller than in the XPS process, which is a direct consequence of the charge conservation in the former [54].

4. DES Spectra of Adsorbates

4.A. AUTOIONIZATION VS AUGER DECAY

The Auger sum rule states that the total spectral intensity of a core-valence valence (CVV) Auger spectrum is equal to the total intensity of a spectrum obtained by self-convoluting the Auger initial state (i.e. in the presence of the core hole) valence density of states, using the Cini–Sawatzky formula [80, 81]. The valence electron distribution is influenced by the presence of the electrostatic potential created by the core hole in the initial state. This potential is screened effectively and the screening charge can take part in the Auger process. This will result in an enhancement of the component of the Auger spectrum which involves final state holes of the screening electron orbitals; so called back-bonding peaks [82]. As the CT screening in the core ionized state is hampered in the resonantly excited core states, the ML coupling strength is different between the core ionized and the core excited states. Then between the normal Auger decay and the spectator Auger decay one expects spectral feature differences in the final states which involve the π bonding orbital. Moreover, for all CO adsorption systems studied, the autoionization spectra are nearly identical to the corresponding Auger spectra and no participant decay has been observed in the autoionization spectra [41–45, 83]. Upon comparison between the autoionization and the AES spectra, the spectra line up on the same K.E. scale. So on a calibrated B.E. scale, the two spectra are shifted from each other by the XPS ionization and XAS resonance energy difference. The first explanation for this spectral behavior was that the core excited state in the photoabsorption process is a $^1\Pi$ state which may then decay into the energetically more favourable $^3\Pi$ state (1.46 eV lower in the case of free CO) by exchange of electrons with the metal. The core excited state decays into the fully screened hole state before de-excitation process with the excess energy given up to the substrate [84]. A more acceptable recent explanation is as follows; in the case of physisorption where the adsorbate and substrate are eletronically isolated, unless the system becomes chemisorbed in the presence of a core hole, there is no CT from the substrate to screen the core hole and therefore there is no relationship between the resonant core excitation and the core ionization. For the chemisorption systems, the energy onset in the XAS spectrum should correspond to the XPS B.E. The XAS onset corresponds to the creation of a final state where a core electron is excited to the lowest unoccupied state, i.e. at the Fermi level. In the completely screened XPS final state a charge redistribution takes place where one electron is taken from the Fermi level to locally screen the core hole. Then those two final states should be indistinguishable [67]. The excited electron delocalizes from the adsorbate to the substrate so that the resonantly excited state results in a state that could be considered as locally identical to the XPS lowest energy state before the Auger core hole decay starts [42–44, 83].

The following is a summary of the theoretical work by the author [49]. We consider a quasidiscrete molecular resonant excitation from a core level (c) to a localized (discrete) molecular empty level (a) which is coupled to a continuum of extended states, e.g. of a substrate metal. As the resonantly excited state ($c^{-1}a$) is slightly above the core ionization limit (c^{-1}) in chemisorption systems, we can consider strong CI between the $c^{-1}a$ and $c^{-1}\epsilon_k$ (here ϵ_k denotes a photoelectron in the continuum of extended states of a substrate metal) configurations which are coupled by the CT mechanism. We focus particularly on a competition between decay of the electron in the broadened affinity level (hopping into the continuum, charge delocalization) and decay of the core hole, leading to competition for intensity between an autoionization decay from the core excited neutral resonant state ($c^{-1}a$) and a normal Auger decay from the core ionized ionic state. The intermediate resonant state ($c^{-1}a$) may decay via a different decay channel; autoionization (participant decay) which leads primarily to single ionization and Auger decay which couples to decay of the electron in the affinity level (delocalization, ionization), leading to the doubly ionized states. The latter consists of the Auger decay of the core level which takes place in the absence (normal Auger decay) or presence of an electron in the affinity level (spectator Auger decay) which eventually leads to the same final doubly ionized state.

At or near the resonance, we consider only the resonant amplitude by neglecting the direct ionization amplitudes. Moreover, we neglect the interference between the spectator decay and the normal Auger decay, the ultimate final states of which represent the doubly ionized states. Then the resonant emission spectrum (as a function of the kinetic energy ϵ and the photon energy ω) becomes

$$I(\epsilon; \omega) \sim |Z_{ca}(\omega)|^2 A_a(\omega) \left\{ \frac{\Gamma_v}{\Gamma_t} \delta(\epsilon - \omega - \epsilon_v) \right.$$
$$\left. + \sum_i \frac{\Gamma_s^i}{\Gamma_t} A_s(\epsilon - \epsilon_s^i - \omega) + \frac{\Gamma_{CT}}{\Gamma_t} \sum_j \frac{\Gamma_A^j}{\Gamma_A} A_c(\epsilon - \epsilon_A^j) \right\} \quad (1)$$

The first term in Equation (1) describes the resonant valence (v) photoemission main line. The second term is the spectator decay spectrum and the third term is the normal Auger decay spectrum. $Z_{ca}(\omega)$ is the effective dipole excitation matrix element given by

$$Z_{ca}(\omega) = Z_{ca} + \sum_\epsilon \frac{V_{va} Z_{ve}}{\epsilon_v - \epsilon + \omega - i\delta} \quad (2)$$

The second term in Equation (2) describes the CI between the resonantly core excited discrete state and the singly valence ionized continuum state. The coupling matrix element is V_{va}. ϵ_v is the valence ionization energy. $A_a(\omega)$ is the spectral function of the hole-particle excited ($c^{-1} \to a$) state given by

$$A_a(\omega) = \frac{1}{\pi} \frac{\Gamma_t/2}{(\omega - \epsilon_a + \epsilon_c + U_{ca})^2 + (\Gamma_t/2)^2} \quad (3)$$

Here ϵ_a, ϵ_c and U_{ca} are the affinity of empty level a, the ionization energy of core level c and the effective core hole-particle Coulomb interaction. Γ_t is the total decay width of the hole-particle excited state, given by

$$\Gamma_t = \Gamma_s + \Gamma_v + \Gamma_{CT} \quad (4)$$

Here Γ_v is the participant decay width, Γ_s is the total spectator decay width ($=\Sigma_i \Gamma_s^i$) and Γ_{CT} is the decay width of the resonantly excited neutral state to the core ionized ionic state by a hopping of the excited electron into the metal substrate empty continuum states. $A_s(\epsilon - \epsilon_s^i - \omega)$ is the spectral function of the final $2h1p$ state created by the spectator decay channel i; $c^{-1}a \to v_1^{-1} v_2^{-1} a\epsilon$, the decay width of which is given by Γ_s^i. The spectral function is given by

$$A_s(\epsilon - \epsilon_s^i - \omega) = \frac{1}{\pi} \frac{\Gamma_{CT}/2}{(\epsilon - \epsilon_s^i - \omega)^2 + (\Gamma_{CT}/2)^2} \quad (5)$$

Here ϵ_s^i is the ionization energy of the final $2h1p$ state, given by

$$\epsilon_s^i = \epsilon_{v_1} + \epsilon_{v_2} - U_{v_1 v_2} - \epsilon_a + U_{v_1 a} + U_{v_2 a} \quad (6)$$

Here U is the effective hole–hole (particle) Coulomb interaction. $A_c(\epsilon - \epsilon_A^j)$ is the core hole spectral function given by

$$A_c(\epsilon - \epsilon_A^j) = \frac{1}{\pi} \frac{\Gamma_A/2}{(\epsilon - \epsilon_A^j)^2 + (\Gamma_A/2)^2} \tag{7}$$

Here Γ_A is the total Auger decay width ($=\Sigma_j \Gamma_A^j$). ϵ_A^j is the Auger energy for the decay channel j; $c^{-1} \to v_1^{-1} v_2^{-1} \epsilon_A^j$, given by

$$\epsilon_A^j = \epsilon_{v_1} + \epsilon_{v_2} - U_{v_1 v_2} - \epsilon_c \tag{8}$$

At the resonance, the width of the spectator decay spectrum of adsorbate is larger than that of free molecule by Γ_{CT} (the decay width due to the charge delocalization). At the resonance, the kinetic energy separation between the spectator decay ($\epsilon_s^i + \omega$) and the normal Auger decay (ϵ_A^j) is given by $U_{v_1 a} + U_{v_2 a} - U_{ca}$. In chemisorbed systems such as CO/Ni, the participant and spectator decays are negligible and the normal Auger decay dominates. Thus there is no decay channel which leads to a final state the same as the direct singly ionized valence photoemission main line. As a result there is no interference effect which may lead to Fano-type resonance or antiresonance behavior for the final state. In this case the XAS spectrum is given by $(Z_{ac})^2 A_a(\omega)$, representing the absorption from a discrete level broadened into a Lorentzian predominantly by the CT decay (Γ_{CT}), in contrast to the case of free molecule for which the width is given by the spectator decay width (in the case when the spectator decay dominates much more than the participant decay). The stronger the hybridization between the empty affinity level and the substrate band, the faster is the delocalization of the excited electron from the adsorbate atomic site to the substrate. The rapid delocalization is manifested as a reduction of the core excited state lifetime (a broader core resonance width by an increase of Γ_{CT}).

Γ_s is often very close to Γ_A because the spectator decay energy is large enough so that the presence of a resonantly excited electron does not much influence the Auger electron wavefunction. However, this may not be the case for the Coster–Kroning (CK) decay because CK decay energy is often very small and consequently the CK electron wavefunction may be influenced by the screening of the final state two hole potential by the presence of an extra spectator electron [85].

The CT due to presence of substrate renders a decay of the core excited state into the most favourable core hole state on the time scale of the core hole decay. The CT time depends on a choice of substrate. When the ML coupling is strong, Γ_{CT} is much larger than Γ_s and Γ_v, so that the core hole decay spectrum is dominated by normal Auger decay (this is the case with CO/Ni, CO/Pd etc.). When Γ_{CT} and $\Gamma_s(\Gamma_v)$ are comparable, the resonant spectrum splits into the spectral feature due to spectator Auger (or participant) decay from a neutral core excited state and that due to normal Auger decay from an ionic state with the total spectral intensity ratio of Γ_s (or $\Gamma_v)/\Gamma_{CT}$. Such examples recently observed, are the autoionization spectra of Ar adsorbed on a number of substrates; graphite, Pt(111), Cu(100), Au(110) and Ag(110), observed at the Ar $2p$–$4s$ resonance [46, 48]. The presence of the substrate renders the decay of the core excited neutral state to an ionic core hole state, on the time scale of the core hole lifetime, by the decay of the excited 4s electron to the Fermi level of the substrate. In the autoionization spectrum, this is manifested in the increase of the portion of Auger decay feature (from 0.54 to 0.85 in the order of Pt, Au, Cu and Ag). So the CT probability (Γ_{CT}) can be determined from the portion of the Auger decay in the spectrum, namely $\Gamma_{CT}/(\Gamma_s + \Gamma_{CT}) \simeq \Gamma_{CT}/(\Gamma_A + \Gamma_{CT})$ [48, 49]. The ionic (XPS) and neutral core hole state (XAS) energy separation decreases in the order of Pt, Au, Cu and Ag (from 3.9 to 2.5 eV). The admixture of the CT configuration becomes more dominant as a result of a decreasing CT energy. So there is a correlation between the energy separation and Γ_{CT} [48, 49].

An indication of chemisorption in the core excited state could be an effect of the core hole which could perturb the properties of the adsorption system. The attractive potential of the core hole may bring the empty adsorbate level closer to the Fermi level so that a chemical binding between the adsorbate and substrate will be formed. In that case, if the integrated intensity of the XAS resonance peak does not change in comparison to the gas-phase data, this is an indication of chemisorption in the core excited state, despite the physisorption in the ground state [48, 49].

So far we have considered the case when the lowest core hole state is an ionic state, not a neutral core excited state. We consider the case when the lowest core hole state is a neutral core excited state (or CT screened state), not an ionic state. In this case there will be competition between the normal Auger decay and the CT core hole screening process, the latter of which leads to spectator and participant decay features in the Auger spectrum. In this case, above the XPS ionization limit the emission will be given by [49].

$$I(\epsilon) \propto |Z_{ck}|^2 \left\{ \sum_i \frac{\Gamma_A^i}{\Gamma_t} A_c(\epsilon_{v_1}^i + \epsilon_{v_2}^i - U_{v_1 v_2}^i - \epsilon) \right.$$
$$\left. + \frac{\Gamma_{CT}}{\Gamma_t} \sum_j \frac{\Gamma_s^j}{\Gamma_a} A_a(\epsilon - \epsilon_s^j) + \frac{\Gamma_{CT}}{\Gamma_t} \sum_l \frac{\Gamma_v^l}{\Gamma_a} A_a(\epsilon - \epsilon_v^l) \right\} \quad (9)$$

Here E is the core hole energy parameter. The first term describes the normal Auger decay and A_c is the core hole spectral function, the width of which is given by the total width Γ_t (XPS width). Γ_t is the sum of the normal Auger decay ($\Sigma \Gamma_A^i = \Gamma_A$) and the decay of the core ionized state to the neutral screened state by the CT screening (Γ_{CT}). The second term describes the spectator decay. A_a is the spectral function of the CT state with the width Γ_a (instead of Γ_t in Equation (3)). Here Γ_a is the decay width of the CT state which is a sum of the spectator decay width ($\Gamma_s = \Sigma_j \Gamma_s^j$) and the participant decay width ($\Gamma_v = \Sigma_l \Gamma_v^l$). So above the XPS ionization limit, the autoionization spectrum width is given by Γ_a. The last term describes the participant decay from a neutral state created by the CT from the substrate. The total spectral intensity is equal to the total spectral intensity of the XPS core hole spectrum. The intensity ratio of the Auger and autoionization decay spectrum is reversed when the relative position of the XPS and XAS maxima is reversed. We can tell whether the screened core hole state (CT shakedown state; XAS like state) can be observed in the XPS spectrum by a cluster model approach. From the relative spectral intensity of the neutral (autoionization) and ionic (Auger) decay processes in the Auger spectrum which is given by Γ_{CT}/Γ_A, the neutralization rate (Γ_{CT}) can be determined. Recently such a case was studied for N_2 physisorbed on graphite ($\Gamma_{CT} = 0.07$ eV, $\Gamma_c = 0.12$ eV and $\delta E = -2.9$ eV) [47]. Here δE is the energy separation between XPS and XAS maxima. As $\Gamma_{CT} \ll |\delta E|$, we obtain [49]

$$\frac{I_S}{I_M} \simeq \frac{\Gamma_{CT}}{2\pi(\delta E)^2} \tag{10}$$

Here I_S is the spectral intensity of the shakedown state and I_M is that of the KT state. The probability one observes the CT shakedown state (screened core hole state) in the XPS spectrum is very small (the intensity is 0.001). Indeed the N 1s XPS spectrum of N_2/graphite does not show such a state [47].

4.B. SPECTRAL FEATURE CHANGES WITH PHOTON ENERGY VARIATIONS

Wurth et al. [42a,b] observed a spectral intensity decrease of the π back-bonding peaks in the O 1s DES spectra of different CO adsorption systems when the photon excitation energy is increased 6 eV above the resonance energy (which is considered to be the excitation to the π character orbital). They interpreted this as ampering of π CT from the metal by the electron resonantly excited to the Rydberg level. Recently the Uppsala group also carefully studied the DES spectra of several CO adsorption systems [83]. In the O 1s DES spectra of the CO/Ni(100) system they observed a notable spectral intensity dependence on the excitation energy, already at the π resonance maximum in comparison to the XPS ionization limit; however, no corresponding changes are observed in the C 1s DES spectra. The spectral feature changes with photon excitation energy variations indicate that despite a rapid delocalization of the excited electron, the decay spectrum is influenced by a temporary existence of the excited electron. In other words the time scale of relaxation of valence orbitals due to the delocalization of the excited electron is comparable to that of the core hole decay. The spectral changes are much more notable in the O 1s excitation spectrum. As the electron resonantly excited to π^* and Rydberg derived levels tends to be more localized in the O 1s excitation than in the C 1s one, there are more significant spectral intensity changes in the O spectrum than in the C one [40].

With an increase of the ML coupling strength the 1π orbital (which is orthogonal to 2π orbital) will be polarized more toward the O and the spectral intensity of the $1\pi^{-2}$ two-hole state peak increases dramatically in the O 1s DES spectrum. So the ML coupling strength in the resonantly excited state can be monitored by observing the spectral intensity changes of the backbonding peak and $1\pi^{-2}$ peak in the O 1s DES spectrum [40]. In the O 1s DES spectrum of CO/Ni(100), with photon energy increase from the XPS ionization limit, the spectral intensity of $1\pi^{-2}$ peak decreases relative to that of the $4\sigma^{-1}1\pi^{-1}$ peak. This is because the ML coupling strength in the core excited state is weaker than that in the core ionized state due to the hindrance of the polarization of the bonding orbital towards the ligand by the resonantly excited electron [40]. The O 1s DES spectra of CO/H/Ni show significant spectral feature changes with different geometries, namely top, bridge and hollow sites [83]. With an increase of the coordination number, the spectral intensity of $1\pi^{-2}$ peak increases relative to the $4\sigma^{-1}1\pi^{-1}$ peak and the back-bonding peak intensity also increases in the O 1s DES spectra; the chemisorption energy of the core excited state increases with an increase of the coordination number. This is also in accord with the increase of the XAS width (ML hybridization width) and the decrease of the f-value (more polarization of the bonding orbital toward the ligand) with an increase of the coordination number [56, 83]. The chemisorption energy of the ground state can be estimated from the chemisorption energy data of both ground state and core excited state of the systems whose DES spectra resemble to those of CO/H/Ni systems [40]. It was shown that the chemisorption energy of the ground state increases with an increase of the coordination number and coadsorption of H weakens the bonding [40].

4.C. XAS SPECTRUM VS DES SPECTRUM [49]

As Equation (1) shows, near or at the resonance, the total intensity of core hole decay spectrum is the XAS spectrum intensity. Thus the total intensity of decay spectra recorded at several photon energies can be used to monitor the XAS spectrum. When the XAS spectrum consists of more than two resonant excitation spectra (e.g. N_2/Ni due to the inequivalent N atoms [43, 79]), the XAS spectrum can be separated into different excitation components (e.g. different atomic sites), if the decay spectrum recorded at different photon energies can be decomposed into the spectral profiles due to each resonant excitation.

By collecting the emitted electrons at a constant binding energy (photon energy minus kinetic energy), one obtains the constant ionic (final) state spectroscopy (CISS) spectrum. The CISS spectrum integrated over the whole B.E. range gives the XAS spectrum in total electron yield (in the case when both autoionization and normal Auger decay channels are possible, the autoionization part of the XAS spectrum). Thus CISS can be interpreted as partial electron yield at constant binding energy. Thus by the CISS spectrum one can determine also the autoionization and the Auger spectral intensity ratio. One can also assign spectral features of the XAS spectrum by studying the CISS spectrum of a specific final state which can be reached only from a particular initial state.

5. Concluding Remarks

The concept of cooperative core hole screening mechanism was introduced to explain the strong spectral feature changes of adsorbate core hole spectra with variations of adsorbate/substrate systems, in a systematic manner, in terms of local metal $s-d$ populations. The near degeneracy of local metal configurations such as $d^{n-1}s^1$ and d^n plays such an important role that even the one-electron picture of the core ionization breaks down. The change of local $s-d$ population from metal surfaces to alloy surfaces and coadsorbed metal surfaces will lead to substantial spectral feature changes. From this viewpoint, the core hole spectra can be used for the study of local electronic structures of adsorbate/substrate systems. The XAS spectrum provides information on both neutral ground state, as well as resonantly excited state. The former is reflected in the spectral intensity (sum rule), while the latter lies in the spectral features such as the spectral width and the excitation energies (final state rule). The relaxation of the occupied orbitals in the resonantly core excited neutral state is modest so that the local s–d population in the resonantly excited state is almost the same as that in the neutral ground state. The charge delocalization of the resonantly excited electron from the adsorbate atomic site to the substrate metal band empty states, triggers the s–$d\sigma$ mechanism and subsequently the cooperative core hole screening mechanism so that the resonantly excited state relaxes to the lowest energy core ionized state. When the metal-ligand coupling is strong (chemisorbed), the charge delocalization occurs before the core hole decay starts. This results in the domination of the Auger spectral features in the de-excitation spectra of adsorbates. On the other hand when the core excited neutral state is lower than the core ionized state (physisorbed systems), upon the ligand core ionization, there is CT from the metal substrate to the ligand core hole site. The CT time is much longer than the photoionization time scale, however, it is comparable with the time scale of the core hole decay. As a result the autoionization spectral features due to the decay from the neutral CT screened hole state appear in the Auger spectrum measured at the excitation energy higher above the XPS ionization limit. The CT time can be determined from the portion of Auger (autoionization) spectral features in the autoionization (Auger) spectra.

References

1. J. C. Fuggle, T. E. Madey, M. Steinkiberg, and D. Menzel: *Chem. Phys. Lett.* **33**, 233 (1975).
2. J. C. Fuggle, E. Umbach, D. Menzel, K. Wandelt, and C. R. Brundle: *Solid State Commun.* **27**, 65 (1978).
3. P. R. Norton, R. L. Tapping, and J. W. Goodale: *Surf. Sci.* **72**, 33 (1978).
4. C. R. Brundle, P. S. Bagus, D. Menzel, and K. Hermann: *Phys. Rev.* **B24**, 7041 (1981).
5. E. Umbach: *Surf. Sci.* **117**, 482 (1982).
6. E. W. Plummer, W. R. Salaneck, and J. S. Miller: *Phys. Rev.* **B18**, 1673 (1978).
7. H. J. Freund, E. W. Plummer, W. R. Salaneck, and R. W. Bigelow: *J. Chem. Phys.* **75**, 4275 (1981).
8. K. Schöhammer and O. Gunnarsson: *Solid State Commun.* **23**, 691 (1977); **26**, 147 (1978); **26**, 399 (1978).
9. B. Gumhalter and D. M. Newns: *Phys. Lett.* **53A**, 137 (1975).
10. B. Gumhalter: *J. Phys.* **C10**, L219 (1977).
11. N. D. Lang and A. R. Williams: *Phys. Rev.* **B16**, 2048 (1977).
12. A. Kotani and Y. Toyozawa: *Jpn. J. Phys.* **35**, 1073 (1973); **35**, 1082 (1973); **37**, 912 (1974).
13. O. Gunnarsson and K. Schönhammer: *Phys. Rev. Lett.* **41**, 1608 (1978).
14. P. S. Bagus and K. Hermann: *Surf. Sci.* **89**, 588 (1979); *Solid State Commun.* **38**, 1257 (1981).
15. P. S. Bagus and M. Seel: *Phys. Rev.* **B23**, 2065 (1981)
16. R. P. Messmer, S. H. Lamson, and D. R. Salahub: *Solid State Commun.* **36**, 265 (1980).
17. R. P. Messmer, S. H. Lamson, and D. R. Salahub: *Phys. Rev.* **B25**, 3576 (1982).
18. D. Saddei, H.-J. Freund, and G. Hohlneicher: *Surf. Sci.* **102**, 359 (1981) and references therein.
19. C. M. Kao and R. P. Messmer: *Phys. Rev.* **B31**, 4835 (1985).
20. H.-J. Freund, R. P. Messmer, C. M. Kao, and E. W. Plummer: *Phys. Rev.* **B31**, 4848 (1985) and references therein.
21. E. J. Baerends and P. Ros: *Int. J. Quant. Chem.* **S12**, 169 (1978).
22. G. Loubriel: *Phys. Rev. B* **20**, 5339 (1979).

23. J. C. Fuggle and D. Menzel: *Vakuum-Tech.* **27**, 130 (1978); in *Proc. 7th IVC and 3rd ICSS* (Vienna 1978), p. 1003.
24. E. Umbach: *Solid State Commun.* **51**, 365 (1984).
25. W. F. Egelhoff, Jr.: *Surf. Sci.* **141**, L324 (1984).
26. A. Nilsson, H. Tillborg, and N. Mårtensson: *Phys. Rev. Letters* **67**, 1015 (1991).
27. D. Lovric, B. Gumhalter, and K. Wandelt: *Surf. Sci.* **278**, 1 (1992).
28. A. Nilsson and N. Mårtensson: *Phys. Rev.* **B40**, 10249 (1989).
29. C. W. Bauschlicher Jr., S. R. Langhoff, and L. A. Barnes: *Chem. Phys.* **129**, 431 (1989) and references therein.
30. M. R. A. Blomberg, P. E. M. Siegbahn, T. J. Lee, A. P. Rendell, and J. E. Rice: *J. Chem. Phys.* **95**, 5848 (1991) and references therein.
31. E. W. Plummer and W. Eberhardt: *Adv. Chem. Phys.* **49**, 533 (1982).
32. H.-J. Freund and E. W. Plummer: *Phys. Rev.* **B23**, 4859 (1981) and references therein.
33. M. Ohno and W. von Niessen: *Phys. Rev.* **B42**, 7370, (1990); **B44**, 1896 (1991), **B45**, 1851 (1992); **B45**, 9382 (1992); *J. Chem. Phys.* **95**, 373 (1991); *Surf. Sci.* **269**, 258 (1992); *J. Chem. Phys.* **97**, 2767 (1992); **97**, 6953 (1992); *Chem. Phys.* **158**, 1 (1991); *J. Electron Spectr. and Rel. Phenom.* **58**, 219 (1992).
34. M. Ohno and P. Decleva: *Surf. Sci.* **258**, 91 (1991); **269/270**, 264 (1992); *Chem. Phys.* **156**, 309 (1991).
35. P. Decleva and M. Ohno: *Chem. Phys.* **160**, 341 (1992).
36. P. Decleva and M. Ohno: *Chem. Phys.* **160**, 353 (1992); **164**, 73 (1992); *J. Chem. Phys.* **96**, 8120 (1992).
37. M. Ohno and P. Decleva: *Chem. Phys.* **169**, 173 (1993).
38. J. Stöhr: *NEXAFS Spectroscopy*, Springer-Verlag, Heidelberg (1992).
39. J. Stöhr and R. Jaeger: *Phys. Rev.* **B26**, 4111 (1982).
40. M. Ohno and P. Decleva: *Surf. Sci.* **284**, 372 (1993); *Chem. Phys.* **171**, 9 (1993); *J. Chem. Phys.* **98**, 8070 (1993); *Surf. Sci.* **296**, 87 (1993).
41. C. T. Chen, R. A. DiDio, W. K. Ford, E. W. Plummer, and W. Eberhardt: *Phys. Rev.* **B32**, 8434 (1985).
42(a). W. Wurth, C. Schneider, R. Treichler, E. Umbach, and D. Menzel: *Phys. Rev.* **B35**, 7741 (1987).
42(b). W. Wurth, C. Schneider, R. Treichler, D. Menzel, and E. Umbach: *Phys. Rev.* **B37**, 8725 (1988).
43. O. Björneholm, A. Sandell, A. Nilsson, N. Mårtensson, and J. N. Andersen: *Phys. Scrip.* **T41**, 217 (1992).
44. M. Ohno: *Phys. Rev.* **B45**, 3865 (1992).
45. G. Illing, T. Porwol, I. Hemmerich, G. Dömötör, H. Kuhlenbeck, H.-J. Freund, C.-M. Liegner, and W. von Niessend: *J. Electron. Spectr. Rel. Phen.* **51**, 149 (1990).
46. W. Wurth, P. Feulner, and D. Menzel: *Phys. Scipta.* **T41**, 213 (1992).
47. O. Björneholm, A. Nilsson, A. Sandell, B. Hernnäs, and N. Mårtensson: *Phys. Rev. Letters* **68**, 1892 (1992).
48. A. Sandell, A. Nilsson, O. Björneholm, B. Hernnäs, P. Bennich, and N. Mårtensson: to be published. Published in A. Sandell, Ph.D. thesis, Uppsala University 1993, Almqvist and Wiksell, International, Stockholm, Sweden.
49. M. Ohno: *Phys. Rev.* **B50**, 2566 (1994).
50. G. Blyholder: *J. Chem. Phys.* **68**, 2772 (1964); *J. Vac. Sci. Technol.* **11**, 865 (1974).
51. M. R. A. Blomberg, U. Brandemark, P. E. M. Siegbahn, K. Broch-Mathisen, and G. Karlström: *J. Phys. Chem.* **89**, 2171 (1985).
52. C. W. Bauschlicher. Jr., P. S. Bagus, C. J. Nelin, and B. O. Roos: *J. Chem. Phys.* **85**, 354 (1986) and reference therein.
53. G. van der Laan, B. T. Thole, H. Ogasawara, Y. Seino, and A. Kotani: *Phys. Rev.* **B46**, 7221 (1992).
54. M. Ohno: *J. Electron Spectr. and Rel. Phenom.* **69**, 225 (1994).
55. P. J. Bassett and D. R. Lioyd: *J. Chem. Soc. Dalton Trans.* **1**, 248 (1972).
56. H. Tillborg, A. Nilsson, and N. Mårtensson: *Surf. Sci.* **273**, 47 (1992).
57. J. G. Lore, S. Hag, and D. A. King: *J. Chem. Phys.* **97**, 8789 (1992) and references therein.

58. G. Wendin and M. Ohno: *Phys. Scrip.* **14**, 148 (1976); M. Ohno: *Phys. Scrip.* **21**, 589 (1980); *J. Phys.* **C13**, 447 (1980); M. Ohno and R. E. LaVilla: *Phys. Rev.* **A38**, 3479 (1988); **B37**, 10915 (1988); **B39**, 8845, 8852 (1989).
59. L. S. Cederbaum, J. Schirmer, W. Domcke, and W. von Niessen: *Intern. J. Quantum Chem.* **14**, 593 (1978); L. S. Cederbaum, W. Domcke, J. Schirmer, and W. von Niessen: *Adv. Chem. Phys.* **65**, I. Prigogine and S. A. Rice (eds.), Wiley, New York (1986), and references therein.
60. C. N. R. Rao and G. R. Rao: *Surf. Sci. Rpts.* **13**, 221 (1991).
61. F. U. Hillebrecht, J. C. Fuggle, P. A. Bennett, Z. Zolnierek, and Ch. Freiburg: *Phys. Rev.* **B27**, 2179 (1983).
62. K. Siegbahn, C. Nordling, G. Johansson, J. Hedman, P. F. Hedman, K. Harmin, U. Gelius, T. Bergmark, L. O. Werme, R. Manne, and Y. Baer: *ESCA Applied to Free Molecules*, North-Holland, Amsterdam (1969).
63. A. P. Hitchcock and C. E. Brion: *J. Electr. Spectr. Rel. Phen.* **18**, 1 (1980).
64. O. Björneholm, A. Nilsson, E. O. F. Zdansky, A. Sandell, B. Hernnäs, H. Tillborg, J. N. Andersen, and N. Mårtensson: *Phys. Rev.* **B46**, 10353 (1992).
65. E. Shigemasa, K. Ueda, Y. Sato, T. Sasaki, and A. Yagishita: *Phys. Rev.* **A45**, 2915 (1992).
66. C. T. Chen, Y. Ma, and F. Sette: *Phys. Rev.* **A40**, 6737 (1989) and references therein.
67. A. Nilsson, O. Björneholm, E. O. Zdansky, H. Tillborg, N. Mårtensson, J. N. Andersen, and R. Nyholm: *Chem. Phys. Lett.* **197**, 12 (1992).
68. A. Sandell, A. Nilsson, and N. Mårtensson: *Surf. Sci.* **241**, L1 (1991).
69. C. W. Bauschlicher, Jr. and P. S. Bagus: *J. Chem. Phys.* **80**, 944 (1984).
70. W. L. Jolly: in C. R. Brundle and A. D. Baker (eds), *Electron Spectroscopy; Theory, Technique and Applications*, Vol. 1., Academic Press, New York (1977).
71. P. S. Bagus, A. R. Rossi, and Ph. Avouris: *Phys. Rev.* **B31**, 1722 (1985).
72. I. Kojima, A. K. Srivastava, E. Miyazaki, and H. Adachi: *J. Chem. Phys.* **84**, 4455 (1986).
73. W. F. Egelhoff, Jr.: *Phys. Rev.* **B29**, 3681 (1984).
74. H. Conrad, G. Ertl, J. Küppers, and E. E. Latta: *Surf. Sci.* **50**, 296 (1975).
75. J. C. Tracy: *J. Chem. Phys.* **56**, 2736 (1972).
76. U. von Barth and G. Grossmann: *Solid State Commun.* **32**, 645 (1979).
77. U. von Barth and G. Grossmann: *Phys. Rev.* **B25**, 5150 (1981).
78. E. Rühl and A. P. Hitchcock: *J. Am. Chem. Soc.* **111**, 2614 (1989).
79. A. Sandell, O. Björneholm, A. Nilsson. E. O. F. Zdansky, H. Tillborg, J. N. Andersen and N. Mårtensson: *Phys. Rev. Lett.* **70**, 2000 (1993).
80. P. Hedegård and F. U. Hillebrecht: *Phys. Rev.* **B34**, 3045 (1986).
81. M. Cini: *Solid State Commun.* **24**, 681 (1977); *Phys. Rev.* **B17**, 2788 (1978); G. A. Sawatzky: *Phys. Rev. Lett.* **39**, 504 (1977).
82. D. R. Jennison, G. D. Stucky, R. R. Rye, and J. A. Kelber: *Phys. Rev. Lett.* **46**, 911 (1981).
83. A. Sandell, O. Björneholm, A. Nilsson, B. Hernnäs, J. N. Andersen and N. Mårtensson: *Phys. Rev.* **B49**, 10136 (1994).
84. W. Eberhardt, E. W. Plummer, C. T. Chen, and W. K. Ford: *Aust. J. Phys.* **39**, 853 (1986).
85. M. Ohno: *Phys. Rev.* **Aug**, 4430 (1994).

Part II: Structure and Reactivity of Radicals on Surfaces

Electron Magnetic Resonance of Aromatic Radicals on Metal Oxide Surfaces

R. B. CLARKSON[1],* KAREN MATTSON[2], WENJUN SHI[2], WEI WANG[2], and R. L. BELFORD[2]
Departments of Veterinary Clinical Medicine[1] and Chemistry[2] and the Illinois EPR Research Center, University of Illinois, Urbana, Illinois 61801, U.S.A.

(Received: 15 December 1993; accepted: 5 July 1994)

Abstract. Multi-frequency electron magnetic resonance (EMR) methods provide a powerful approach to the study of radicals adsorbed on metal oxide surfaces. The structure, adsorption characteristics, surface environment, and mobility of surface species often can be determined. In this review, EMR studies of radicals produced on oxide surfaces from polynuclear aromatic hydrocarbons, nitroaromatics, and sulfur-containing aromatics are considered. Intra- and intermolecular spin interactions are probed by techniques which emphasize Zeeman or non-Zeeman interactions, and couplings between unpaired electrons and nuclei such as ^{1}H, ^{14}N, ^{17}O, ^{25}Mg, and ^{27}Al are discussed.

Key words. Electron magnetic resonance, EPR, ENDOR, ESE, ESEEM, aromatic radicals, nitroaromatics, thiophenes, metal oxide surfaces, alumina, silica–alumina, silica, magnesia, titania, powder lineshape, orientation selection, exact cancellation, EPR simulation, ENDOR simulation, ESEEM simulation.

1. Introduction

Electron Magnetic Resonance (EMR) is a family of spectroscopic methods that includes continuous-wave electron paramagnetic (or spin) resonance (EPR or ESR) and electron-nuclear double resonance (ENDOR), as well as pulsed techniques such as electron spin echo (ESE), electron spin echo envelope modulation (ESEEM), and pulsed ENDOR. EMR methods have been useful in studying the formation, structure, and reactivity of aromatic radicals on metal oxide surfaces since EPR instrumentation became commercially available in the late 1950s. (See O'Reilly's review for an excellent overview of early work [1]). Advances in instrumentation since then have enhanced spectral resolution for both the Zeeman and non-Zeeman interactions of these chemical systems, providing much more detailed information; advanced theoretical tools to analyze this additional information from orientationally disordered samples (powders, glasses, frozen solutions) have enabled researchers to utilize the new spectroscopic information to answer questions concerning the atomic and electronic structure of the organic radicals, as well as to probe their environment and mobility at the surface. We hope that a brief description of some EMR methods used in this work will provide a useful, if cursory overview of the rationale for their application, as well serve as a guide to the numerous acronyms by which they are conventionally identified. Basic

---

* Author for correspondence.

references for each technique also are provided. Applications of EMR to aromatic radicals on metal oxide surfaces not only provide a wealth of information concerning the radical and its surface environment, they also deepen our understanding of magnetic resonance by providing challenging experimental and theoretical problems.

2. Electron Magnetic Resonance Methods

Continuous-wave EPR was the first EMR method used to study aromatic radicals. The usual environment of these radicals on surfaces restricts their mobility, and they often produce powder-like EPR spectra. Important magnetic interactions which may contribute to the lineshapes of such spectra (depending on the structure of the radical species and the surface to which it is adsorbed) include electronic and nuclear Zeeman, nuclear hyperfine, and quadrupole terms. These often are represented in the form of an effective spin Hamiltonian, which for dilute paramagnetic samples may be written as:

$$\mathcal{H} = \mathcal{H}_S^{Zeeman} + \sum_{i=1}^{n} (\mathcal{H}_I^{Zeeman} + \mathcal{H}_{SI}^{Hyperfine} + \mathcal{H}_I^{Quadrupole})_i \tag{1}$$

The electronic Zeeman interaction in purely hydrocarbon π-radicals is nearly isotropic, and the extent of anisotropy in nitrogen, oxygen, and sulfur-containing aromatic radicals still is very modest. Since the Zeeman interaction is magnetic-field-dependent (e.g., $\mathcal{H}_S^{Zeeman} = \beta S_1 \cdot \mathbf{g} \cdot \mathbf{B}_0$), a careful investigation of field-dependent interactions often benefits from taking spectra at several microwave frequencies, particularly very high frequencies (VHF, above 35 GHz), which recently have become available to the experimentalist as commercially available components for the construction of VHF-EPR instruments. This multi-frequency EPR approach has been reviewed by Belford and Clarkson [2,3], and has proved very important in our work with all classes of aromatic radicals. VHF-EPR studies of aromatic radicals provide enhanced spectral resolution in many cases; the magnetic field separation ΔB of two slightly different g-values, g_1 and g_2, increases with the frequency in a linear fashion:

$$\Delta B = \left(\frac{h}{\beta}\right)\left(\frac{1}{g_1} - \frac{1}{g_2}\right)\nu \tag{2}$$

In addition to its enhanced spectral resolution, VHF-EPR also has excellent sensitivity ($<10^8$ unpaired spins at 95 GHz [3]), which can be useful in studying samples of limited size, and provides an extended range of sensitivity to motion-induced effects.

Electron-nuclear hyperfine interactions often give enough spectral information to make possible an unambiguous identification of the radical under investigation. In low-viscosity liquids, where rapid tumbling averages away the anisotropic components, well-resolved, isotropic hyperfine information gives a detailed account of molecular structure as it is reflected in the wide range of interactions that are possible between unpaired electrons and nuclei with magnetic moments ($I \neq 0$). The information content is even richer when hyperfine interactions are observed in solids, since then the full range of isotropic and anisotropic interactions may be investigated to provide information on the distances and orientations of specific dipole–dipole electron–nuclear couplings, as well as on isotropic contact terms – information dependent on *both* the atomic arrangement and the electronic structure of the radical. This can be seen in the expressions for a single hyperfine interaction:

$$\mathcal{H}_I^{Hyperfine} = \mathcal{H}_{Contact} + \mathcal{H}_{Dipolar} \quad (3)$$

$$\mathcal{H}_{Contact} = \frac{8\pi}{3} g\beta g_N \beta_N |\psi(0)|^2 \mathbf{S} \cdot \mathbf{I} \quad (4)$$
$$= hA_0 \mathbf{S} \cdot \mathbf{I}$$

$$\mathcal{H}_{Dipolar} = g\beta g_N \beta_N \left[\frac{\mathbf{S} \cdot \mathbf{I}}{r^3} - \frac{3(\mathbf{S} \cdot \mathbf{r})(\mathbf{I} \cdot \mathbf{r})}{r^5} \right] \quad (5)$$

Here, A_0 is the isotropic hyperfine coupling, and r is the distance between the nucleus and unpaired electron. In the case where solid samples are single crystals, particular orientations of the crystal in the external magnetic field result in spectra with features that can be related to the angles (θ, ϕ in spherical polar coordinates) which describe the orientation. By rotating the crystal and taking spectra at intervals of (θ, ϕ), one can determine the anisotropic Zeeman and hyperfine interactions, providing a detailed set of data which can be analyzed to determine atomic and electronic structure.

Aromatic radicals on surfaces usually appear as disordered solids (frozen solutions or powders) rather than as single crystals. In this case, the EMR spectra consist of simultaneous contributions from all possible orientations ($\Sigma\Sigma\theta$, ϕ), and are called *powder spectra*. All of the orientation-dependent information from Zeeman and hyperfine interactions is convoluted into a single powder lineshape, which often appears as a broad, asymmetric resonance with several turning points. While Zeeman anisotropy in many aromatic radicals is small (particularly for hydrocarbon species), there remains substantial hyperfine anisotropy, which holds important information concerning the structure, orientation, and environment of the surface species, and which in powdered samples usually cannot be analyzed accurately by EPR alone. A desire to improve resolution in powders led Hyde *et al.* to first apply *ENDOR* spectroscopy to frozen solutions of triphenylmethyl radicals [4], and this method has proven very useful for radicals on surfaces as well. Good introductions to the method of continuous-wave (cw) ENDOR in powders may be found in references [5] and [6], while the technique applied to radicals in solution is beautifully treated in reference [7]. Pulsed ENDOR, a more recent methodology (reviewed in reference [8]), has additional advantages for the study of surface radicals. ENDOR spectra from suitably prepared samples of aromatic radicals adsorbed on metal oxide surfaces often give well-resolved and complex spectra with linewidths of ca. 50–100 kHz, an improvement in resolution of better than two orders of magnitude over conventional cw EPR.

Electron Spin Echo (ESE) methods provide information on the weakest hyperfine couplings, often reporting on electrons and nuclei separated by more than 1.0 nm [9–11]. ESE is particularly well-suited to examine the environment of radicals adsorbed on surfaces, in part because it gives information on the distances, orientations, and number of neighboring nuclei ($I \neq 0$), with a sensitivity that does not decrease for low-g_n nuclei, as it does for cw ENDOR. Thus, it is possible to examine the immediate surroundings of a surface radical and study hyperfine interactions with nuclei like ^{17}O, ^{25}Mg, ^{27}Al, and ^{29}Si *from the vantage point of the adsorbed paramagnetic species*. Furthermore, in the multi-frequency approach to ESE work, it sometimes is possible to find a value of the magnetic field for which the nuclear Zeeman interaction just equals the isotropic portion of the hyperfine coupling. Then, if the nucleus under investigation has $I = 1$ (e.g. ^{14}N), a cancellation of terms occurs which leaves the nuclear quadrupole coupling as the only effective splitting mechanism for the nuclear spin levels in one electronic spin mainfold, producing electron spin echo envelope modulation (ESEEM) from the nucleus that reports on the pure nuclear quadrupole resonance (NQR) [12,13]. Not only does this enhance spectral resolution for nitrogen-containing radicals, it also provides NQR information which may be useful in interpreting the surface environment of adsorbed species.

Regardless of the EMR spectroscopic technique that is used, spectra from disordered materials present a challenge to interpret. The convolution of spectral intensity from all orientations produces powder lineshapes, which require very careful simulation before reliable determinations of the **g**, **A**, and **Q** matrices can be made. In the cases of ENDOR and ESE, additional attention must be paid to the effect of Zeeman anisotropy in producing 'orientation selection', which can produce spectra that have single crystal characteristics. Approaches that are used in the simulations described in this paper have been reported for EPR [14,15], ENDOR [16,17,18], and ESEEM [19,20,21]. These simulation techniques have become less time consuming since the introduction of automated fitting strategies, which can begin with a set of reasonable but somewhat inaccurate parameters for the spin Hamiltonian, and converge onto an optimized best fit, with good reliability. Among the most robust of the optimization methods, the multi-variable simplex approach is one which we frequently employ [22,23].

3. Hydrocarbon Polynuclear Aromatic Radicals

The adsorption of polynuclear aromatic compounds (PAs), from the gas phase or from solution, onto suitably activated metal oxide surfaces is known to produce radical ions. Some of the very first surface chemistry studied by EPR involved the production of perylene radicals on activated alumina and silica-alumina powders [24,25,26]. Unlike other radical species which can be formed on MO_x surfaces by one-electron transfer reactions (e.g., O^-, O_2^-, CO_2^-), many PA radicals are easily prepared, remain stable for hours or days at room temperature, and can be formed in concentrations suitable for all EMR spectroscopic methods (ca. 10^{12}–10^{14} radicals per cm^2 of surface). Ultraviolet or gamma irradiation may be used to enhance the extent of electron transfer if higher radical densities are desired. Alternative synthetic routes for the production of cation and anion radicals (e.g. $PA^{\bar{\cdot}}$ produced with Na or K metal reduction; $PA^{\dot{+}}$ formed in conc. H_2SO_4 [27,28], or by UV irradiation in boric acid glass [29]) provide more homogeneous preparations, which are important for comparison purposes in order to identify surface-related spectroscopic effects. The overall chemistry of PA radicals on oxide surfaces offers a rich and flexible system, with opportunities to investigate many interactions important for surface science and catalysis, as well as a synthetic route for the formation and stabilization of all classes of aromatic radicals.

Many EPR studies involving PA radicals on metal oxides have been designed to inform about the nature of one-electron donor and acceptor sites on model catalysts and catalyst supports. Work on Al_2O_3 [30], SiO_2/Al_2O_3 [31], and Mo/Al_2O_3 [32], for example, showed that the ease of formation of $PA^{+\cdot}$ radicals was a linear function, inversely proportional to the ionization potential (I.P.) of the parent PA compound. Richardson further demonstrated that the extent of PA radical formation on metal-ion exchanged faujasite was related to the difference between the I.P. of selected PA molecules and the electron affinities of the corresponding cations ($PA^{+\cdot}$) [33]. Dollish and Hall first observed that on γ-alumina, this one-electron transfer process required the presence of molecular oxygen [34], and Lewis and Singer demonstrated that in an inert solvent, mild thermal treatment with oxygen alone was sufficient to create radicals from a variety of PA molecules, including anthracene, tetracene, pentacene, perylene, and pyrene [35]. The role of oxygen as a catalyst or participant in the surface electron transfer process still is under investigation. Photochemical reactions of PA molecules at oxide surfaces provide additional insights into the formation of cation radical species [36].

Because the loss of spectral resolution when PA radicals are observed on surfaces makes an unambiguous assignment of the paramagnetic species difficult or impossible *by EPR alone*, several models have been put forth concerning the exact nature of the adsorbed species. For perylene (Pe) radicals formed on activated *γ-alumina*, for example, it has been suggested that the surface species is the cation ($Pe^{+\cdot}$) [34], the anion ($Pe^{-\cdot}$) [37], an aryloxy radical which develops from $Pe^{+\cdot}$ over time [38], and a pair of radicals, one of which is formed by electron transfer ($Pe^{+\cdot}$) while the other is the result of hydrogen addition from Brønsted acid sites ($PeH^{+\cdot}$) [39]. Each of these possible structures has different EMR spectral features, and it is possible to distinguish between them by suitable spectroscopic methods. Because of unfavorable linewidths in these systems, it also has been impossible to resolve by EPR the weak hyperfine interactions between the PA surface radicals and neighboring magnetic nuclei at the surface – interactions which undoubtedly do exist, and which can help to describe the nature of the donor/acceptor sites.

In order to address the many important scientific questions which are posed by PA radicals on oxide surfaces, it is advantageous to adopt the broader EMR approach, making use of the complementary information available from multi-frequency EPR, ENDOR, and ESE. To demonstrate this multiple spectroscopic approach, it is useful to begin with the spectrum of $Pe^{+\cdot}$ in concentrated H_2SO_4, shown in Figure 1(a). Here, rapid tumbling of the cation radical averages away most of the anisotropy in the hyperfine interactions, **A**, leaving a well-resolved narrow line spectrum, from which the isotropic coupling constants ($A_0(i)$, see Equation 4) may be determined. Slight anisotropy in **g** also is averaged away by the tumbling. The 60 hyperfine lines and three isotropic coupling constants in the $Pe^{+\cdot}$ solution spectrum form an overall lineshape that unambiguously can be assigned to the radical cation.

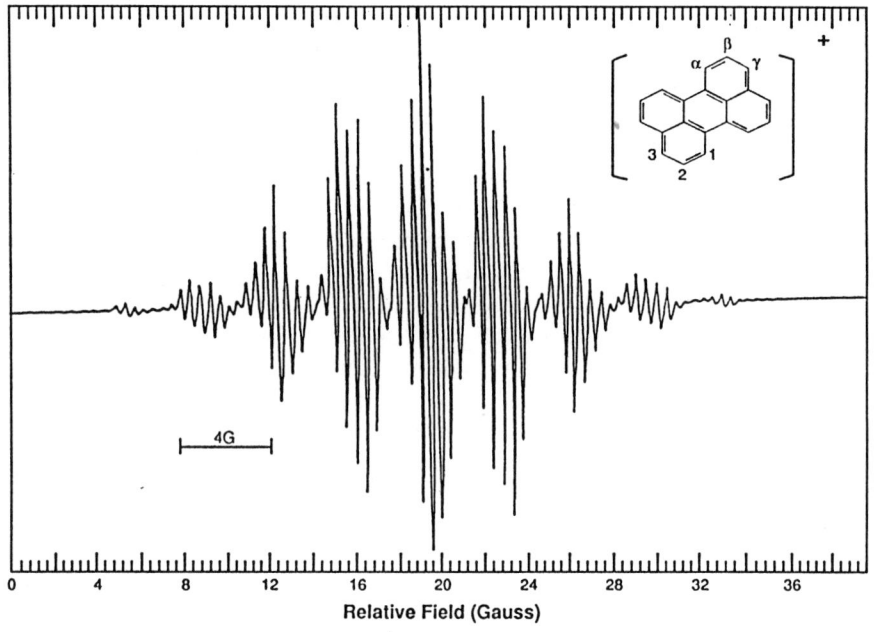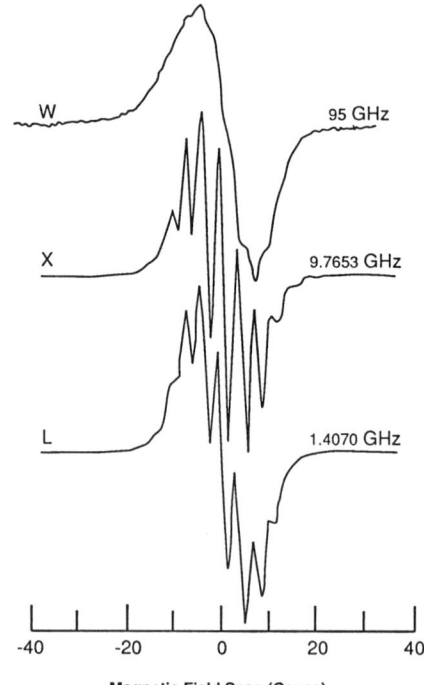

Fig. 1. (a) Solution spectrum of Pe^{+} in conc. H$_2$SO$_4$ taken at 9.5 GHz, room temperature. (b) Spectra of perylene radicals on alumina at 1 GHz, 9.5 GHz, and 95 GHz, room temperature.

Figure 1(b) shows EPR spectra of the radical formed when Pe is adsorbed on activated alumina, measured at three different magnetic fields and microwave frequencies. At all frequencies, there is an obvious loss of spectral resolution when compared to the solution spectrum. Several possible mechanisms, individually or in combination, could account for this change in the spectrum. Anticipating results which will be presented later, we judge it most likely that the major sources of lineshape change in this system are: (i) slower tumbling on the surface, resulting in incomplete averaging of **A**, and to a much lesser extent, **g** anisotropy, and (ii) shorter transverse relaxation (T_2) due to surface interactions. It is now apparent that the loss of EPR resolution on the oxide surface has obscured some vital spectral (hyperfine) information that is needed in order to make a firm assignment of the structure being observed. This lineshape change is the source of much of the ambiguity concerning the nature of the species, and clearly suggests the use of other EMR methods to improve resolution.

Before addressing the problem of hyperfine resolution for species partially immobilized on surfaces, one should consider the other type of resolution, namely Zeeman. Zeeman terms are field-dependent, and higher B_0 fields result in better resolution, as Equation (2) illustrated. The further loss of spectral resolution for Pe radicals as the field/frequency is raised from 0.33T/9.5 GHz to 3.3T/95 GHz in Figure 1(b) is due in part to the greater resolution of electron Zeeman interactions, some of which are random variations in **g** due to inhomogeneous radical environments, and others are truly characteristic of the slight but real anisotropy in the **g** matrix of this species. Random variations in **g** due to environmental inhomogeneities has been termed 'g-strain broadening' (see Froncisz and Hyde [40]); several groups have shown that this frequency-dependent broadening is not necessarily a limitation to resolution for organic radicals [41,42]. While the magnitude of g-anisotropy in Pe radicals is small, as predicted by Stone [43], there is a measurable difference between g_{xx}, g_{yy}, and g_{zz} which must be taken into account when analyzing the VHF-EPR spectrum. Preliminary analyses of the W-band (95 GHz) spectrum shown in Figure 1(b) suggests that the extent of g-anisotropy is well accounted for by $g_{xx} = 2.00223$, $g_{yy} = 2.00278$, and $g_{zz} = 2.00313$ ($\langle g \rangle = \text{Tr } \mathbf{g} = 2.00271$; $\langle g \rangle - g_e = \Delta g = 4 \times 10^{-4}$). The small value of Δg in this case is evidence to support a strictly hydrocarbon species for Pe radicals on alumina; the aryloxy radical (PeO·), for example, is expected to have a value of Δg at least twice this measured amount [35]. VHF-EPR plays an important role in studying such systems, and its potential to identify surface-induced changes in the g-matrix is yet to be explored; applications to heterocyclic aromatic radicals, where Δg is an order of magnitude larger, will be discussed in following sections.

In order to achieve better resolution of hyperfine information from radicals on surfaces, ENDOR spectroscopy may be employed. Figure 2 shows two ENDOR spectra of Pe on activated Houdry-M46, a silica–alumina catalyst [23]. One spectrum is from a sample prepared by contacting the catalyst with a 1 mM solution of Pe in benzene; the other from a sample in which the Pe was introduced onto the catalyst as a vapor. With the exception of a spectral feature at 14 MHz that is present with much greater amplitude in the case of solution deposition and absent for vapor deposition, the two spectra are nearly identical. The 14 MHz peak results from long-range (0.5–1 nm) dipolar interactions between the unpaired

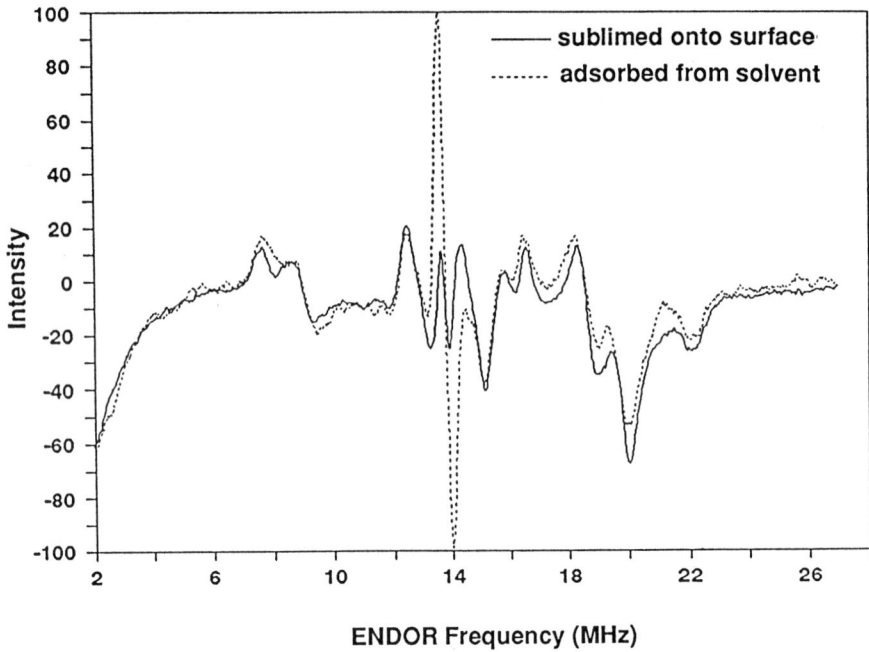

Fig. 2. ENDOR spectra of perylene radicals adsorbed on Houdry M-46 silica–alumina catalyst. Sublimed onto surface (———); adsorbed from benzene (------). $T = 133$ K, $B_0 = 3300$ Gauss [23].

electron and protons (primarily solvent protons in this case), and has been called the 'matrix proton' peak by Hyde [4]; it occurs at the nuclear Larmor frequency ($\nu_n = g_n \beta_n B_0$). Consideration of these long-range interactions will be deferred until later.

The complex lineshape in Figure 2 represents the ENDOR spectrum from protons in the Pe radical. Because of the very small g-anisotropy of this radical and the rapid cross relaxation between different portions of the hyperfine spectrum, the position of the electronic Zeeman excitation has little effect on the signal. Instead, the spectrum represents a powder ENDOR pattern from several magnetically inequivalent protons with anisotropic hyperfine interactions. In order to get a better picture of how this spectrum is related to the ring proton structure in the radical, it is useful to consider the somewhat simpler case of the coronene radical on Houdry M-46. The ENDOR spectrum of this species is shown in Figure 3. The unperturbed molecule can be expected to have 12 equivalent protons, and a simulation of the spectrum based on this assumption also is illustrated in the figure. The excellent agreement between experiment and simulation is no guarantee that this assumption is valid, but it cannot be seriously invalid, or the ENDOR spectrum could not be accounted for with a single anisotropic hyperfine interaction. For coronene on Houdry M-46, one obtains, by simulation, values for the diagonal elements of the **A** matrix of $A_{11} = 1.212$ MHz, $A_{22} = 2.598$ MHz, and $A_{33} = 5.354$ MHz. Since the ground state for a planar coronene molecule would be orbitally degenerate, it has been suggested that the structure undergoes a Jahn–Teller

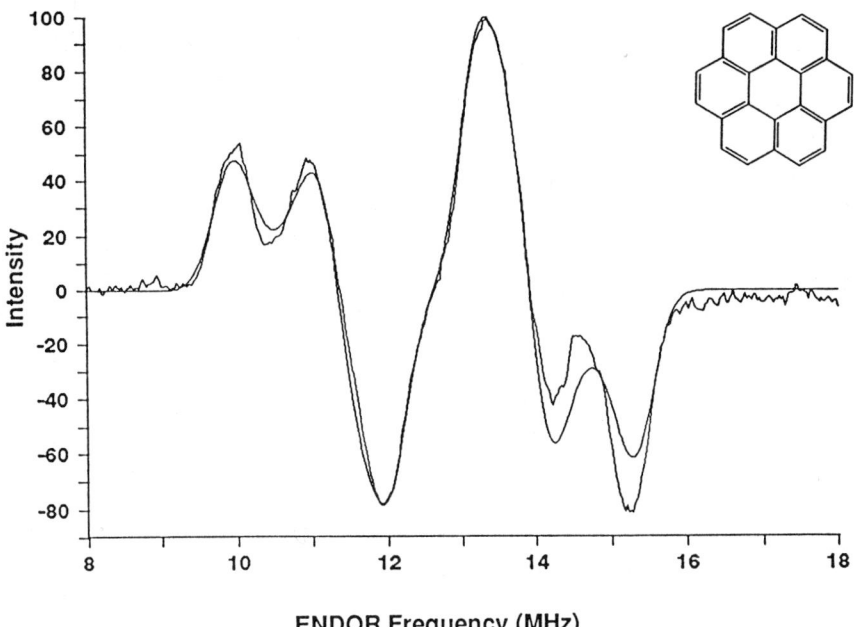

Fig. 3. ENDOR spectrum of coronene radicals adsorbed on Houdry M-46. Experimental spectrum (———); simulated spectrum (------). $T = 133$ K, $B_0 = 3300$ Gauss [23].

distortion, which could make some of the 12 ring protons magnetically inequivalent [44]. If this is the case, the ENDOR results can only be understood if it is assumed that: (i) a dynamic J–T distortion is causing the inequivalent sets of protons to interconvert at a rate fast compared with T_{1n}; or (ii) the J–T distortion preserves the equivalence of the protons; or (iii) inequivalences are too minor to be resolved by ENDOR. While a better understanding of this system will require additional work, it is clear from the current data that coronene radicals on silica–alumina are highly symmetric, with protons whose ENDOR spectrum is well-simulated by assuming that they all are equivalent. An unusual feature unique to the coronene radical on silica-alumina is that the calculated isotropic hyperfine coupling ($A_0 =$ Tr **A** $= 3.055$ MHz) from ENDOR data is much lower than that measured in solution by EPR (4.20 MHz [45]), again suggesting a surface perturbation or dynamical process which requires further study.

The ENDOR lineshape in Figure 3 is the result of a single anisotropic proton hyperfine interaction. In all experimental work from our laboratory, frequency-modulated ENDOR is used, so the data is obtained as the first derivative of an absorption line. Figure 4 illustrates both the absorption and first-derivative lineshapes for the powder ENDOR spectrum of coronene, and notes the positions of the canonical A-values. In general, a powder ENDOR pattern like that for coronene is contributed *by each inequivalent proton* in a PA radical, and thus the analysis of a lineshape like that for $Pe^{+\cdot}$ requires three anisotropic proton hyperfine interactions. Figure 5(a) shows the ENDOR spectrum from $Pe^{+\cdot}$ together with a

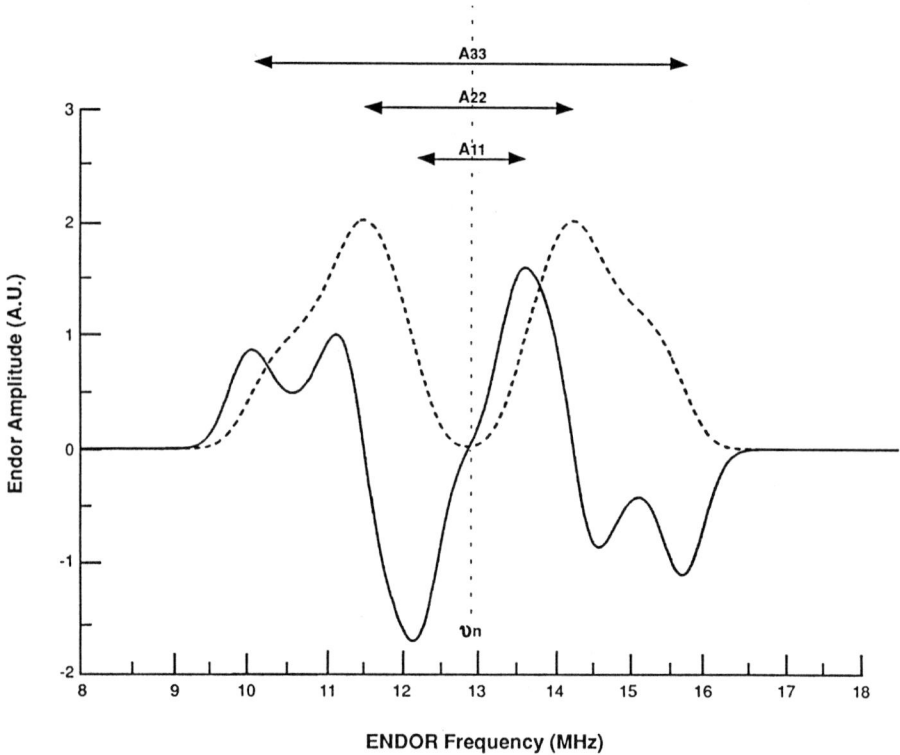

Fig. 4. Simulated ENDOR spectra for coronene. Both absorption (------) and first derivative (——) lineshapes are displayed, together with the values and positions of the canonical A-values. For this simulation, $B_0 = 3053$ Gauss, and $A_{11} = 1.212$ Mhz, $A_{22} = 2.598$ MHz, and $A_{33} = 5.354$ MHz.

simulation; Figure 5(b) shows how the sum of contributions from the three inequivalent perylene protons yields the observed spectrum. Details of the analysis of this species are given in Table I [46], together with comparison data and theoretical predictions [47,48]. By comparing the A-values obtained from a simulation of the powder ENDOR data with theoretical predictions and isotropic couplings measured in solution, and remembering the low value of Δg measured by VHF-EPR, one finds it most reasonable that the perylene radical formed on alumina and silica–alumina is indeed Pe^+. Adsorption onto the surface does not observably alter the symmetry of either Pe or coronene, and no direct evidence for the role of oxygen in the electron transfer process is detected by these measurements. Similar ENDOR measurements have been made for a wide variety of hydrocarbon and heteroatomic (O, N, S) PA radicals on alumina, silica–alumina, as well as in more homogeneous preparations with H_2SO_4 or boric acid glasses (for example, see references [49–51]). ENDOR of perylene radicals has been observed with Pe adsorbed on activated forms of Al_2O_3, SiO_2/Al_2O_3, SiO_2, decatonated NH_4-X zeolite, TiO_2, and MoO_3, and effects of differing surface interactions with the adsorbed species were detected by variations in the β-proton hyper-

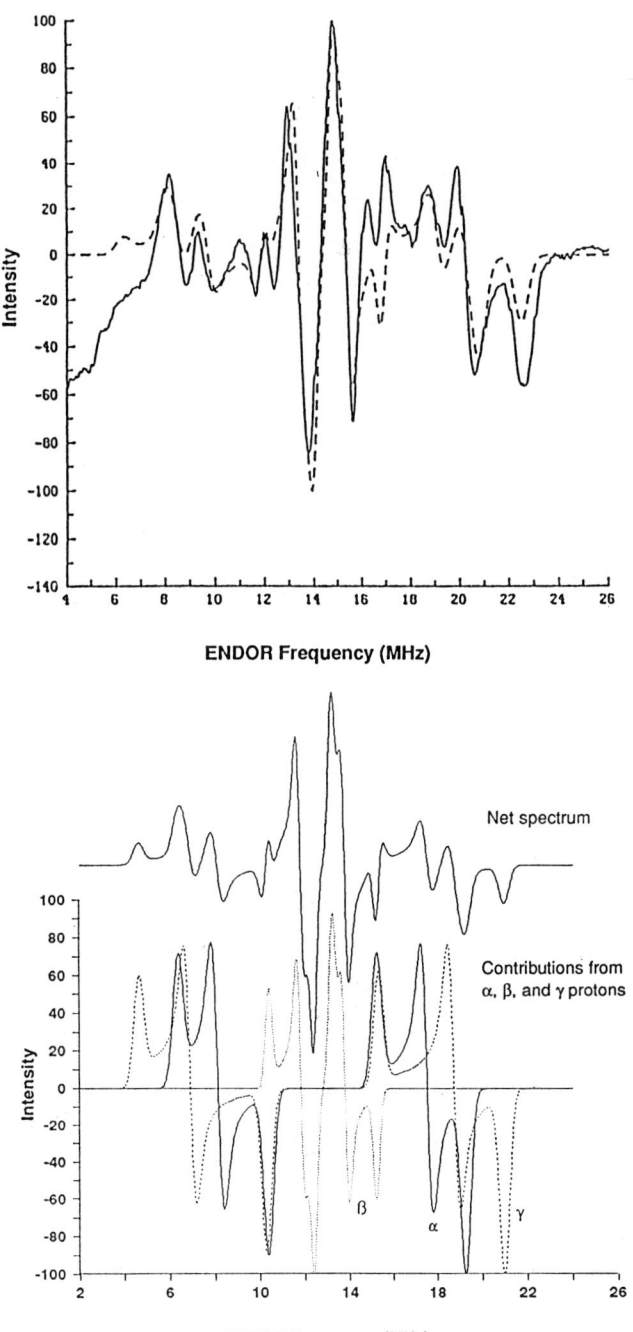

Fig. 5. (a) ENDOR spectrum of perylene radicals on Houdry M-46. Experimental (———) and simulation (------). Simulation parameters are given in Table I [46]. (b) Convolution of contributions from the three inequivalent proton groups, α, β, and γ, which combine to form the powder ENDOR spectrum of perylene cation radical [23].

TABLE I
Hyperfine data for perylene radicals. All values given in MHz

Proton		$|A_{xx}|$	$|A_{yy}|$	$|A_{zz}|$	$|\langle A \rangle|$ or A_0	Comments	
γ	SiO_0/Al_2O_3	4.7	12.2	16.7	11.2	ENDOR, on Houdry M-46 [46]
	Al_2O_3	4.8	12.2	16.8	11.3	ENDOR, on alumina [46]
	Theory	−6.0	−11.7	−15.9	−11.2[a], −12.4[b]	References [46] and [47]
	Liquid solution (+)				−11.44	TRIPLE, cation in liquid TFA [47]
	Liquid solution (−)				9.8	ENDOR, anion in DME [48]
β	SiO_0/Al_2O_3	0.9	2.3	4.8	2.7	ENDOR, on Houdry M-46 [46]
	Al_2O_3	1.0	2.3	4.8	2.7	ENDOR, on alumina [46]
	Theory	−0.3	2.9	4.5	2.4[a], 4.1[b]	References [46] and [47]
	Liquid solution (+)				1.26	TRIPLE, cation in liquid TFA [47]
	Liquid solution (−)				1.2	ENDOR, anion in DME [48]
α	SiO_0/Al_2O_3	5.2	9.7	12.6	9.2	ENDOR, on Houdry M-46 [46]
	Al_2O_3	1.35	9.7	12.6	9.2	ENDOR, on alumina [46]
	Theory	−5.4	−9.9	−12.2	−9.2[a], −8.7[b]	References [46] and [47]
	Liquid solution (+)				−8.6	TRIPLE, cation in liquid TFA [47]
	Liquid solution (−)				8.6	ENDOR, anion in DME [48]

[a] Elementary calculations based on SCF-MO π-electron spin densities; values for cation radical [46].
[b] INDO calculations [47].

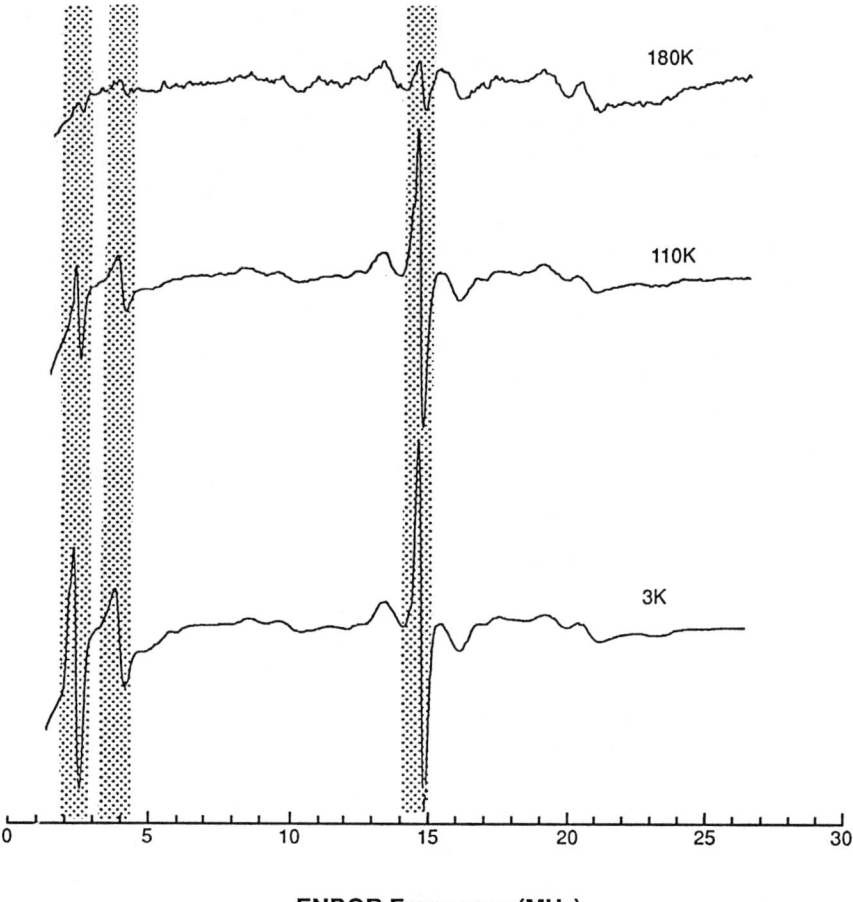

Fig. 6. ENDOR spectra of Pe$^{\pm}$ radical on γ-alumina at 3 K, 110 K, and 180 K. Samples were prepared by contacting activated alumina with a 2 mM solution of perylene in d_6-benzene. Spectrometer gain is identical for each spectrum. Adapted from [55], with permission.

fine coupling constants [52,53]. A clearer understanding of the effects of electrostatic perturbations on the β-proton couplings, a question first considered by Reddoch [54], is a subject for future work.

If unpaired electrons in surface radicals can be thought of as *observation posts* of their environment, then it is certain that they observe magnetic nuclei other than those chemically bound to the paramagnetic molecule. Potential candidates for observation by EPR are limited due to the weak character of such dipole-dipole interactions (usually <1 MHz) and the customary EPR linewidths ($\Delta H_{pp} > 1$ Gauss = 2.8 MHz), but coupling to nuclei like ^{1}H, ^{2}H, ^{13}C, ^{11}B, ^{17}O, ^{27}Al, and ^{29}Si can be detected by ENDOR or ESE techniques under appropriate conditions. As an example, consider the spectra in Figure 6, all of which represent

ENDOR of Pe$^{\pm}$ on activated alumina [55]. At 180 K, a weak signal symmetric about 14.3 MHz represents the now familiar ENDOR from perylene ring protons. At the proton Larmor frequency, $\nu_n = 14.3$ MHz ($B_0 = 3359$ Gauss), a sharp matrix proton resonance (due to nonbonded protons from neighboring Pe molecules and the alumina surface) is seen. Very weak signals also are seen at lower frequencies. As the temperature is lowered, mobility of the surface radicals is more and more restricted, and there is a dramatic growth in the intensity of three peaks at 14.3, 3.7, and 2.30 MHz. Rothenberger et al. suggested that these resonances represented dipolar couplings with ^{1}H, ^{27}Al, and ^{2}H respectively, and they showed by isotopic substitution that protons associated with the alumina surface are important contributors to the ^{1}H signal [53]. Similar ENDOR measurements of Pe$^{\pm}$ on a silica–alumina (Houdry M-46; 88% SiO_2, 12% Al_2O_3) failed to detect the resonance ascribed to aluminum, suggesting that the surface sites occupied by the radicals on the two surfaces are different. On alumina at 3K, the perylene cation radical appears to be relatively immobile at a surface site whose composition includes protons and aluminum, a situation seemingly well-described by the Al^{+3} (cus) site on partially dehydroxylated alumina proposed by Burwell [56] and reviewed by Knözinger [57]. Simulation of ENDOR peaks from nonbonded nuclei can yield quantitative information concerning the environment of unpaired electrons, as Kevan and coworkers have shown [58].

The fact that conventional cw ENDOR becomes less sensitive for low-g_n nuclei clearly is a drawback if one is interested in studying the interactions of surface radicals with elements like oxygen or aluminum. Operating at higher fields/frequencies can help to increase sensitivity, as Möbius has shown [59]. Alternatively, information on dipolar coupling can be obtained through an analysis of electron spin echo envelope modulation (ESEEM) data. This approach was pioneered by Kevan and coworkers, who have reviewed the literature through 1987 [60]. Because the primary data from this experimental method are collected in the time domain, it is necessary either to simulate the ESEEM pattern or Fourier transform the data and continue analysis in the frequency domain. Utilizing a three pulse (stimulated echo) technique at 8.886 GHz, Snetsinger et al. examined Pe$^{\pm}$ radicals on alumina [61]; ESEEM data and Fourier-transformed data are illustrated in Figures 7(a) and 7(b). The simulation of the ESEEM pattern assumed one ^{27}Al nucleus at a distance of 0.33 nm from the unpaired electron and three protons at a distance of 0.70 nm. In their study of ·CH$_2$OH radicals produced by radiolysis on A-type zeolites, Dikanov et al. also saw ESEEM patterns which included modulation from aluminum [62]. Their simulation of the time-domain data indicated that a single ^{27}Al atom was located at a distance greater than 0.4 nm, corresponding to a more loosely bound species than Pe$^{\pm}$ on alumina. Thus it seems that the identity of the 3.7 MHz peak seen by ENDOR for perylene on alumina is confirmed by ESE measurements to be aluminum. Additionally, ESEEM data has provided a more detailed picture of the coordinationally unsaturated Al^{+3} site. In ESE experiments at microwave frequencies *lower* than X-band, stronger ESEEM often can be observed for weakly interacting nuclei like those seen on oxide surfaces by adsorbed radicals [63]; this approach seems likely to provide even more detailed information concerning active sites responsible for the

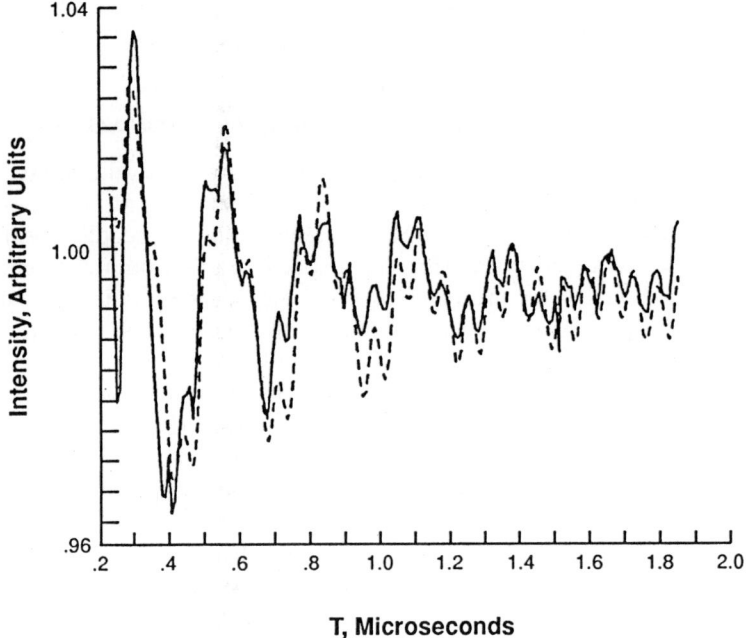

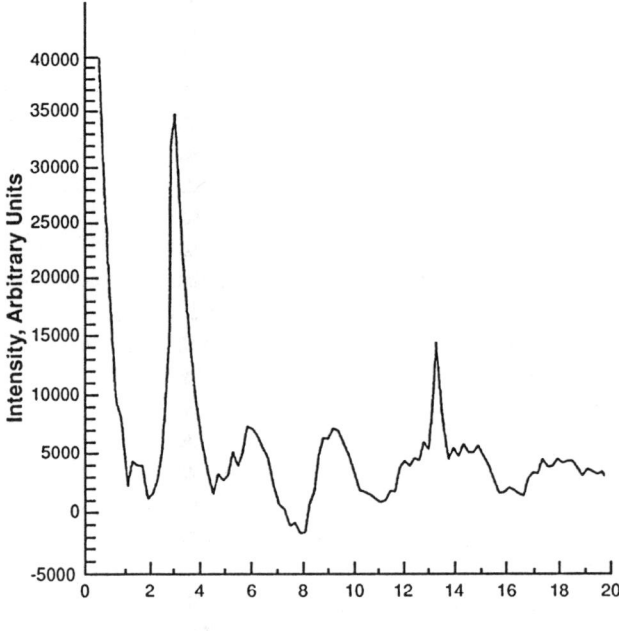

Fig. 7. (a) Simulated (------) and experimental (———) ESEEM of perylene cation radical in benzene on alumina. B_0 = 3160 Gauss; three-pulse stimulated echo sequence; T = 4.2 K [61]. (b) Frequency domain representation of data in Fig. 7(a). The peak at 3.4 MHz is attributed to ^{27}Al, while that at 13.5 MHz is due to matrix protons. Adapted from [61], with permission.

formation of radicals on surfaces. Examples of the application of low-frequency ESE (S-band, 2–4 GHz) will be given in following sections.

4. Polynuclear Aromatic Radicals Containing Heteroatoms

Heteroatoms such as oxygen, nitrogen, and sulfur introduce additional complexity into the EMR spectra of polynuclear aromatic radicals on oxide surfaces. Radicals of furans, nitroaromatics, thiophenes, and similar species readily form on many activated oxides by electron transfer. Such species exhibit both Zeeman and non-Zeeman effects; it is the presence of the heteroatoms, which usually can be neglected in the discussion of spectra from purely hydrocarbon PA radicals, which cause the significant Zeeman effects. In addition to the usual hyperfine interactions, which already have been encountered, heteroatomic species can exhibit a substantial g-anisotropy, brought about by spin–orbit (SO) coupling of the unpaired electron with the heteroatom. This source of anisotropy is almost negligible in hydrocarbon radicals, because of the small SO coupling constant (λ) of carbon (e.g., $\lambda = 29 \text{ cm}^{-1}$ (C), 76 cm^{-1} (N), 151 cm^{-1} (O), and 382 cm^{-1} (S) for valence p electrons). Furthermore, ^{14}N, ^{17}O, and ^{33}S all have $I > 1/2$, and thus possess a nuclear quadrupole moment which influences EMR spectra in several ways, including providing additional modulation mechanisms for ESE observations.

4.1. NITROAROMATIC ANION RADICALS

The anion radicals of many nitroaromatic compounds form by electron transfer on activated metal oxide surfaces. From initial experiments performed by Tench and Nelson on nitrobenzene/ZnO [64], *anion radicals of dinitrobenzene, trinitrobenzene*, and other nitrobenzene derivatives have been detected by EPR on other surfaces, including alumina [65]. Figure 8(a) shows a typical X-band (ca. 9.5 GHz) EPR spectrum of the anion radicals formed when γ-alumina that had been thermally activated *in vacuo* at 650 °C for 24 h, treated with one atmosphere of pure oxygen for one hour, and sealed in a quartz tube with a septum cap and allowed to cool to room temperature, is contacted with a 1 mM solution of m-dinitrobenzene (m-DNB) in benzene. The main spectral features are the result of the ^{14}N hyperfine coupling of the strongly coupled nitrogen ($I = 1$, three hyperfine lines). In contrast, effects of the weakly coupled nitrogen are much too small to be resolved by cw EPR. Clearly, the two equivalent nitro groups of the parent neutral molecule have become very inequivalent in the surface-bound radical form of m-DNB [66]. Figure 8(b) shows the same sample observed at W-band (95 GHz) [66]. The higher field/frequency has resolved the full g-anisotropy in the radical, which is known to be the anion, m-DNB$^{\bar{}}$. For the first time, details of the effects of nitrogen on the electronic Zeeman interaction clearly can be seen and studied. Table II summarizes results of VHF-EPR studies of dinitrobenzene and m-dinitrobenzene on MgO and γ-alumina.

Differences in g, and particularly in A-values for m-DNB$^{\bar{}}$ on alumina and magnesia suggest that the nature of the surface has an effect on these parameters, and current work is focussing on developing a better understanding of this effect. The invariance of g_{zz} in the three systems studied (and its near equivalence to the

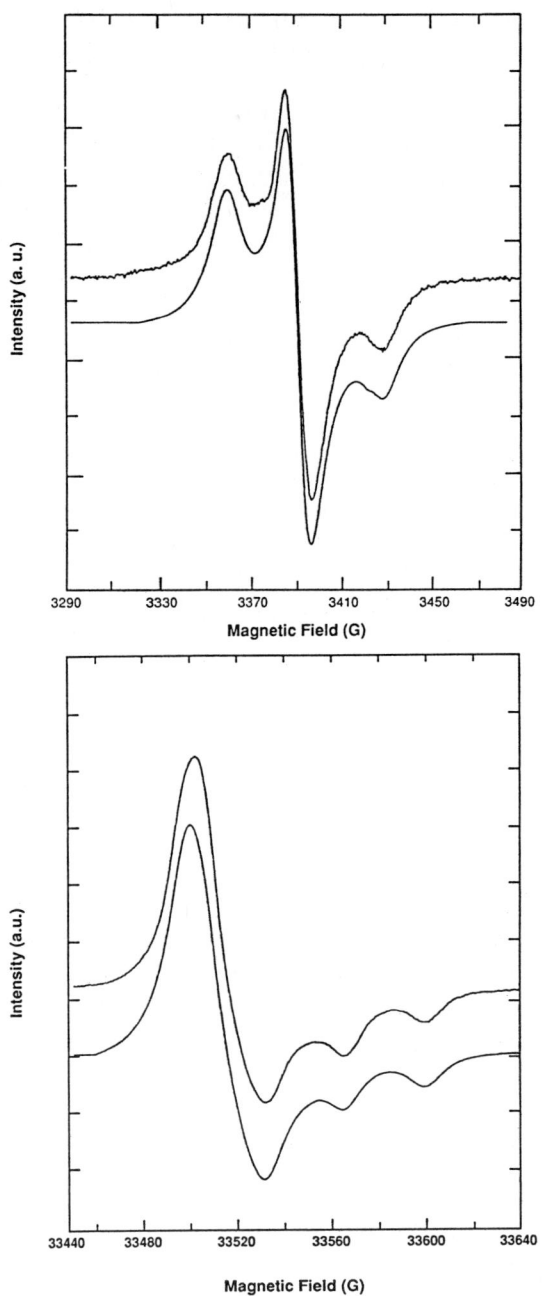

Fig. 8. (a) X-band (9.5226 GHz) spectrum of m-DNB$^{\bar{\cdot}}$ on γ-alumina. Upper trace is experimental, lower trace is simulation. (b) W-band (94.0492 GHz) spectrum of the sample used in Figure 8(a). Upper trace is experimental, lower trace is simulation.

TABLE II
EPR parameters for NB⁻ and m-DNB⁻ on alumina and magnesia. Values obtained by simulation of VHF-EPR data taken at W-band (95 GHz)

Species/Metal Oxide	NB/γ-alumina	m-DNB/γ-alumina	m-DNB/MgO
g_{xx}	2.00480	2.00515	2.00553
g_{yy}	2.00562	2.00585	2.00601
g_{zz}	2.00212	2.00212	2.00212
$\langle g \rangle$	2.00418	2.00437	2.00455
A_{xx}*	29	17	14
A_{yy}	21	16	15
A_{zz}	100	97	86

* A values in MHz. See reference [64] for experimental and theoretical details.

g-value of a free nitrogen atom, 2.0021 [67]) may imply that p_{xy} orbitals participate in the formation of weak bonds between the radical and the surface, while the p_z remains largely uninvolved, as would be the case if the nitrogen atom displayed sp^2 hybridization. The symmetry of the radical is nonaxial, a fact which could not be appreciated at lower frequencies. Formation of nitrobenzene and dinitrobenzene radicals on ZnO, TiO_2, and Houdry M-46 also was observed, although only the three samples in Table II were studied in detail.

Hyperfine coupling to the weakly-interacting nitrogen in m-DNB⁻ cannot be resolved by EPR; Singel and coworkers first used ESE to study this interaction for the radical on alumina [12,13]. Additionally, they postulated that for a correct choice of magnetic field strength B_0, the nuclear Zeeman interaction in this weakly coupled nitrogen (^{14}N) could be made to cancel exactly the isotropic hyperfine interaction, leaving only the nuclear quadrupole interaction as an effective mechanism to provide some of the nuclear state splittings producing ESEEM, as seen in Equation (6).

$$\mathcal{H}_n = U \cdot \mathbf{I} \left(\pm \frac{a_{\text{isot}}}{2} - g_n \beta_n B_0 \right) + \mathcal{H}_Q \qquad (6)$$

In this expression, $U = \mathbf{B}_0/B_0$ is a unit vector in the direction of the static magnetic field. At 'exact cancellation' in the electron spin manifold where it occurs, $\mathcal{H}_n = \mathcal{H}_Q$, and a simplification of the ESEEM pattern is predicted, allowing the direct observation of NQR transitions. Figure 9 shows Fourier-transformed ESEEM from m-D^{14}NB⁻ on γ-alumina; the four spectra are taken at four different field/frequency values. Three of the peaks in these spectra do not move with field, marking them as NQR transitions. The fourth peak increases in frequency as the field strength increases, indicating it belongs to the other electron spin manifold, where cancellation will not occur.

Experimental and theoretical work suggests that the most complete cancellation effect will occur at a frequency near 3 GHz in these samples. Considerable discussion has centered on whether 'exact cancellation' has any real significance in systems with g and A-anisotropy, and to what extent NQR frequencies are 'tainted' by nuclear Zeeman and superhyperfine contributions resulting from incomplete cancellation [68]. The work of Shi and coworkers clearly demonstrates that for

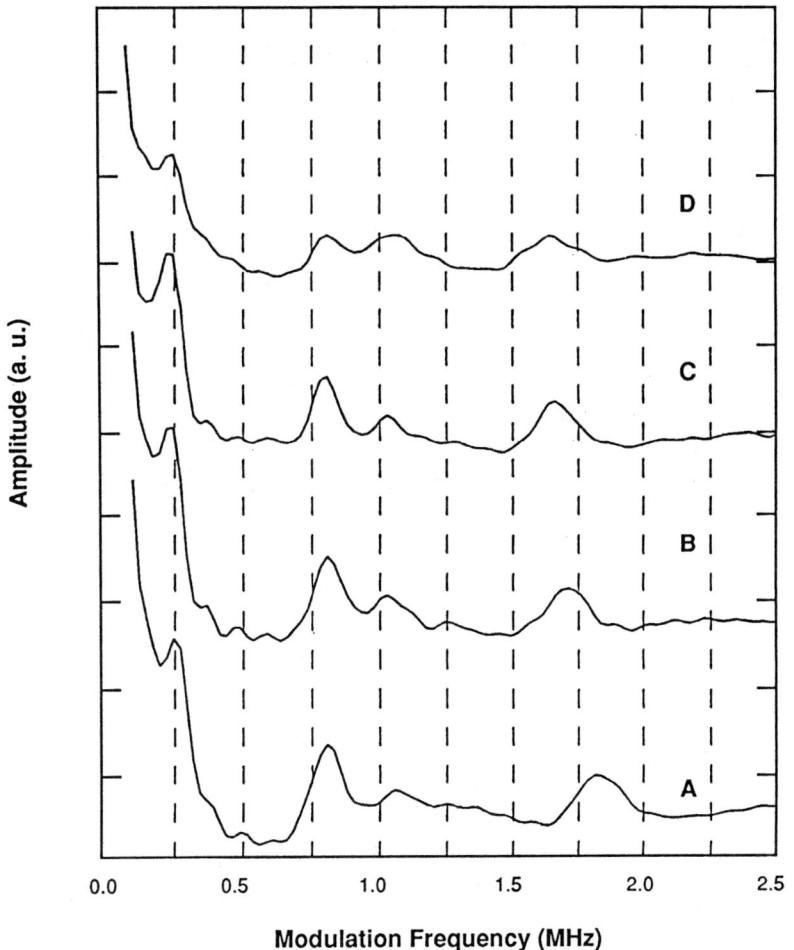

Fig. 9. Fourier transformed ESEEM of m-DNB$^{\top}$ on γ-alumina. $A = 3.85$ GHz, $B = 3.24$ GHz, $C = 3.01$ GHz, and $D = 2.81$ GHz. All spectra taken at 7 K, using stimulated echo (three-pulse) sequence with phase cycling.

radicals with modest anisotropies like m-D^{14}NB$^{\top}$, there is a range of field/frequency values over which NQR transitions can be observed without a measurable admixture of nuclear frequencies [66,69]. Thus, if the appropriate frequencies are available, ESE provides a unique experimental route to probe the environments of ^{14}N-containing radicals by studying nuclear quadrupole interactions.

Several interesting conclusions can be reached by investigating the ESEEM of m-DNB$^{\top}$ on alumina. Firstly, the simulation of the envelope modulation obtained at S-band (2–4 GHz) frequencies requires one weakly coupled ^{14}N nucleus; if more nitrogens are included in the simulation, an obvious mismatch between simulation and experiment is seen. This result indicates that the radicals on alumina are isolated, and not in the form of dimers or polymeric clusters. Second, Flockhart

TABLE III
ESEEM Parameters for weakly coupled nitrogen in m-D^{14}NB$^{\bar{}}$ on activated metal oxide surfaces

Oxide			a_{iso}*	e^2qQ*	η
γ-alumina	freshly made	exp.		1.243	0.36
		cal.	0.65	1.245	0.36
	one year old	exp.		1.219	0.37
		cal.	0.65	1.220	0.37
MgO	freshly made	exp.		1.324	0.38
		cal.	0.66	1.321	0.38

* Units of MHz. See reference [64] for experimental and theoretical details.

et al. observed that the intensity of the EPR signal from this species on alumina grew with time [70]. If low-frequency ESE measurements are made over time, this increase in radical concentration is observed, and small but real changes in the quadrupole coupling constant are noted. Table III summarizes the results from ESEEM simulations of data taken over one year. While it is still too early to assess the precision of e^2qQ measurements by ESEEM techniques, changes of a few percent in the quadrupole coupling constant and the asymmetry parameter do significantly alter the pattern of echo envelope modulation that is observed. From these observations, combined with VHF-EPR data on the g and A-matrix values for the strongly coupled ^{14}N, a picture emerges of measurably different surface interactions of the m-dinitrobenzene radical anion on different activated oxide surfaces. This picture may provide new information concerning the active sites and surface environments of adsorbed species.

In addition to probing intramolecular hyperfine couplings, low frequency ESE experiments prove very sensitive to couplings between unpaired electrons in nitroaromatic anion radicals and neighboring magnetic nuclei. For example, if γ-alumina is activated in the manner previously described, but if it is treated for one hour with oxygen gas enriched 50% in ^{17}O, then a three-pulse ESE measurement will show, in addition to modulations from nitrogen and protons, a low-frequency modulation which is ascribed to ^{17}O, as shown in Figure 10. Simulation with one or two oxygens produced nearly equivalent agreement with the data, and thus the exact form of the elusive oxygen species that may promote electron transfer at oxide surfaces still is unknown. Table IV gives best simulation parameters for the data. This ESEEM result is, however, the first spectroscopic evidence that an oxygen species remains closely associated with surface radicals, and further experiments are in progress to better identify its structure.

Finally, three-pulse echo experiments on the m-DNB anion radical adsorbed on alumina and magnesia also have observed modulations from ^{25}Mg (on MgO) and ^{27}Al (on alumina); simulation parameters also are given in Table IV. The distance from the radical to the surface metal ion in both of these systems is larger than that measured for the Pe$^{\dot{+}}$—Al distance on alumina (0.33 nm). Future work will compare different radical surface species on the same substrate in greater detail, in order to better understand the relationships between radical molecular structures, surface site geometry, and ESEEM-derived parameters.

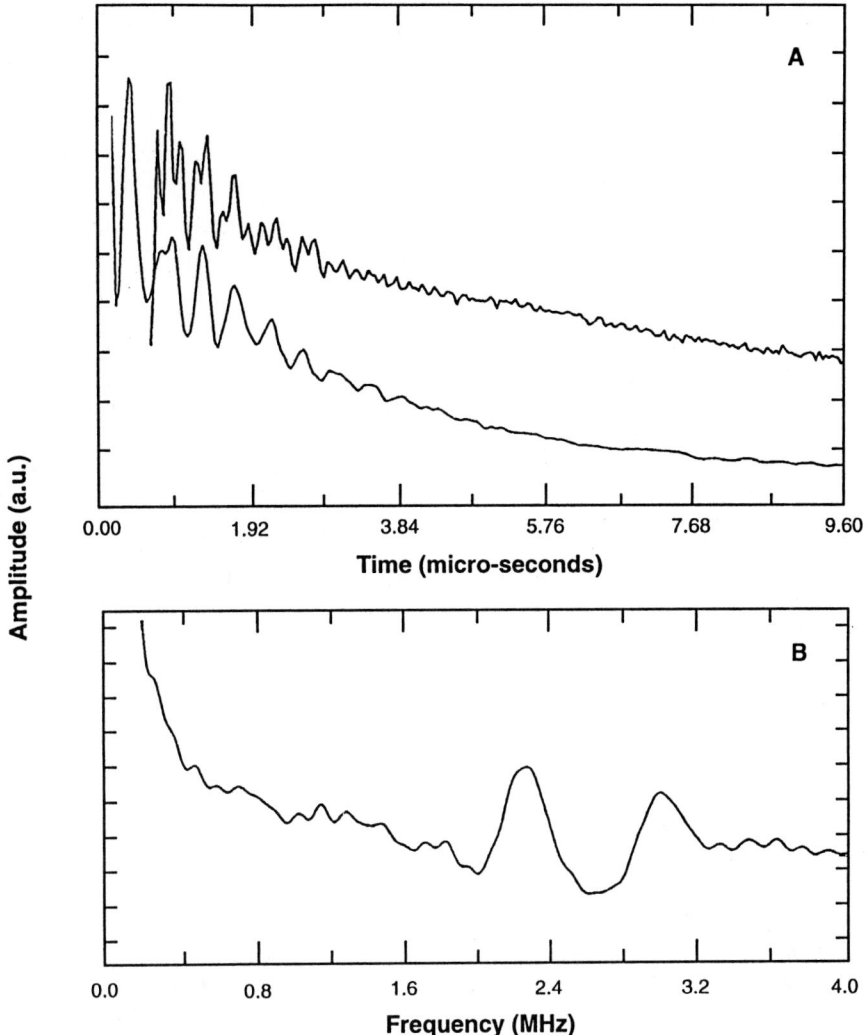

Fig. 10. ESEEM from m-DNB$^-$ on alumina pretreated with ^{17}O-enriched O_2. Stimulated echo (three-pulse) sequence with phase cycling. Frequency = 3.8496 GHz, T = 4.2 K. Simulation parameters: r_{eff} = 0.24 nm, a_{iso} = 0.02 MHz, e^2qQ = 2.7 MHz, η = 0.9.

TABLE IV
Simulation parameters for ESEEM from m-DNB$^-$ on various substrates

	r_{eff} (nm)	a_{iso} (MHz)	e^2qQ (MHz)	η
^{17}O on γ-alumina	0.24	0.02	2.7	0.9
^{27}Al (γ-alumina)	0.49	0.02	†	
^{25}Mg (MgO)	0.45	0.03	‡	

† One or two oxygen atoms.
‡ Very small.

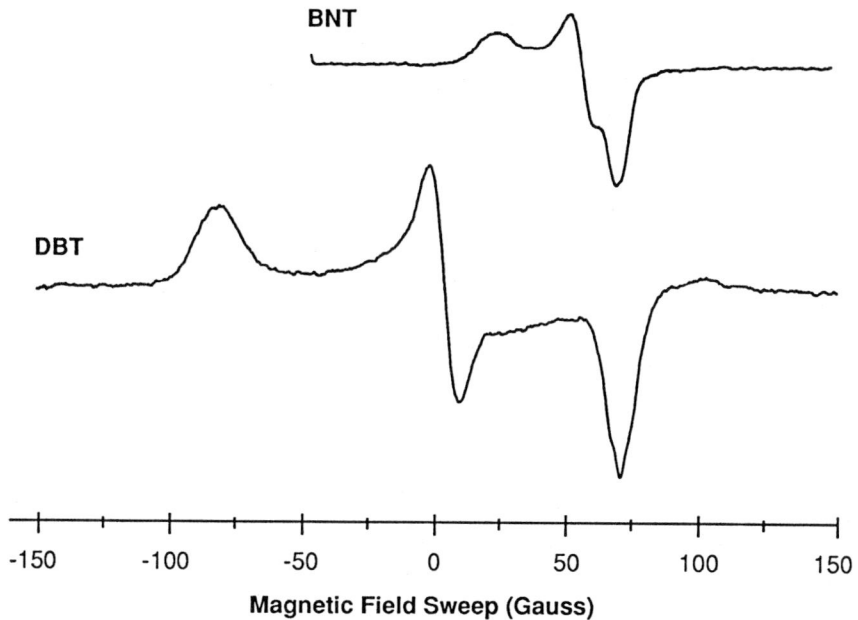

Fig. 11. W-band EPR spectra of DBT and BNT radicals on alumina. The overall magnetic field sweep range is 300 Gauss. Magnetic field sweep centered at 3.35T.

4.2. SULFUR-CONTAINING AROMATIC RADICALS

Sulfur-containing aromatic radicals constitute another interesting class of compounds which form by electron transfer on metal oxide surfaces. Among these, the thiophenes have proven especially interesting for EMR investigation, and radicals from many members of this class (e.g., thiophene (THI), *benzothiophene* (BTH), *dibenzothiophene* (DBT), *benzonaphthothiophene* (BNT), *dinaphthothiophene* (DNT), etc.) have been studied on alumina and silica-alumina, as well as in homogeneous boric acid glass [71,72]. Our laboratory has been especially interested in these systems as models for thiophenes found in fossil fuels, as well as for their rich chemistry [73,74].

Perhaps the most striking feature of the EPR spectra of thiophenes is the influence of sulfur on the g-matrix; the larger spin–orbit coupling of sulfur shifts g-values away from the free electron value, g_e, in a manner which can be represented simply by:

$$g_{x,y,z} = g_e + \lambda/K_{x,y,z} \tag{7}$$

where λ is the spin–orbit coupling constant, and $K_{x,y,z}$ is a function dependent on the electronic configuration of the molecule. Figure 11 illustrates the effect of the larger λ of sulfur on VHF-EPR spectra of DBT and BNT radicals on alumina. Well-resolved spectral features correspond to the powder patterns for systems with $g_{xx} \neq g_{yy} \neq g_{zz}$. While these features cannot be resolved at conventional X-band (9.5 GHz) frequencies, VHF-EPR provides data which can be simulated accu-

TABLE V
Summary of g-matrix values for selected thiophene radicals on alumina

Species	g_1	g_2	g_3
THI	2.00506 ± 0.00050	2.00506 ± 0.00050	2.00234 ± 0.00010
BTH	2.00642 ± 0.00038	2.00451 ± 0.00012	2.00226 ± 0.00013
DBT	2.01130 ± 0.00026	2.00624 ± 0.00015	2.00236 ± 0.00010
BNT	2.00583 ± 0.00012	2.00495 ± 0.00012	2.00224 ± 0.00011
DNT	2.00518 ± 0.00052	2.00326 ± 0.00047	2.00224 ± 0.00051

rately, providing the first opportunity to study these radicals in detail. Table V summarizes the g-matrix data for selected thiophene radicals formed on an alumina surface.

The data in Table V show a trend in g-values as the number of carbon atoms in the aromatic ring cluster increases from 4 in Th to 20 in DNT. Similar results are obtained when these radicals are formed by uv irradiation in a boric acid glass matrix, suggesting that surface effects have a small but measurable effect on the observed g-matrices of the radicals on alumina. Figure 12 illustrates the variation of the canonical g-values of the radicals with the size of the fused-ring system. It is observed that the canonical g-values do not follow a monotonic trend within the thiophenes. Rather, the g-shift starts from THI, rises up, peaks at DBT, then decreases to DNT. Also, the lower field g-matrix components g_1 and g_2 behave similarly while g_z stays constantly near the free-electron g value. The latter behavior is expected because all these radicals are presumably planar π radicals, for which the delocalized unpaired electron is in p-orbitals along a direction perpendicular to the molecular plane, and the g-shift along this direction is expected to be zero in the first order approximation. The observation that the principal values of the g-matrices are not a simple monotonic function of the number of benzene rings fused to thiophene indicates that the size of a particular thiophene molecule is not the sole factor affecting the g-matrix of thiophenic radicals.

VHF-EPR information makes possible an investigation of the relationships between g values in these radicals and their electronic structure. Based on notions including 'frozen orbitals' and the additive property of g-shifts, Stone [75] proposed that the g-shifts in aromatic hydrocarbon radicals would behave as $\Delta g = b + \lambda c$, where λ is the coefficient in the simple Hückel Molecular Orbital (HMO, or SHMO) energy expression $E = \alpha + \lambda\beta$, and α and β are the Coulomb and resonance integrals, taken for this series as 1.2 and 0.65, respectively [76]; b and c are empirical parameters. Figure 13(a) shows the measured g-shift (plotted as $\Delta g = g_e - \langle g \rangle$) against λ from HMOs. Four out of five points fall into a straight line in this particular sulfur heterocyclic series. Similarly, g-shifts correlate well with simple HMO calculations of unpaired spin density on the thiophenic sulfur, as seen in Figure 13(b). More detailed *ab initio* calculations of the electronic structure of these radicals reveal a complexity in the relationships between molecular and electronic structure and the g-matrix which further work certainly needs to clarify; the simple HMO correlations should be viewed merely as one indication that details of adsorbed radical structure can be probed effectively by VHF-EPR

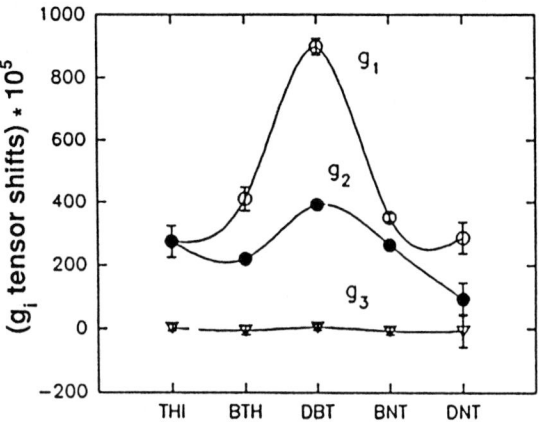

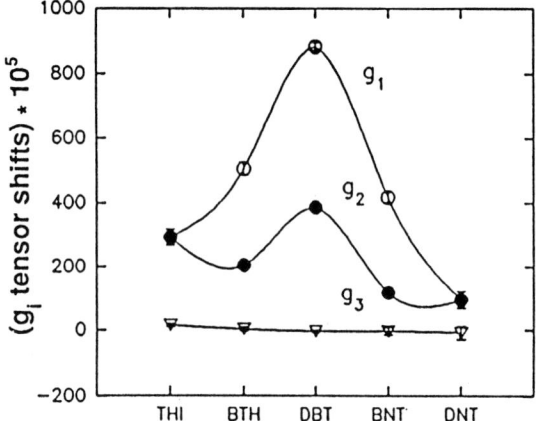

Fig. 12. Variation in g-shifts for matrix elements from a series of thiophenes. Upper graph represents radicals on alumina; lower graph gives data in boric acid glass matrix.

methods. Yet even at this level of approximation, key points such as the relationship between g-shifts and the unpaired electron density on sulfur, and the small but measurable effects of surface interactions on the g-values for thiophene clearly are indicated.

While proton hyperfine splittings in thiophenic radicals can be resolved by EPR at X and W-band frequencies when samples are prepared in a homogeneous solid matrix like boric acid glass, linewidths for the species on alumina are too broad for the weaker interactions to be seen. ENDOR and ESE methods do make possible such measurements, and the interpretation of such data requires a careful consideration of 'orientation selection' caused by the significant g-anisotropy of sulfur-containing aromatics. Measurements at magnetic field increments which

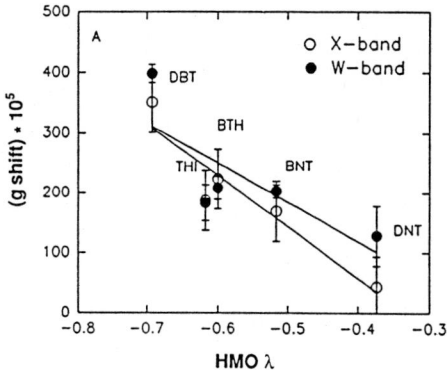

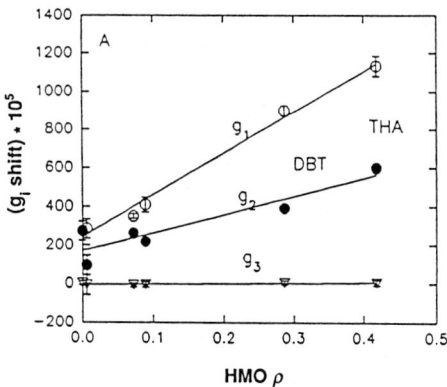

Fig. 13. (a) g-shifts for a series of thiophene radicals plotted against HMO λ. (b) g-shifts for a series of thiophene radicals plotted against HMO ρ.

span the EPR linewidth must be made, as a contour plot of the proton ENDOR from the two-sulfur aromatic radical of thianthrene on a silica–alumina surface reveals in Figure 14, where field-dependent variations in the measured hyperfine couplings are seen. An interesting asymmetry in the data from this system, together with an analysis of the g-matrix obtained from EPR, can be interpreted as arising from the radical adsorbed with one edge down on the surface and the other lifted at an angle to it. Future ENDOR and ESE work with sulfur-containing aromatics will investigate both the orientation of adsorbed species and the surface environments with which they interact.

5. Conclusions

EMR methods provide a powerful approach to the study of radicals adsorbed on disordered materials. Experimental and theoretical advances have made possible the analysis of powder spectra with good precision, allowing investigators to probe

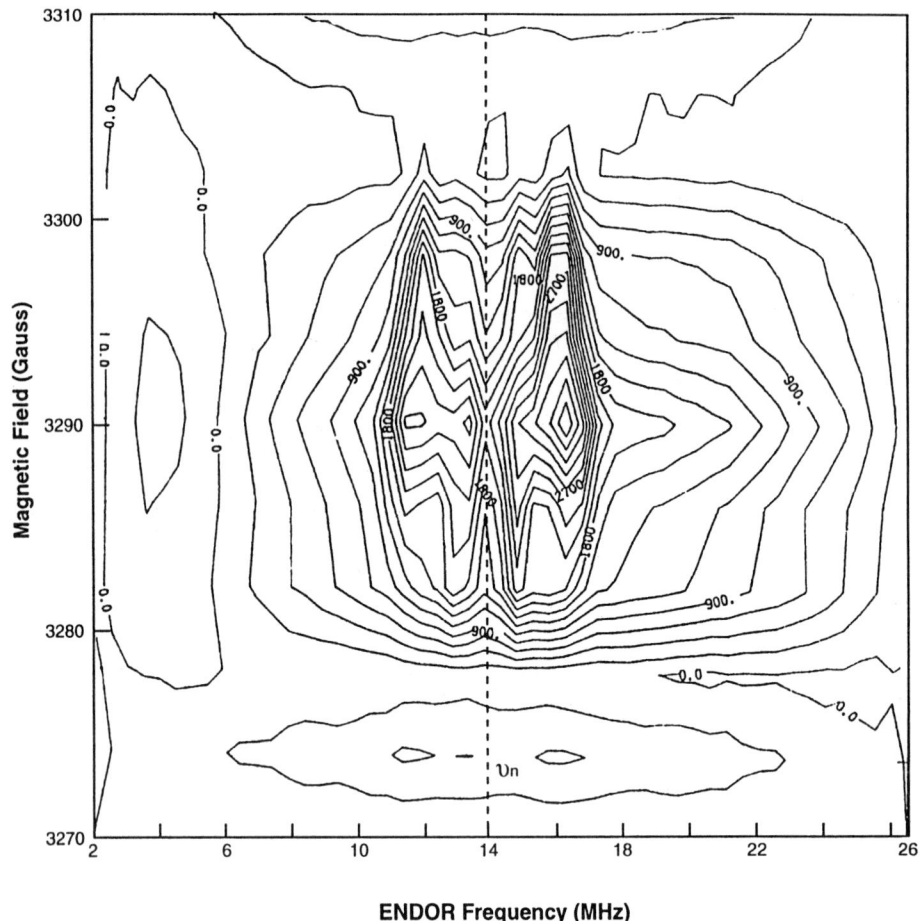

Fig. 14. Two-dimensional ENDOR (ν, B) of thianthrene radical on alumina. The variation of measured hyperfine coupling with field is seen in the contour plot.

details of radical structure, adsorption, and surface environment and mobility, with resolution often approaching that usually associated with single crystal systems. Developments in magnetic resonance thus provide the opportunity to study important chemical events, such as surface electron transfer, in systems similar, if not identical to, functioning catalysts or solid-state devices. Multi-frequency EPR, ENDOR, and ESE provide complementary information which, taken together, often allows definitive characterization of surface radicals.

Acknowledgements

Partial support for this work was provided by the National Institutes of Health (GM42208, RR01811), the U.S. Department of Energy (through the Illinois Clean Coal Institute, ENR ICCI RC ANTIC), and the scientific resources of the Illinois

EPR Research Center. The opinions expressed are those of the authors, and do not necessarily reflect those of the NIH, DOE, or ICCI.

References

1. P. E. O'Reilly: *Adv. Catal.* **12**, 31 (1960).
2. R. L. Belford, R. B. Clarkson, J. B. Cornelius, K. S. Rothenberger, M. J. Nilges, and M. D. Timken: 'EPR Over Three Decades of Frequency: Radiofrequency to Infrared', in J. A. Weil (Ed.), *Electron Magnetic Resonance of the Solid State*, Chemical Institute of Canada, Ottawa, p. 21 (1987).
3. R. L. Belford and R. B. Clarkson: 'Multifrequency Electron Paramagnetic Resonance Spectroscopy', in R. E. Botto and Y. Sanada (Eds.), *Magnetic Resonance of Carbonaceous Solids*, Advances in Chemistry Series 229, American Chemical Society, Washington, p. 107 (1993); W. Wang, R. L. Belford, R. B. Clarkson, P. H. Davis, J. Forrer, M. J. Nilges, M. D. Timken, T. Walczak, M. C. Thurnauer, J. R. Norris, and Y. Zhang: *Appl. Magn. Res.* **6** (1994).
4. J. S. Hyde, G. H. Rist, and L. E. G. Eriksson: *J. Phys. Chem.* **72**, 4269 (1968).
5. L. Kevan and L. D. Kispert: *Electron Spin Double Resonance Spectroscopy*, John Wiley, New York (1976).
6. L. Kevan and P. A. Narayana: 'Disordered Matrices', in M. M. Dorio and J. H. Freed (Eds.), *Multiple Electron Resonance Spectroscopy*, Plenum Press, New York, p. 229 (1979).
7. H. Kurreck, B. Kirste, and W. Lubitz: *Electron Nuclear Double Resonance Spectroscopy of Radicals in Solution*, VCH Publishers, New York (1988).
8. C. Gemperle and A. Schweiger: *Chem. Revs.* **91**, 1481 (1991).
9. L. Kevan and R. N. Schwartz: *Time Domain Electron Spin Resonance*, John Wiley, New York (1979).
10. L. Kevan and M. K. Bowman: *Modern Pulsed and Continuous-Wave Electron Spin Resonance*, John Wiley, New York (1990).
11. S. A. Dikanov and Yu. D. Tsvetkov: *Electron Spin Echo Envelope Modulation (ESEEM) Spectroscopy*, CRC Press, Boca Raton (1992).
12. H. L. Flanagan and D. J. Singel: *J. Chem. Phys.* **87**, 5606 (1987).
13. S. A. Cosgrove and D. J. Singel: *J. Phys. Chem.* **94**, 2619 (1990).
14. M. J. Nilges: *Ph.D. Dissertation*, University of Illinois, Urbana, Illinois (1979).
15. R. L. Belford, P. H. Davis, G. G. Belford, and T. M. Lenhardt: *ACS Symposium Series No. 5*, p. 40 (1974).
16. L. R. Dalton and A. L. Kwiram: *J. Chem. Phys.* **57**, 1132 (1972).
17. C. G. Hurst, T. A. Henderson, and R. W. Kreilick: *J. Am. Chem. Soc.* **107**, 7294 (1985).
18. Fa. S. Jiang, T. M. Zuberi, J. B. Cornelius, R. B. Clarkson, R. B. Gennis, and R. L. Belford: *J. Am. Chem. Soc.* **115**, 10293 (1993).
19. J. B. Cornelius, J. McCracken, R. B. Clarkson, R. L. Belford, and J. Peisach: *J. Phys. Chem.* **94**, 6977 (1990).
20. J. B. Cornelius: *Ph.D. Dissertation*, University of Illinois, Urbana, Illinois (1987).
21. A. V. Astashkin, S. A. Dikanov, and Yu. D. Tsvetkov: *Chem. Phys. Lett.* **136**, 204 (1987).
22. W. Spendley, G. R. Hext, and F. R. Himsworth: *Technometrics* **4**, 441 (1962).
23. K. J. Mattson: *Ph.D. Dissertation*, University of Illinois, Urbana, Illinois (1990).
24. A. J. Rooney and R. C. Pink: *Proc. Chem. Soc.* 70 (1961).
25. D. M. Brauer: *Chem. Ind. (London)* 177 (1961); *J. Catal.* **1**, 372 (1962).
26. W. K. Hall: *J. Catal.* **1**, 53 (1962).
27. A. Carrington, F. Dravnicks, and M. C. R. Symons: *J. Chem. Soc.* 947 (1959).
28. A. J. Bard, A. Ledwith, and H. J. Shine: *Adv. Phys. Org. Chem.* **13**, 155 (1976).
29. G. Vincow and P. M. Johnson: *J. Chem. Phys.* **39**, 1143 (1963).
30. G. M. Muha: *J. Phys. Chem.* **71**, 633 (1967); B. D. Flockhart, J. A. N. Scott, and R. C. Pink: *Trans. Faraday Soc.* **62**, 730 (1966).
31. G. M. Muha: *J. Phys. Chem.* **82**, 1843 (1978).
32. L. Petrakis, P. L. Meyer, and T. P. Debies: *J. Phys. Chem.* **84**, 1020 (1980); L. Petrakis, P. L. Meyer, and G. L. Jones: *J. Phys. Chem.* **84**, 1029 (1980).

33. J. T. Richardson: *J. Catal.* **9**, 172 (1967).
34. F. R. Dollish and W. K. Hall: *J. Phys. Chem.* **71**, 1005 (1967).
35. I. C. Lewis and L. S. Singer: *J. Phys Chem.* **85**, 354 (1981).
36. Y. Mao and J. K. Thomas: *Langmuir* **8**, 2501 (1992); X. Liu and J. K. Thomas: *Langmuir* **9**, 727 (1993).
37. G. M. Muha: *J. Catal.* **58**, 470 (1979).
38. B. D. Flockhart and M. A. Salem: *J. Colloid Interface Sci.* **103**, 76 (1985).
39. T. Wozniewski, E. Fedorynska, and S. Malinowski: *J. Colloid Interface Sci.* **87**, 1 (1982).
40. W. Froncisz and J. S. Hyde: *J. Chem. Phys.* **73**, 3123 (1980).
41. O. Y. Grinberg, A. A. Dubinskii, and Ya. S. Lebedev: *Russ. Chem. Rev. Engl. Transl.* **52**, 850 (1983).
42. E. Haindl, K. Möbius, and H. Z. Oloff: *Z. Naturforsch.* **40a**, 169 (1985).
43. A. J. Stone: *Mol. Phys.* **7**, 311 (1964).
44. L. C. Snyder: *J. Phys. Chem.* **66**, 2299 (1962).
45. H. Ohya-Nishiguchi: *Bull. Chem. Soc. Jpn.* **52**, 7 (1979).
46. R. B. Clarkson, R. L. Belford, K. S. Rothenberger, and H. C. Crookham: *J. Catal.* **106**, 500 (1987).
47. R. Makela and M. Voulle: *Finn. Chem. Lett.* **3**, 66 (1984).
48. F. W. Heineken and T. C. Christidis: *Proc. Congress Ampère 18th*, 503 (1974).
49. H. Sang, H. Wang, and K. P. Such: *Magn. Res. Chem.* **30**, 150 (1992).
50. R Erickson, M. Lindgren, A. Lund, and L. Sjöqvist: *Collolds and Surfaces A* **72**, 207 (1993).
51. R. I. Samoilova, A. V. Astashkin, S. A. Dikanov, D. Goldfarb, and E. V. Lunina: *Colloids and Surfaces A* **72**, 29 (1993).
52. R. B. Clarkson: 'ENDOR Studies of Radicals Adsorbed on Metal Oxide Powders', in J. P. Fraissard and H. A. Resing (Eds.): *Magnetic Resonance in Colloid and Interface Science*, D. Reidel, Dordrecht, p. 425 (1980).
53. N. M. Atherton and C. E. Oliver: *J. Chem. Soc. Faraday Trans.* **84**, 10 (1988).
54. A. H. Reddoch: *J. Chem. Phys.* **43**, 225 (1965).
55. K. S. Rothenberger, H. C. Crookham, R. L. Belford, and R. B. Clarkson: *J. Catal.* **115**, 430 (1989).
56. R. L. Burwell: *J. Catal.* **86**, 301 (1984).
57. H. Knözinger: *Catalysis by Acids and Bases*, Elsevier, Amsterdam, p. 11 (1985).
58. P. A. Narayana, M. K. Bowman, D. Becker, L. Kevan, and R. N. Schwartz: *J. Chem. Phys.* **67**, 1990 (1977).
59. O. Burghaus, M. Rohrer, T. Gotzinger, M. Plato, and K. Möbius: *Meas. Sci. Technol.* **3**, 765 (1992).
60. L. Kevan: *Acc. Chem. Res.* **20**, 1 (1987).
61. P. A. Snetsinger, J. B. Cornelius, R. B. Clarkson, M. K. Bowman, and R. L. Belford: *J. Phys. Chem.* **92**, 3696 (1988).
62. S. A. Dikanov, V. F. Yudanov, R. I. Samiolova, and Yu. D. Tsvetkov: *Chem. Phys. Lett.* **52**, 520 (1977).
63. R. B. Clarkson, M. D. Timken, D. R. Brown, H. C. Crookham, and R. L. Belford: *Chem. Phys. Lett.* **163**, 277 (1989); R. B. Clarkson, D. R. Brown, J. B. Cornelius, H. C. Crookham, W.-J. Shi, and R. L. Belford: *Pure & Appl. Chem.* **64**, 893 (1992).
64. A. J. Tench and R. L. Nelson: *Trans. Faraday Soc.* **63**, 2254 (1967).
65. M. Branca and V. Indovina: *J. Chem. Soc. Faraday Trans.* **86**, 403 (1990).
66. Wenjun Shi: *Ph.D. Dissertation*, University of Illinois, Urbana, Illinois (1993).
67. C. J. Ultee: *J. Phys. Chem.* **64**, 1873 (1960).
68. H. L. Flanagan, G. J. Gerfen, and D. J. Singel: *J. Chem. Phys.* **88**, 20 (1988).
69. W. Shi, W. Wang, M. J. Nilges, R. B. Clarkson, and R. L. Belford: 'Nitroaromatic Anions and Copper Sites: 2–4 GHz ESEEM and 95 GHz EPR', 35th Rocky Mt. Conference on Analytical Chemistry-EPR Symposium, Denver, CO, July 25–29 (1993).
70. B. D. Flockhart, I. R. Leith, and R. C. Pink: *Trans. Faraday Soc.* **66**, 469 (1969).
71. G. Hwang and H. Chon: *Zeolites* **10**, 101 (1990).
72. Wei Wang: *Ph.D. Dissertation*, University of Illinois, Urbana, Illinois (1993).

73. R. B. Clarkson, Wei Wang, D. L. Brown, H. C. Crookham, and R. L. Belford: *FUEL* **69**, 1405 (1990).
74. R B. Clarkson, W. Wang, D. L. Brown, H. C. Crookham, and R. L. Belford: 'Electron Magnetic Resonance of Standard Coal Samples at Multiple Microwave Frequencies', in R. Botto and Y. Sanada (Eds.): *Techniques in Magnetic Resonance for Carbonaceous Solids*, ACS Advances in Chemistry Series **229**, American Chemical Society, Washington, p. 507 (1993).
75. A. J. Stone: *Mol. Phys.* **6**, 509 (1963).
76. P. D. Sullivan: *J. Am. Chem. Soc.* **90**, 3618 (1968).

ESR Studies of Organic Radical Cations in Zeolites

CHRISTOPHER J. RHODES* and CHANTAL S. HINDS
Department of Chemistry, Liverpool John Moores University, Byrom St., Liverpool L3 3AF, U.K.

(Received: 15 December 1993; accepted: 5 July 1994)

Abstract. An overview is presented of the formation of organic radical cations in zeolites. Attention is paid firstly to their deliberate production by radiolysis and then to the spontaneous oxidation by activated H-exchanged zeolites of adsorbed organic substrates. The nature of the oxidation site arising from thermal pretreatment in oxygen is considered, and possible mechanistic details are outlined for the frequently observed oligomerisations of simple substrates. The broad conclusions are that radical cations are formed from that component of the organic product mixture with the lowest ionisation potential, and that the dominant molecular transformations are driven by proton (Brønsted) catalysis: it appears that there is little evidence that radical cations are involved particularly in heterogeneous catalysis by zeolites, but they may well precede the formation of neutral radicals which are implicated as reaction intermediates. The reorientational dynamics of alkene radical cations in zeolites, as determined by ESR spectroscopy, are also considered.

Key words. Zeolites, radical cations, ESR.

1. Introduction

Numerous reports have been published, spanning a period of nearly 30 years, of the formation of radical cations when organic substrates interact with activated zeolites. In general, it is necessary to first activate the zeolite by heating to 700–1100 K in air or oxygen and to then admit the organic species from the gas phase onto the zeolite; but in the case of certain molecules with low ionisation potentials prior activation beyond drying is sometimes unnecessary, although a higher spin concentration is always obtained after thermal oxygen treatment.

The first clue that open-shell species have been formed is provided by the sudden development of colour in the zeolite – this being reinforced by the observation of electron spin resonance (ESR) spectra that may be assigned to particular radical cation structures derived from transformations of the substrate. This is an important point, since it is usually only in the case of aromatic molecules that a simple ionisation to form the *primary* radical cation occurs as these are stable energy minima; in the case of simple alkenes, extensive transformations usually occur so that the radical cation observed is that from a more stable oligomer. Exceptions to this would appear to be tetramethylethylene (TME) and 9-octalin which both give their monomer radical cations on H-mordenite, but these species may also be formed from simpler (lower carbon number) organic substrates and so are themselves energy minima.

In our own work, we have focused on this 'spontaneous' formation of radical

---

* Author for correspondence.

cations in activated, acidic, H-exchanged forms of zeolites (mordenite, ZSM-5, X, Y) since it appears to us that this is most relevant to heterogeneous catalysis by these materials; however, reports have appeared which show that radical cations may be produced in a variety of non-acidic zeolites by X- or γ-radiation. This provides a method for studying the properties of radical cations over a much wider range of temperature than is possible by conventional matrix isolation methods; nonetheless, it is probably no coincidence that we have seen a recent resurgence of interest in the topic of radical cations in zeolites, because a very extensive literature has been compiled during the past decade using freon matrices.

As a result of this, it is relatively easy to obtain 'fingerprint' ESR spectra of authentic radical cations in order to confirm or otherwise assignments made in zeolite media: in consequence, much earlier work has now been refuted and it appears likely that more revision of the often rather ambiguous literature will transpire as the subject is thus placed upon firmer ground.

2. Generation of Monomer Radical Cations by Radiolysis

It is instructive to consider first the formation of radical cations by radiolysis in various media since this involves a simple electron loss from the substrate monomer – without extensive structural modification or oligomer formation occurring, which we discuss subsequently. In both freon [1–3] and zeolite [5–9] media, the essential principle is that electron loss centres (holes) ($C^{+\cdot}$) are created by the radiation and to which electron transfer may occur from the substrate molecule (S) so long as its ionisation potential is less than that of the pre-ionised centre (C): this may all be summarised in Equations (1)–(2). It is of course necessary that, whatever the medium, the electrons produced in (1) are eliminated from the scene, otherwise they would either simply recombine with the hole centres ($C^{+\cdot}$) or be captured by the radical cations ($S^{+\cdot}$). In freon matrices [1–3], e.g. $CFCl_3$, electron capture by the medium is facile, and while the nature of the $CFCl_3/e^-$ centre remains controversial [10, 11] it is unequivocal that below ca. 150 K its ESR signature is extremely broad and does not contaminate appreciably the sharper signal from the radical cation. Above 150 K, dissociation of the centre is possible (5) since signals from $CFCl_2^\cdot$ radicals appear [12].

$$C \rightarrow C^{+\cdot} + e^- \tag{1}$$

$$S + C^{+\cdot} \rightarrow S^{+\cdot} + C \tag{2}$$

$$e^- + X \rightarrow X/e^- \tag{3}$$

$$e^- + CFCl_3 \rightarrow [CFCl_3]^- \tag{4}$$

$$[CFCl_3]^- \rightarrow CFCl_2 + Cl^- \tag{5}$$

In zeolite media, it is not quite so clear how the electrons are removed, but undoubtedly they are, since the spectra reported by the groups of Iwasaki [4] and of Trifunac [5–9] show exclusively the radical cation component. The scheme in Equations (6)–(9) is proposed [5], in which the initial step is ionisation of the zeolite matrix (Z): as in freon matrices, the positive hole must be mobile so that substrate molecules evenly but dilutely located within the microporous structure can be ionised (7). Quite reasonably, it is considered that likely electron trapping sites are cations [5]; however, nothing akin to a sodium cluster, e.g. Na_4^{3+}, was observed, even though Na^+ cations are in the majority and paramagnetic centres of this kind have been observed previously in Na–Y zeolite [13]. It has therefore been assumed that Fe^{3+} ions present oxidise these initially formed clusters, as in (9): in any event, the electrons are deeply trapped, since they could not be released by exposure either to visible or UV light [5], as judged by the insensitivity of the radical cation signals to this treatment.

$$Z \rightarrow Z^{+\cdot} + e^- \quad (6)$$

$$Z^{+\cdot} + S \rightarrow Z + S^{+\cdot} \quad (7)$$

$$e^- + Na_4^{4+} \rightarrow Na_4^{3+} \quad (8)$$

$$Na_4^{3+} + Fe^{3+} \rightarrow Na_4^{4+} + Fe^{2+} \quad (9)$$

2.1. SPECIFIC STUDIES

2.1.1. *Saturated Alkanes*

The first report of organic radical cations generated by radiolysis and stabilised in a zeolite which provides a direct comparison with freon matrix work is due to the Iwasaki group [4] – this being an extension of their comprehensive work on alkane radical cations. The motivation was to test the site of deprotonation in the reaction between a primary alkane radical cation and a neutral alkane molecule; previous work by this group [14, 15] strongly indicated that deprotonation occurs at the position of highest spin and therefore charge density. This is a topic of great significance to the fundamental radiation chemistry of hydrocarbons and has been the subject of considerable activity [14–19], but appears to depend on the conformation of the alkane chain, since *gauche* isomers show maximum spin population at the 2-position, from which deprotonation takes place preferentially; this contrasts with the linear conformation in which the 1-position is activated. In the linear channels of the (Na)ZSM-5 zeolite, only the latter conformation is stabilised, leading exclusively to 1-alkyl radicals [4].

More subtle influences of zeolite media have been discovered in the case of *cis*- and *trans*-decalin radical cations, both of which may be switched between closely lying electronic states with different coupling constants [6]. Matrix effects of this kind are harder to explain, but it is most significant that these cationic species are stabilised in silicalite with almost zero aluminium, and hence cation, content. One might ask, therefore: how are the radiolytically produced electrons trapped? It is well known that defect centres are produced by radiolysis of silica [20], arising from electron loss (holes) and electron gain: the implication is that holes can oxidise organic substrates present, and their counterpart centres are probably stable dislocations of the silicalite structure arising from electron addition. While some Fe^{3+} impurities may be present in very low amount, the overall scheme of Equations (6)–(9) would surely be less effective than for zeolites with their relatively high cation component. Given that the intensity of the radical cation signals in silicalite is no less than in zeolite media, no compelling evidence is provided that cations are dominant electron traps in these matrices. We propose that the radiation chemical properties of zeolites (silicalite) and solid freons are rather similar in that both are 'self trapping' in regard to electron removal, and this is the reason for their ability to stabilise radical cations. Further stability must be provided in zeolite media in which radical cations are stabilised against the negatively charged zeolite framework and along with shape selectivity will restrict their diffusion through the microporous network.

In NaX and NaY zeolites *cis*-decalin radical cations are proposed [6] to deprotonate, giving the '*cis*-decalin neutral radical' which it is proposed has couplings of $a(2H) = 40$ G, $a(3H) = 9$ G, and a simulation based on these values is given which, curiously, is said to be in 'good agreement' with the experimental spectrum. However, the latter clearly comprises a dominant *quartet* with *ca.* 40 G coupling while the simulation is principally a *triplet*. While the authors are correct from a *chemical* viewpoint, they have been misled by the previous mis-assignment of the *cis*-decalin radical by Lloyd and Williams [21] who quote the above parameters. Recent work by Roberts [22], shows that the correct values for the 9-decalinyl radical are $a(2H) = 41.10$ G, $a(1H) = 38.7$ G, $a(2H) = 9.25$ G (+smaller couplings) which explains the *quartet* which is *actually* present, since all three β-protons will be equivalent within the solid state linewidths. It is noteworthy that Roberts obtained the same decalinyl radical by chlorine abstraction from both *cis*- and *trans*-chlorodecalins and so the proposal by Lloyd and Williams [21] that 9-decalinyl can exist in distinct *cis* and *trans* forms is incorrect: this is supported by another study using muon spectroscopy [23].

The tetramethylcyclopropane (TMC) radical cation has been studied in freon and in xenon matrices, but was also observed by Trifunac in a NaX matrix [5]. In the highly constrictive $CFCl_3$ medium, the ring-closed structure was observed [24], but above 110 K in the less restricting $CF_2ClCFCl_2$ matrix a ring-opened structure was adopted [24] with the unpaired electron localised to a single Me_2C— centre: the latter was also observed in solid xenon and ascribed to a neutral radical formed by an ion-molecule reaction [25]. Despite the more open supercage structure of NaX, the ring-closed cation was stable up to 200 K without reacting with neutral TMC molecules, and is probably a consequence of cation interactions with the negative zeolite framework which impedes their diffusion to encounter neutral molecules.

Interesting results were obtained with *trans*-2,3-dimethyloxirane which we include here as a hetero-cyclopropane analogue. In freon media a definite matrix effect was discovered [26]: namely, that a ring-opened, allylic, MeCH—O—CHMe$^{+\cdot}$ structure (**I**) was adopted in $CFCl_3$ [26], while in $CF_2ClCFCl_2$ this symmetrical cation was transformed at 95 K to a distonic radical cation, MeCH·—O—CHMe$^+$, (**II**), in which the spin and charge are essentially localised to separate ends of the molecule [27]. In NaY [5], a quintet pattern, $a(4H) = 22.5$ G, was observed as the signature of the species (**II**) – this being stable up to 200 K – while (**I**) was not observed at any temperature between 77 and 200 K. It may well be that (**I**) is able to reorient with respect to the matrix in both $CF_2ClCFCl_2$ and NaX so that a nucleophilic coordination by a Cl or a O(Al, Si) atom leads to structures of type MeCH·—OCH(Me)—X, where X = Cl or O, so that the spin is necessarily localised. This may not be possible in the more restrictive $CFCl_3$ environment.

2.1.2. Alkenes

Tetramethylethylene (TME) has proved a useful probe of mobility and reactivity in zeolites [7]. Depending on the concentration of TME and on the microporous structure (supercages or channels) of the host, monomer or dimer $(Me_2C\!\!=\!\!CMe_2)_2^{+\cdot}$ radical cations may be observed, or the neutral radical $Me_2\!\!=\!\!C(Me)CH_2$ resulting from deprotonation within the dimer when the necessary reacting configuration can be achieved. It is interesting to note that neutral radicals are not observed in H-zeolites because the highly protic environment will tend to favour the conjugate acid (radical cation) form. This is similar to the explanation for the stability of Me_3C^+ carbenium ions to deprotonation in superacid media [28].

At high concentrations (ca. 7%) of TME, the spectrum of the dimer is present exclusively, but solely the monomer at ca. 0.1% of TME. In silicalite, at 5% TME loading, the $TME^{+\cdot}$ monomer is present at 77 K, but at and above 100 K the deprotonated neutral radical is observed. Spectra for 7% TME in Na-Ω-5 (mazzite) following γ-irradiation at 77 K show, at 77 K, the $TME^{+\cdot}$ radical cation, but, at 175 K, there is considerable overlap of this with the neutral species. In the latter case, it is noteworthy that the resolved fine structure as observed in silicalite was absent. It is proposed that the difference between the spectra of the $TME^{+\cdot}$ radical cation in ZSM-5 and in Na-Ω-5 arises from both chemical and physical interaction with the zeolite framework in regard particularly to the mobility of the radical cations, although the chemical processes involved must be the same in all media. In the straight channels of ZSM-5 and silicalite, the encounter and aggregation of a monomer radical cation are allowed even at low temperatures when the concentration range is favourable, but at 50–90 K the dimers dissociate. At higher temperatures, >110 K, the reactive encounter involves deprotonation. In Na-Ω-5, where the cages are larger, the neutral radicals start to form above 110 K, but are probably present below this. Differences in spectral resolution are similarly accounted for in terms of the rotational freedom permitted by the differing framework structures.

2.1.3. Allenes

Tetramethylallene (TMA) belongs to the D_{2d} point group and possesses two degenerate π-orbitals. Thus. following ionisation, the resulting $TMA^{+\cdot}$ radical cation is subject to Jahn–Teller distortion. A substantial variation in the (averaged) methyl proton coupling constants with the twist angle (θ) is predicted by AM1-UHF calculations – increasing as (θ) decreases. Experimentally, it is found that the cation exhibits different couplings in different media [8]: this points to a sensitivity of the Jahn–Teller distortion to environmental influences. We feel that it is significant that the allene radical cation is believed to possess high positive charge density at the central carbon atom, while the unpaired electron density is concentrated at the termini of the allene unit [29]. Since these distributions vary with (θ), we consider the reverse tenet, that a strong electrostatic field from the negatively charged zeolite framework, behests the further localisation of positive charge at C2 which increases (θ): thus it is in NaY that the coupling constant is largest. The electrostatic influence of the freon medium will be very low in the absence of actual chemical bonding. In the NaY medium [8], the broad lines relative to those in solid $CFCl_3$ or fluid hydrocarbon media, over parallel temperature ranges, strongly support an appreciable surface interaction and slower motional averaging of the hyperfine anisotropy of $TMA^{+\cdot}$ in the zeolite. Up to 298 K, the absence of neutral $Me_2C=C=C(Me)CH_2$ radicals accords with a very low rate of diffusion of $TMA^{+\cdot}$ cations in NaY.

2.1.4. Bicycloalkenes

The radical cation of hexamethyl(Dewar benzene) (HMD) was first studied by Rhodes using a solid $CFCl_3$ matrix [30] and the symmetrical ESR pattern with $a = 9.5$ G assigned to the 12 equivalent protons of four methyl groups: thus the $^2B_2$ state was implicated, wherein the SOMO is distributed between the two alkene units. On increasing the temperature to 140 K, a sudden and irreversible spectral change occurred which was a direct observation of the highly thermodynamically favoured ring-opening to the hexamethylbenzene isomer with $a(18H) = 6.7$ G. Below 120 K, NaY is similarly successful [5] in stabilising the ring-closed $^2B_2$ form, and is probably a consequence of (i) the electrostatic interaction with the zeolite framework, and (ii) the fact that HMD is of similar dimensions to the 0.7 nm apertures of the 1.25 nm Y supercages. The latter is likely responsible for low diffusion of HMD molecules and thus no products of ion–molecule reaction were observed. It should also be noted that the alternative $^2A_1$ state of $HMD^{+\cdot}$ was not observed in NaY, despite its implication as being present simultaneously with that $^2B_2$ in a photo-CIDNP experiment [31]; this is therefore, clearly, an *excited state* of the radical cation.

It is pertinent that another CIDNP study [31], this time of quadricyclane (**IV**) and norbornadiene (**III**), provided evidence that their radical cations are distinct species, in accord with results from pulse radiolysis [32]; however, on radiolytic oxidation in freon matrices [2], only (**III**)$^{+\cdot}$ was obtained from either substrate. Similarly, only (**III**)$^{+\cdot}$ was obtained from (**III**) and (**IV**) in a NaZSM-5 matrix, but its chemistry is more complex [9]; namely that, after annealing to 200 K, the cyclopentadiene radical cation was observed, thus implicating the thermally forbidden cycloreversion (10). It is proposed that this transformation occurs via excited states of (**III**)$^{+\cdot}$ which are stabilised by the zeolite. This requires implicitly that such states are thermally accessible from the originally observed *ground state*. The requirement that excited states are involved in (10) assumes that the reaction is concerted and it may be that the interaction of (**III**)$^{+\cdot}$ with the zeolite is sufficiently asymmetric to lift this condition.

At 260 K, the neutral cyclopentenyl radical (**V**) is formed, so a formal transfer of H^- to the cyclopentadiene radical cation must occur, but a mechanism for this is not obvious.

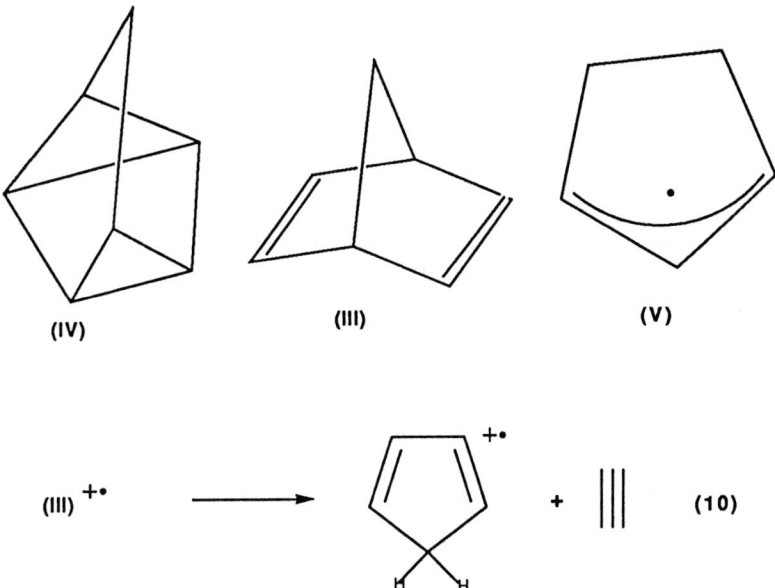

3. Spontaneous Generation of Radical Cations by Zeolites

Having covered the generation of radical cations in zeolites by radiolysis, we now move to the aspect of certain zeolites spontaneously to oxidise organic substrates to their radical cations. Very useful coverage of the field up to *ca.* 1975 is provided in a review by Loktev and Slinkin [33], which focuses particularly on oxidation but also on reduction by zeolites, and concerns mainly aromatic radical cations. It is concluded that substrates with ionisation potentials >9 eV do not form radical cations on X- or Y-type zeolites, but benzene and some derivatives are apparently oxidised by a CeY zeolite. Substrates of lower ionisation potentials, however, such as polycyclic aromatics (anthracene, perylene, coronene, naphthalene) are quite readily oxidised, and their *primary* radical cations are observed; (stilbene and 1,3-diphenylbutadiene were very recently shown to be oxidised on NaZSM-5)[34].

Among the deductions arising from these investigations are that more than one site of oxidation is present, and that some of those less powerfully oxidising are located on the external zeolite surface, since the larger substrates cannot penetrate the zeolite micropores but nonetheless give ESR signals. The evidence is strong that prior heat treatment of H-exchanged zeolites in oxygen (air) significantly enhances radical cation yields in most cases, especially when the substrate is small enough to be admitted to the zeolite channels and shows that, under these conditions of activation, strongly oxidising sites are formed at the internal surface. In the case of H-mordenite [33,35,36] and HZSM-5 [36], sites are formed which can oxidise certainly benzene (I.P. = 9.25 eV) and possibly dimethylacetylene (2-butyne (I.P. = 9.8 eV) [37,38]. It has been proposed that when H-zeolites are heated in air, Brønsted sites are converted into half as many Lewis sites, and it

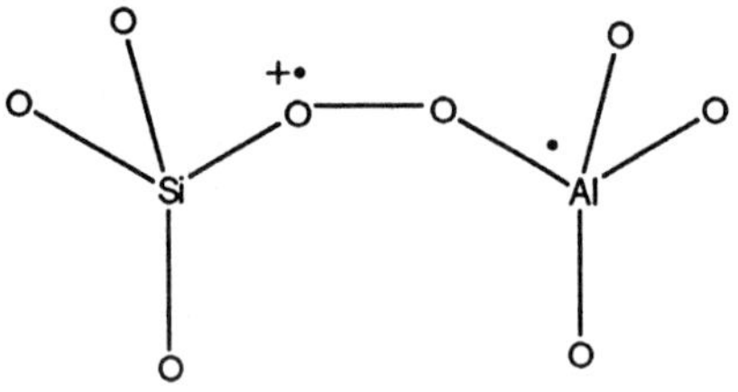

(VI)

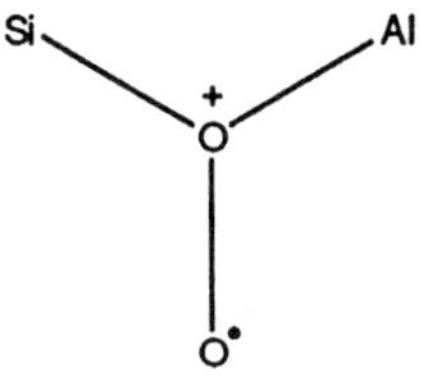

(VII)

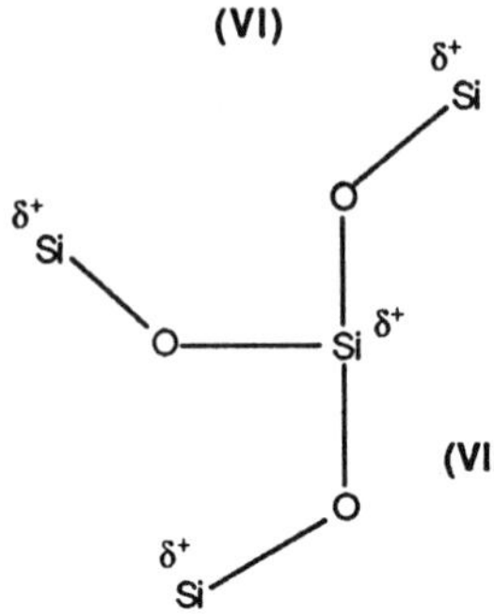

(VIII)

is these that are oxidising [33]. In our own work, we have not observed radical cation signals using Na-zeolites [39], which would appear to support this; but heating in a pure N_2 atmosphere was similarly unsuccessful, demonstrating a chemical role for oxygen in the activation process. An enhanced role for oxygen in radical cation formation has been shown in a number of studies [33], but no precise chemical structure has been elucidated. In our view it is likely that Lewis sites coordinate with molecular oxygen to form powerfully oxidising species (**VI**). We proposed an alternative mechanism previously [39], leading to similar sites (**VII**), and indeed addition of an electron to (**VI**) could lead to (**VII**) (we reserve judgement as to whether this might be mechanistically significant).

We have interpreted [39] the well known 'invisibility' of the electron-gain centres in terms of a very broad line which represents an 'envelope' from a range of SiO----Al coordination strengths and thus varying g-components; we now argue further that such a range of sites provides a range of oxidising power. The unmodified Lewis sites are probably the most weakly oxidising and can be formed on both internal and external surfaces but their own natural structural diversity must give a range of oxidising strength; when they involve coordinated oxygen these are much more strongly oxidising, but appear to be formed with the greatest difficulty. Work by Kucherov and Slinkin [40] demonstrates that the presence of SO_2 either enhances the oxidising power of certain potential oxidation sites, or, it might be argued, unblocks sites that have already acquired electrons; these are probably closer to the unmodified Lewis variety. The conventional depiction of Lewis sites as involving a formal 'Si^+' unit [41] evokes a sense of unease, but the positive charge must be delocalised over its silyloxy substituents, as in (**VII**), thus providing stability and a rationale for their formation. If these sites are of such

$$C_6H_6^{+\bullet} + C_6H_6 \rightleftharpoons (C_6H_6)_2^{+\bullet} \qquad (11)$$

an extended nature one might indeed expect their redox power to be somewhat variable according to differences in the overall structure of larger associated 'silica' units. The proximity of tetracoordinate 'Al$^-$' units within this would be expected to reduce electron demand and might explain why it is the 'silica-rich' zeolites that are highly efficacious in radical cation production from adsorbed organics. Redox variability of this kind would also apply to 'Si$^+$' units modified by O_2 association.

We note that Shih [38] had formulated the redox sites in oxygen activated HZSM-5 as 'solid state defects', Si—O$^+$·—Al, to which he assigned a single sharp line at $g = 2.01$. In our view this is unlikely, for the following reasons: if the structure is essentially localised in this way, a ^{27}Al coupling is to be expected but no structure vas observed. For any lower O----Al coordination (i.e. closer to Si—O·----Al) substantial line broadening is expected – also at odds with the experiment. It is further significant that this sharp line is still present along with the spectra of radical cations and so is not involved in the oxidation. We agree with Shih that the signal is from a spin-bearing surface oxygen species, given the g-value, but not of the above formulation; nor that is the true redox centre.

3.1. SPECIFIC STUDIES

3.1.1. *Aromatic Radical Cations*

As noted before, aromatic substrates more complex than benzene usually form only their monomer radical cations in zeolites. While the benzene radical cation has similarly been observed, more complex chemistry is further evident and which provides a probe of how radical cation reactivity is mediated by zeolite media. Of especial interest is the early study by Corio and Shih [42] which may be regarded as the first attempt to interpret the ESR spectrum resulting from the adsorption of benzene (and several methylated derivatives) on activated H-mordenite. Following the interaction of liquid benzene with this zeolite, the ESR spectrum of the benzene monomer radical cation was observed, along with a central component with a splitting of 2.2 G assigned correctly to the $(C_6H_6)_2^{+\bullet}$ dimer cation. The loss after one hour of the dimer signal leaving only that with a splitting of 3.8 G was assigned to the sole presence of the monomer cation, and therefore that the dimer had dissociated, meaning that the reaction (11) is completely reversible. One can speculate that dissociation of the dimer might be assisted by interaction of the more charge localised $C_6H_6^{+\bullet}$ monomer with the zeolite surface. We have shown that even neutral cyclohexadienyl radicals are impeded in their motion in NaX zeolite and so are strongly surface associated [43]. In another study [44], it was concluded that the biphenyl radical cation was formed rather than the benzene monomer cation since it was shown that authentic biphenyl adsorbed on H-mordenite gives the *same* spectrum as does benzene.

Thus there is a controversy which is hard to resolve, and indeed a very recent study [45] has shown the presence of two 'dimer' species, in addition to the monomer, on HY zeolite, using ENDOR spectroscopy. We suggest that an ENDOR study of the benzene/mordenite system could be decisive in this matter because only two resonances would be present for the benzene monomer radical cation, but six for the biphenyl radical cation.

Durene, pentamethylbenzene and hexamethylbenzene apparently give only their monomer cations on H-mordenite, although there is the hint of secondary, possibly π-dimer, species after some hours [44].

3.1.2. Alkene Adsorption on H-Zeolites

(i) *Open-chain alkenes*. The adsorption of C3—C5 1-alkenes on H-mordenite was first reported by Leith [46]: all alkenes gave analogous spectra which consisted of at least 19 components with a spacing of 8.1 G. The spectrum was assigned, for the case of propene, to the conformationally fixed allyl radical MeCHC=CH—CHCH$_2$Me. However, it is most unlikely that the Me groups would not be freely rotating, even at very low temperatures, and the $(-)8.1$ G splitting is too large for the nodal proton of an allyl radical. Nevertheless, Leith's notion that the propene is dimerised after adsorption is a very salient one. In our recent study of the propene/mordenite system [39], we obtained spectra almost identical with those of Leith, but arrived at a different conclusion, since we showed that almost identical patterns could be produced by γ-irradiation of TME in CFCl$_3$ matrices: at low concentrations a pattern of 13 lines with $a(12H) = 17.6$G was observed, from the TME$^{+\cdot}$ monomer radical cation, while at increased TME concentrations a second set of lines appeared from the (TME)$_2^{+\cdot}$ dimer. Thus we assigned the spectrum from the propene/mordenite system to TME$^{+\cdot}$/(TME)$_2^{+\cdot}$ radical cations. In regard to the mechanism by which TME could arise from propene it is very significant that when the H-mordenite sample was exchanged by deuterium the ESR spectra were barely resolved single peaks, but with a total width compatible with the formation of TME$^{+\cdot}$(TME)$_2^{+\cdot}$ radical cations which had been completely substituted by deuterium. This, coupled with the fact that we could not obtain TME derived spectra from reactions of genuine propene radical cations in freon matrices, lead us to conclude that a prior dimerisation of two propene C3 units occurs via a proton catalysed route [39]. We further propose that the identity of the spectra found for all C3—C8 1-alkenes studied lies in the decomposition of the C4—C8 series into propene prior to its conversion to TME [39].

In a study of butene adsorption on H-mordenite, TME$^{+\cdot}$/(TME)$_2^{+\cdot}$ spectra were again recorded [47]; significantly, a gas chromatographic analysis showed TME to be only a very minor component of the products. This gives rise to the following question, therefore: why should this be the only product detected by ESR? In our view, this is a general feature – that one radical cation component is detected out of the many possible, given the often complex mixtures that are formed within the micropores of zeolite catalysts. It appears reasonable that the cause lies in the relative ionisation potentials of the various molecules present: i.e. the molecule with the lowest ionisation potential is ionised preferentially. This selectivity must result from thermodynamic control since ionisation of any molecule with a substantially lower I.P. than that of the preionised acceptor site would occur. We envisage

$$2 \;\text{>=<}^{+\bullet} \longrightarrow \;\text{>+—|—+<} \quad (12)$$

$$2 \;\text{>=<}^{+\bullet} \longrightarrow \;\triangle\!\!\!\!\!\text{Y}^{+} \;+\; \text{>|—+<}^{H} \quad (13)$$

$$\text{>=<}^{+\bullet} \;+\; \text{>=<} \longrightarrow \text{>+—|—|<}^{\bullet} \quad (14)$$

that electron transfer can occur between product molecules, in an equilibrium cascade, until the minimum energy is achieved (i.e. the species of lowest I.P. is ionised to its radical cation). Thus, while ESR is rather selective in its detection of product species, what it does do is reveal, of all the species present, that product of catalysis which has the lowest I.P.

(ii) *Branched alkenes.* Ichikawa et al. [47] report that all butene isomers, including isobutene (IB), form $TME^{+\bullet}/(TME)_2^{+\bullet}$ on adsorption on H-mordenite. It is interesting and rather curious that, in our own study in collaboration with Roduner, we instead observed the 2,5-dimethylhexa-2,4-diene (DMHD) radical cation, as confirmed by its 'fingerprint' obtained in a solid $CFCl_3$ matrix [48]. We note that Kucherov and Slinkin reported [36] spectra that could readily be assigned to $TME^{+\bullet}$ (and possibly the dimer) by the interaction of both IB and its dimer (2,4,4-trimethylpent-1-ene) with activated H-mordenite.

Roduner's group have studied [49] a variety of branched alkenes in regard to spontaneous radical cation formation on HZSM-5. The case of TME is extremely interesting since the initially observed spectrum shows the presence of a *mixture* of $TME^{+\bullet}$ and $DMHD^{+\bullet}$: furthermore, as a function of time the $TME^{+\bullet}$ signal was seen to diminish, while that from $DMHD^{+\bullet}$ rose rapidly over 30 min before itself decreasing. The kinetics for the decay of $TME^{+\bullet}$ are best described as second-order and implicate a bimolecular rate-limiting step, possibly involving combination of two $TME^{+\bullet}$ radical cations (12): the problem is that it is in no way easy to see how they might lose C_4H_{10} overall, form DMHD. Another possibility is that the two cations undergo a disproportionation via H-atom transfer (13). A rate-limiting step in which a neutral TME molecule is attacked by a $TME^{+\bullet}$ radical cation (14) is not supported by the kinetics since, with neutral TME being in excess, are expected to be pseudo-first order, and a distonic intermediate of this kind was not detected: again a ready route to loss of C_4H_{10} is not obvious.

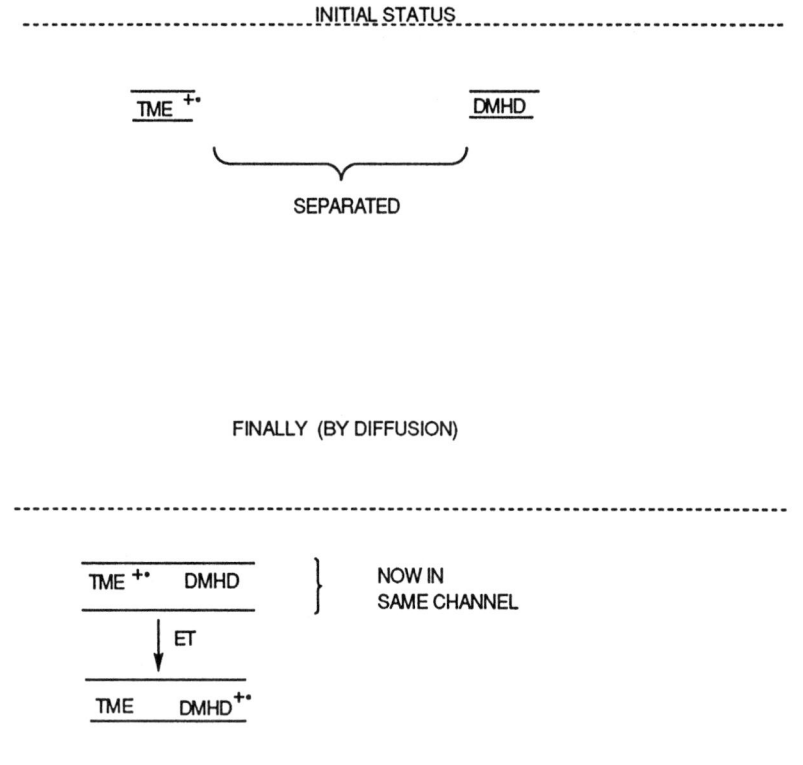

Scheme 1.

We think it more probable that the two events (loss of $TME^{+\cdot}$ and gain of $DMHD^{+\cdot}$) are not connected, and DMHD is formed by a pathway independent of $TME^{+\cdot}$ decay. We envisage that TME molecules are oxidised on encountering redox sites mainly within channels and that DMHD molecules are also formed and are oxidised, hence the initial mix of $TME^{+\cdot}$ and $DMHD^{+\cdot}$. Over a period of time, DMHD molecules are further formed, and, in the proximity of a $TME^{+\cdot}$ cation, undergo an electron transfer: $TME^{+\cdot} + DMHD \rightarrow TME + DMHD^{+\cdot}$. Since this process involves the encounter of two molecules in the same channel as the rate limiting step (the electron transfer being very fast) the kinetics are expected to be close to second order, as observed. This is outlined in Scheme 1.

It is significant that both 2,3-dimethylbut-1-ene and 2,3,3-trimethylbut-1-ene behave similarly to TME both in that they form $TME^{+\cdot}$ and $DMHD^{+\cdot}$ radical cations and the decay kinetics of these. This leads to the conclusion that TME is readily formed as a primary product from these alkenes. We believe that TME is converted to DMHD predominantly by a proton catalysed route: one can speculate on the mechanism for this, but to move in advance of conjecture it will be necessary to determine the nature of counter-molecules formed along with DMHD, i.e. in the process $C_6H_{12}(TME) + C_6H_2(TME) \rightarrow C_8H_{14}(DMHD) + C_4H_{10}$, is C_4H_{10} lost

$$\text{X}=\text{ } \xrightarrow{\text{H}^+} \text{ } \underset{+}{\text{X}}\diagdown_{\text{H}} \longrightarrow \text{X}=\text{X} \quad (15)$$

in one step, e.g. as isobutane, or is the process stepwise, so that the lost C_4H_{10} reflects, e.g. $C_2H_4 + C_2H_6$, or $2CH_4 + C_2H_2$, or $CH_4 + C_3H_8$? Given preliminary gas chromatographic data that the products formed in these reactions are rather complex mixtures, this may prove a difficult task, but attention to the 'volatile' fraction could provide the answer to many mechanistic questions in all these reactions.

In the case of 2-methylbut-2-ene and 2-methylbut-1-ene the DMHD radical cation is not observed but instead a species whose ESR spectrum is satisfactorily simulated with coupling constants of 12.2 G to three equivalent CH_3 groups and 5 G to a unique proton. Results for the primary 2-methylbut-2-ene radical cation in freon matrices [50] show larger couplings than this and eliminate this assignment; however, a diene structure $Me_2C=CH—CH=CHMe^{+ \cdot}$ would account for the observed coupling constants – that for the methyl groups being equal to that measured for $DMHD^{+ \cdot}$. The ESR spectrum recorded following adsorption of 4-methylpent-2-ene is complex [49] and reveals the time-dependent superposition of features from at least two species. While a definite analysis is difficult, the 12 G coupling to at least two methyl groups suggests that a diene radical cation with some terminal methyl substitution is present.

Shih [38] observed that 3,3-dimethylbut-1-ene (DMB) forms the $TME^{+ \cdot}$ monomer radical cation on adsorption on H-mordenite and proposed that the $DMB^{+ \cdot}$ radical cation is first formed but rearranges to $TME^{+ \cdot}$; even at 123 K. In order to determine the stability of $DMB^{+ \cdot}$ [39], we produced it in a solid freon matrix and examined its ESR spectrum in the temperature range 77–160 K, but no feature from $TME^{+ \cdot}$ was detected: thus, $DMB^{+ \cdot}$ is certainly stable at a temperature 40° higher than that used by Shih. While $CFCl_3$ has been known to retard rearrangements of radical cations that involve large changes in geometry [30], this is usually only below ca. 140–150 K where the matrix softens slightly, and the shift of a methyl group (+ a proton) in $DMB^{+ \cdot}$ is not a great change: we therefore believe that this result argues against a rearrangement of $DMB^{+ \cdot}$ in mordenite and it is more plausible that there is an initial proton catalysed rearrangement (15) of DMB to TME which is then preferentially oxidised, given that it has the lowest I.P.

(iii) *Cycloalkenes*. The adsorption of cyclopentene on H-mordenite was reported by Shih [51, 52] and later by Kucherov and Slinkin [53]. Shih assigned the spectrum that he observed to the primary cyclopentene radical cation with $a(4H) = 37$ G and $a(2H) = 14$ G. Since his early study, authentic spectra of this species have been obtained in solid freon matrices [50, 54], and it is clear that it is not formed in the zeolite medium. We repeated this work and obtained identical spectra [37], but it became clear that the correct analysis was not a quintet of *triplets* as had been assumed, presumably to fit the cyclopentene structure, but a quintet of quintets. We found that the spectrum was reversibly temperature depen-

(Structure IX: shown with Ha (axial) and He (equatorial) protons on a twisted C=C framework)

$$C_5H_8^{+\bullet} + C_5H_8 \longrightarrow C_5H_9^+ + C_5H_7^\bullet \qquad (16)$$

dent and at 195 K revealed nine equidistant lines ($a = 26$ G) from eight equivalent protons. It occurred to us that this was probably the 9-octalin radical cation, and we were able to reproduce the 77 K spectrum from a sample of authentic octalin in a solid $CFCl_3$ matrix. The 9-octalin cation adopts a structure [55] which is twisted about the C=C bond (IX) with a set of four equivalent pseudo-axial and another set of 4 equivalent pseudo-equatorial protons: the respective couplings are 37 G and 14 G. At elevated temperatures, there is a rapid inversion of the rings which accompanies the alternate twist of the C=C unit – hence the eight protons become indistinguishable at 298 K (fast exchange) but show a classical linewidth alternation in the region of intermediate exchange [37,55]. Simulation of the spectrum at 195 K gave an estimated activation energy for the exchange of 14.4 kJ/mol, and is to be compared with that of 18.8 kJ/mol measured in liquid hydrocarbon solution using an optically detected magnetic resonance technique [55]. While these values agree within the experimental error, comparison of actual spectra taken at the same temperature in both media shows that the inversion really is faster in the zeolite, and that these numbers are significant: we suggest that, in solution, there is a contribution to the activation energy arising from the reorganisation of the solvation shell during the inversion. For a nonpolar hydrocarbon solvent this is expected to be only a small amount, but perhaps of the order of 1 kcal/mol, as implied here.

Since 9-octalin is a formal dimer of cyclopentene, we considered that the octalin radical cation might arise by the reaction between a cyclopentene radical cation and a neutral cyclopentene molecule. We therefore investigated this reaction in a freon matrix [37] but found only the proton transfer reaction (16), the cylopentenyl radical being the sole paramagnetic product. It appears, therefore, that octalin arises via a nonradical cation route and we propose Scheme 2 which is driven by proton catalysis: as with the propene adsorption, discussed earlier, we find support for the extensive involvement of protons when a deuterium exchanged mordenite was used – again a poorly resolved feature was obtained, but whose overall width accords with the formation of octalin cations perdeuterated at the γ-positions.

Crockett and Roduner [56] made a subsequent study of a range of cycloalkenes on H-mordenite, including cyclopentene. Their results are in accord with ours but the spectrum they obtain on heating the cyclopentene sample to 295 K is quite

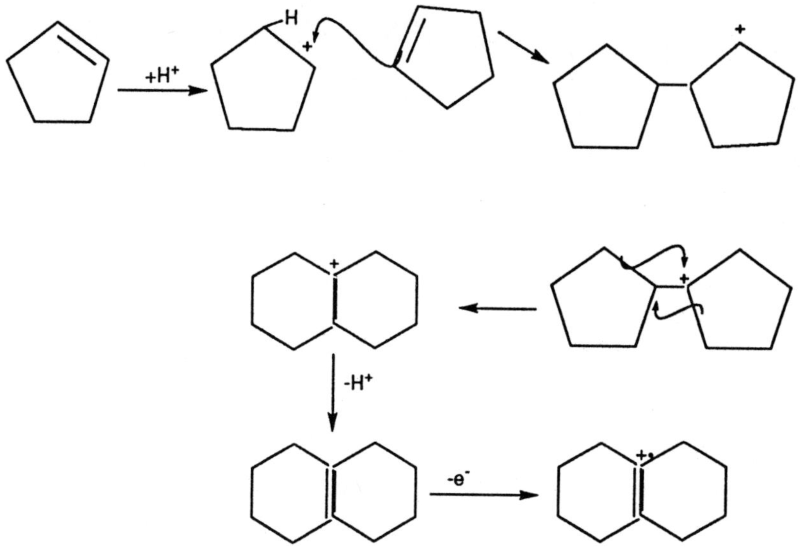

Scheme 2.

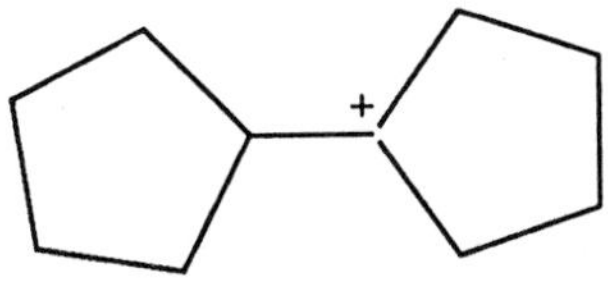

(X)

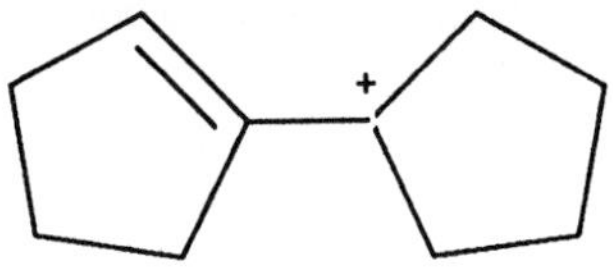

(XI)

different from that which we observed. This they assign convincingly to the cyclopentylidenecyclopentene radical cation, which can be considered to arise by loss of H_2 from the tertiary carbenium ion (**X**) to yield a more stable allylic ion (**XI**), which deprotonates and is oxidised. There is ample precedent for the conversion of tertiary to allylic carbocations by H_2 elimination [57]. In our study [30], we discovered a change in the spectrum at 295 K, but this was fully reversible, and on cooling again to 77 K the original quintet of quintets pattern was restored. We attribute this effect to the localisation of octalin cations to more confined regions of the zeolite structure which impedes intramolecular motion: we could obtain

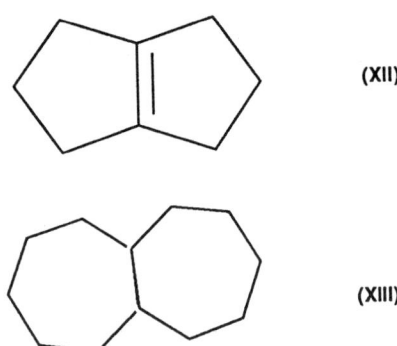

(XII)

(XIII)

good dynamic simulations of the higher temperature spectrum using a smaller rate of inversion, and which led to an estimated activation energy of *ca.* 28 kJ/mol.

It is very interesting that the H-mordenite and HZSM-5 used by the Zurich group seems so much more active toward diene formation than are samples of these zeolites used by other workers (c.f. DMHD vs. TME formation alluded to earlier), and must be catalytically significant: g.c./m.s. analysis would be most revealing of any *overall* differences in product distributions between different batches of a given catalyst. Octalin cations were also formed from cyclodecene on H-mordenite, but decayed at 294 K to form a secondary species which is not assigned but is not the cyclopentylidenecyclopentene (CPC) radical cation [56]. This observation militates against $CPC^{+ \cdot}$ being formed directly from octalin$^{+ \cdot}$, and we propose that CPC is formed via cyclopentene from (**X**)/(**XI**), as discussed earlier, but is then oxidised by existing octalin$^{+ \cdot}$ cations: octalin$^{+ \cdot}$ + CPC$^-$ → octalin + CPC$^{+ \cdot}$; this being in analogy with the TME$^{+ \cdot}$/DMHD oxidation proposed in Scheme 1. Octalin itself initially formed its primary radical cation, but decayed while a secondary species (possibly a conjugated hexalin diene radical cation?) grew in.

A single species was formed when either cyclooctene, cyclooctadiene or bicyclo[3.3.0]oct-2-ene interacted with H-mordenite, which at 120 K displayed couplings of $a(4H) = 36.1$ G, $a(4H) = 13.9$ G: this was ascribed to the bicyclo[3.3.0]oct-1-ene cation (**XII**). Similarly, a species with almost identical splittings [$a(4H) = 35.2$ G, $a(4H) = 13.7$ G] was formed when cyclohexene, 1-methylcyclohexene, 4-methylcyclohexene or cyclododecene were adsorped, proposed to be the cation (**XIII**); we recall that for octalin$^{+ \cdot}$ $a(4H) = 37.0$ G, $a(4H) = 14.0$ G [37]. Given the well established differences in conformational behaviour of 5-, 6- and 7-membered rings, it appears very surprising that these very different structures would show almost identical coupling constants, and is certainly at odds with the ratios of axial/equatorial β-proton couplings [50, 54] for cyclopentene and cyclohexene cations: 1.1 and 2.4 respectively. [It is significant that ratio for cyclohexene$^{+ \cdot}$ agrees closely with that for octalin$^{+ \cdot}$ (2.6) and suggests that the presence of the second ring has little effect on the conformation of each ring.] We feel that such similar coupling constants point to the *same* structural unit being formed in all cases, *namely an octalin radical cation*. Clearly, as there are small differences, both between coupling constants and the apparent propensity to ring inversion, *precisely* the same molecule is *not* being formed in all cases, and we believe that these octalin cations differ in *substitution* by alkyl groups at the γ-positions. For instance, we can envisage that cyclooctene might form, say, 2,7-di-*n*-propyloctalin: depending on the number of carbon atoms that need be accounted for around the C10 moiety, the substitution pattern is expected to change; in our work on the formation of a similar structure from hexa-1,5-diene [58], we proposed the 2,7-dimethyloctalin cation (see section on dienes, below). In such a situation, minor differences in conformational behaviour as are observed would be expected.

An ENDOR study of all these systems would be invaluable in determining remote substitution from long-range small couplings that are not resolved by ESR; equally incisive would be the fingerprinting of authentic (**XII**) and (**XIII**) cations in freon matrices to compare with the zeolite assigned species.

1,3- and 1,4-dimethylcyclohexene gave identical spectra with coupling constants of $a(6H) = 11.7$ G and $a(4H) = 19.4$ G assigned, respectively, to the methyl and β-protons in the 2,2'-dimethyl-1,1'-dicyclohexenyl (**XIV**) radical cation unit: a radical cation mechanism is proposed on the grounds that the tertiary ion (**XV**) would attack from the 1-position; however, loss of H_2 to form the allylic ion (**XVI**) would account for the position of coupling as in (**XVII**). We envisage Scheme 3 to account for the formation of (**XVII**).

α-Pinene, *trans-iso*-limonene and α-terpinene gave identical spectra on mordenite, which are assigned to the 4-isopropyl-1-methylcyclohexa-1,3-diene radical cation (**XVIII**) [59]. What is very exciting is that when α-pinene was adsorbed onto a deuterium-exchanged mordenite no incorporation of deuterium was observed, and it is argued therefore that a Brønsted catalysed route is not responsible for the transformation to (**XVIII**): rather that oxidation of pinene occurs at a Lewis site and the resulting radical cation (**XIX**) rearranges directly. It would be instructive to generate genuine (**XIX**) in a γ-irradiated Na-mordenite and to ascertain its propensity for thermal rearrangement as proposed.

3.1.3. *Alkynes*

To our knowledge, there are three reports of alkyne adsorption on zeolites, being confined to acetylene [40] and dimethylacetylene (DMA) [37,38]. The latter material was studied on HZSM-5 by Shih [38], who concluded that the spectrum consisted of seven lines with a splitting of 10 G which he assigned to a Jahn–Teller distorted form of the monomer $DMA^{+\cdot}$ radical cation. This, however, must be refuted because the splitting is roughly half that measured for $DMA^{+\cdot}$ in a solid freon matrix [60]. We obtained a similar spectrum from DMA in H-mordenite [37] which comprises at least nine lines (as also does the spectrum in HZSM-5 [38] on closer inspection) and which we assigned to a dimeric species: the tetramethylcyclobutadiene (TMCB) radical cation, as confirmed by studies in solid freon matrices [37,60]. In both mordenite and ZSM-5 media, at 350 K, the spectrum from $TMCB^{+\cdot}$ decayed but was replaced by that from hexamethylbenzene (HMB) radical cations [$a(18H) = 6.7$ G]. In this case, a radical cation mechanism cannot be excluded because $DMA^{+\cdot}$ is known to undergo cycloaddition with

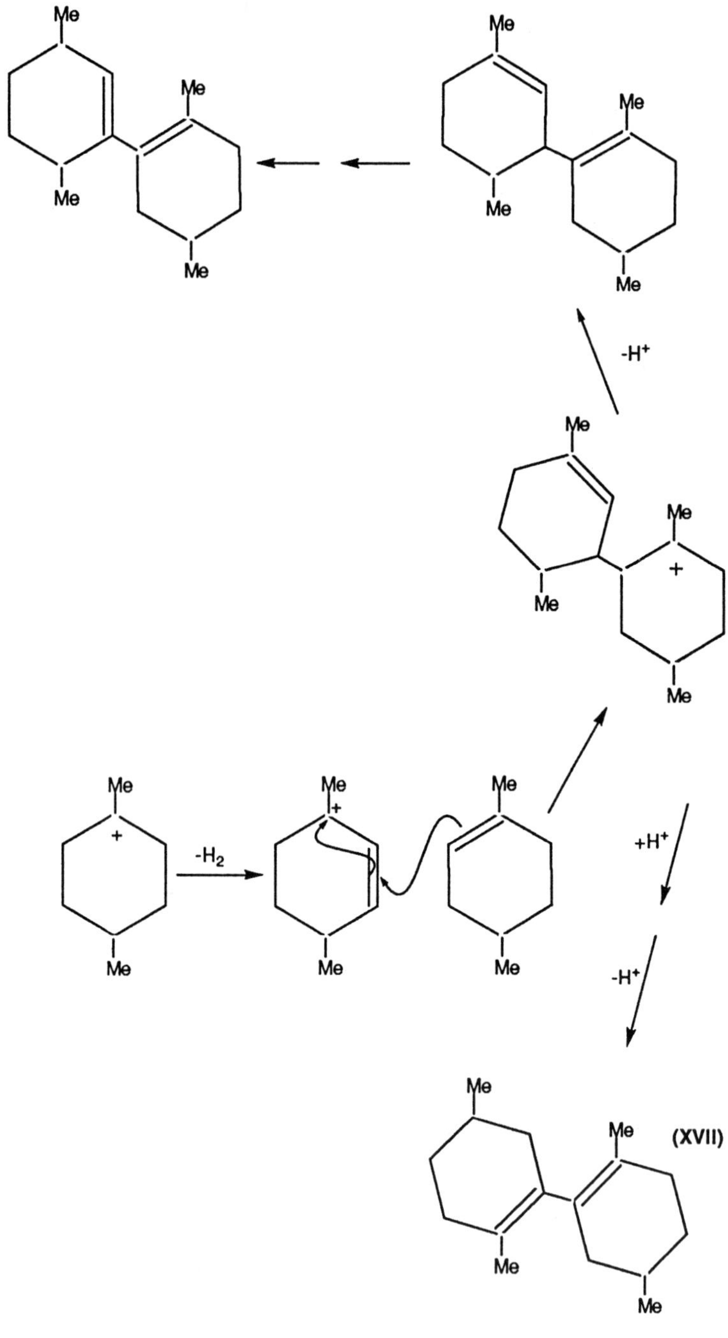

Scheme 3.

a neutral DMA molecule to form TMCB$^{+\cdot}$ even at low temperatures [37,60]: nonetheless, protonated cyclobutadienes are known in superacid media from NMR studies [61], and could be intermediates in an H-zeolite (Scheme 4). Since HMB$^{+\cdot}$ is not formed from TMCB$^{+\cdot}$ in freon matrices, preferring instead to fragment (17), we have no evidence that a stepwise DMA$^{+\cdot}$ → TMCB$^{+\cdot}$ → HMB$^{+\cdot}$ route operates. Thus, while DMA$^{+\cdot}$ → TMCB$^{+\cdot}$ is supported (but not proved, see above) we again propose that HMB is formed via acid catalysis from DMA and is subsequently oxidised, possibly via E.T. to TMCB$^{+\cdot}$: TMCB$^{+\cdot}$ + HMB → TMCB + HMB$^{+\cdot}$.

Kucherov and Slinkin studied the adsorption of acetylene (ethyne) on mordenite [40] but did not assign the spectrum: however, it consisted of at least five and probably seven lines with a spacing of *ca.* 4–5 G, and we propose the trimer – the C$_6$H$_6^{+\cdot}$ radical cation. More rapid diffusion of C$_2$H$_2$ could certainly lead to this if the concentration were high enough, even at lower temperatures than are required for HMB$^{+\cdot}$ formation from DMA.

3.1.4. Dienes

We have found that almost identical spectra are observed following the interaction of hexa-1,5-diene, *cis* or *trans* piperylene (penta-1,3-diene), or penta-1,4-diene with H-mordenite, and must also be assigned to 9-octalin radical cations [58]. The former diene has already been mentioned, and we believe that the product radical

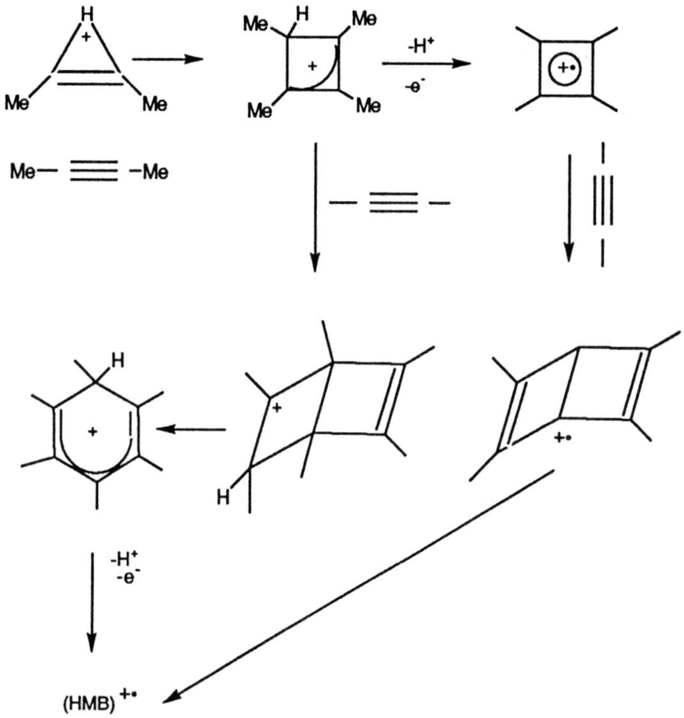

Scheme 4.

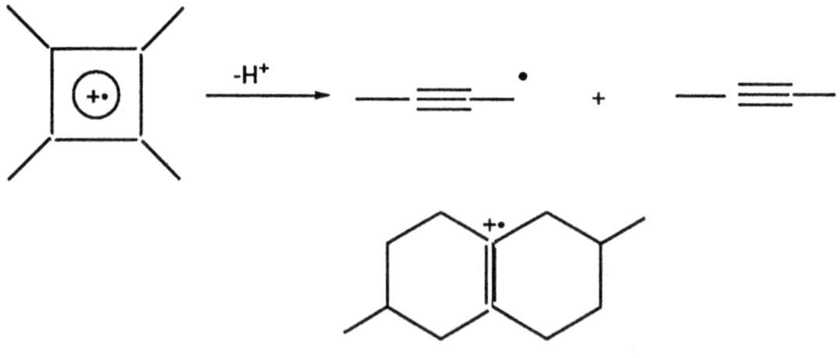

cation is the 2,6-dimethyl derivative (**XX**), most likely with the groups occupying pseudo-equatorial positions. We favour here also a proton catalysed route since this leads smoothly to (**XX**), avoids the non-obvious loss of two carbon atoms to form a C10 structure, and fits with the extensive exchange of protons (deuterons) between the substrate and the zeolite that we observed [58]. In freon matrices, unimolecular cyclisation occurred via the cyclohexane-1,4-diyl radical cation, stabilised in the $CF_2ClCFCl_2$ matrix, leading to the cyclohexene radical cation: neither species was detected in the zeolite medium.

Results for penta-1,4-diene in freon media were inconclusive regarding its reactivity but did not support the involvement of radical cations in mordenite. In the case of the piperylenes, chain-end allyl radicals (**XXI**) and (**XXII**) were formed in $CF_2ClCFCl_2$, in preference to deprotonation, while isomerisation yielded clear features from cyclopentene radical cations in a $CFCl_3$ matrix. In the latter aspect, our results accord with those of Fujisawa *et al.* [62], but differ in that we were able, in part, to stabilise the primary radical cations which are not, therefore, unstable to rearrangement and are energy minima: as with our results for organotin radical cations [63], at 77 K only rearrangement (fragmentation) from excited species occurs. In view of the formation of octalin$^{+\cdot}$ from cyclopentene in H-mordenite, the isomerisation of *cis*- and *trans*-piperylene radical cations to cyclopentene$^{+\cdot}$ might be considered significant; however, as the latter were only found to deprotonate in freon media, no direct mechanism involving their intermediacy is supported.

3.1.5. *Mobility of Alkene Radical Cations in Zeolites*

Given that the framework of a zeolite is negatively charged, it is reasonable that the surface motion of radical cations will be impeded, and we have presented earlier indirect evidence from their observed reactivity. When molecules are adsorbed within zeolites in spherical supercages which are large compared with the molecular dimensions, their motion is expected to be isotropic as we found for neutral cyclohexadienyl radicals at room temperature in NaX [43]. As located in channels, this may not be the case, especially when the radical is of comparable dimensions to its host: in which case, an easier motion about one axis over others will confer anisotropy to the measured *g*- and *A*-tensors. Roduner [64] has discovered that, particularly at low temperatures, the ESR spectra of TME$^{+\cdot}$ in HZSM-5 and of DMHD$^{+\cdot}$ in H-mordenite show pronounced anisotropy and he has succeeded in obtaining excellent simulations which allow certain conclusions to be drawn regarding their molecular motion. Thus it appears that TME$^{+\cdot}$ in ZSM-5 at 293 K does not tumble about its C=C axis, but instead rotates about the plane of the molecule. Since only at the channel intersections is there sufficient room for this to occur, one may conclude that these are where the radical cations are located. Indeed, it is significant that, according to computational modelling studies, this is the preferential site for small hydrocarbons in ZSM-5 [65]. Since rotation of TME$^{+\cdot}$ about its plane-perpendicular axis [(z) in (**XXIII**)], has the largest moment of inertia, it is expected to be the most excited – space permitting. At 100 K, the observed anisotropy is more compatible with a reorientationally frozen species, but now twisted, with a dihedral angle of *ca.* 60° between the

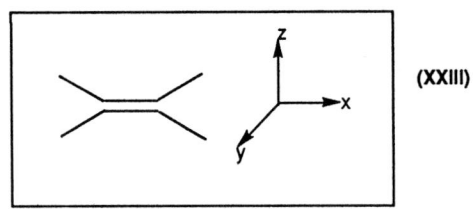

(XXIII)

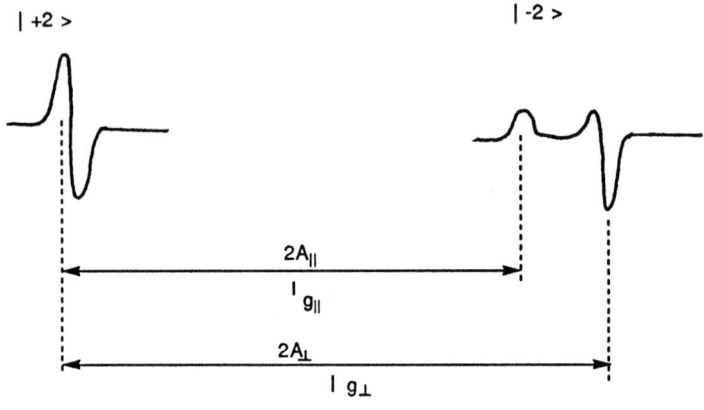

Scheme 5.

Me$_2$C— planes. It is proposed that torsional motion at the higher temperature averages out this distortion.

In the case of DMHD$^{+\cdot}$ in H-mordenite at 149 K, the anisotropy in the g- and A-parameters is relatively small, indicating that the radical is mobile; however, the lack of coaxial symmetry in the apparent g- and A-tensors in the simulation rules out a rotational motion about the long axis of the molecule. It is proposed that the radical undergoes a librational or 'rattling' motion within the zeolite channel which averages the anisotropies. From a closer consideration of our spectrum of 9-octalin$^{+\cdot}$ in mordenite at 195 K [37], we notice that the non-alternate lines are broadened in the direction of high field, and that $|-2\rangle$ shows a clear $\|$, $\perp$ structure. This can most readily be accommodated by a motion about the C=C (x) axis (**XXIII**) which averages the (z), (y) g- and A-anisotropies: this again is a motion which involves the greatest moment of inertia for the molecule, and is probably also librational since the ca. 0.7 nm mordenite channel would not readily permit a free rotation about the C=C axis for this large molecule. It is salient to mention that the existence and nature of preferential motion in alkene radical cations may be ascertained quite readily from the orientation of 'pair-matched' features: i.e. those differing only in the sign of the nuclear spin quantum number. Essentially, if one examines the $|+2\rangle$ lines of the TME$^{+\cdot}$ spectrum, it is clear that they are of the form shown in Scheme 5: this establishes that $A_\| < A_\perp$ and $g_\| < g_\perp$

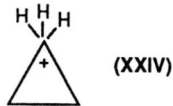
(XXIV)

. For an alkene unit (**XXIII**), assuming bond angles of 120°, which is probably close to the truth, we obtain the following relations for the hyperfine tensor: $A_y = a + 1.25B$, $A_x = a - 0.25B$, $A_z = a - B$, where B is the dipolar coupling. Given that g_z is close to free spin but g_x and g_y are somewhat larger, we obtain the following conditions which signify rotation about each axis shown.

	Rotation axis
$A_\parallel < A_\perp$; $g_\parallel < g_\perp$	z
$A_\parallel > A_\perp$; $g_\parallel > g_\perp$	y
$A_\parallel < A_\perp$; $g_\parallel > g_\perp$	x

This accords with the conclusion that $TME^{+\cdot}$ rotates about its z-axis in mordenite, and since the octalin$^{+\cdot}$ spectrum makes clear (Scheme 5) that $A_\parallel < A_\perp$; $g_\parallel > g_\perp$, we can determine immediately that its reorientational or librational motion occurs about the x-axis.

3.1.6. Saturated Hydrocarbons

The group of Kucherov and Slinkin have reported that radical cations are generated from the adsorption of cyclohexane on both H-mordenite and HZSM-5 [66]: these we are now able to assign once again to octalin radical cations. Chen and Fripiat [67], on the other hand, have assigned ESR spectra arising from methylcyclohexane and 3-methylpentane following adsorption on H-mordenite to neutral radicals. In a very recent paper by Roduner and Crockett [68], these assignments are refuted since the parameters are far better accommodated by radical cation structures: respectively, as 1,2-dimethyl-cyclopentene$^{+\cdot}$ and $TME^{+\cdot}$. It is significant that it is necessary to heat the adsorbed samples for several hours in order to generate radical cations from alkanes, in contrast with the immediate observation of radical cation signals when alkenes are adsorbed. This suggests that the chemistry involved is not driven from the initial formation of the alkane radical cations, but that it is necessary to first generate the alkene species by an overall H_2 elimination, which are then oxidised as usual – hence the observed induction period.

It has been proposed that, on acidic zeolites, protonation of alkanes occurs typically at a tertiary carbon atom leading to H_2 or low molecular weight alkane elimination and alkylcarbenium ions: these can undergo very extensive transformations [57] – probably via corner-protonated cyclopropane intermediates (**XXIV**). For 1,2-dimethylcyclopentene and TME, out of the undoubted mixture of products in each case formed, these are most likely the alkenes of lowest I.P. and are thus detected. We envisage the generation of 1,2-dimethylcyclopentene from methylcy-

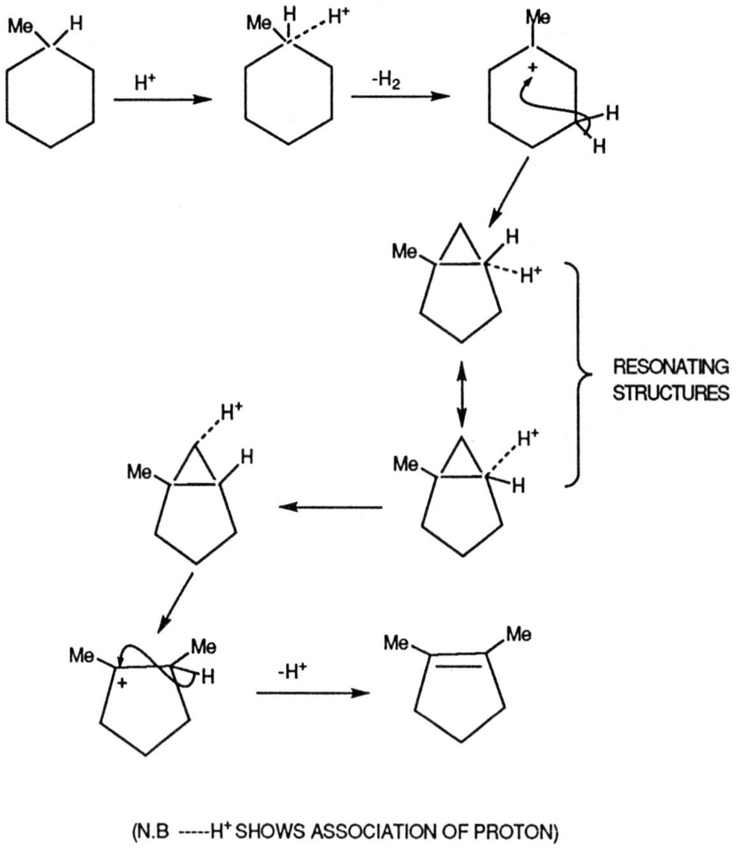

Scheme 6.

clohexane by the proton catalysed route outlined in Scheme 6. In this, we use a formalism (----H^+) to represent the hypervalent (carbonium) ion, i.e. as a structurally implicit bound proton.

4. Conclusions

From the foregoing, it is clear that much of the chemical transformation in acidic zeolites is proton catalysed – radical cations are formed subsequently and selectively from those product species of lowest ionisation potential. Thus, one must ask: are radical cations formed at all as intermediates in zeolite catalysis? There is good evidence that neutral radicals are present in a number of reactions involving heterogeneous catalysts [69–74], including zeolites [72–74], and while it is unlikely that the highly stabilised radical cations that we observe in our ESR studies are directly important to the overall catalytic sequence, we envisage that radical cations of smaller molecules such as $Me_2O^{+\cdot}$ might precede the formation of neutral radicals by deprotonation: i.e. $Me_2O^{+\cdot} \rightarrow H^+ + MeOCH_2$, the involvement of which has been proposed in the methanol-to-gasoline process [37, 72].

In certain instances, it might be possible to distinguish between genuine reactions of radical cations such as dimerisation/elimination, and other processes, by the detection of low molecular weight fragments thus eliminated using techniques such as g.c./m.s. and solid state $^1H/^{13}C$ NMR: thus the ESR studies probably require the support of other physical methods. That ESR assignments may be checked against spectra from genuine radical cations, rather than relying on conjecture, provides a great strength, and the potential use of ENDOR spectroscopy to aid the evaluation of long-range couplings is likely to contribute to a firmer overall framework of knowledge.

Furthermore, it is clear that radical cations can act as probes of molecular motion in zeolites and accurate determinations of the anisotropies present, by computer simulation of spectra, enable an understanding of preferential reorientation and diffusion properties of organic substrates in zeolites – and indeed on solid catalysts in general.

It would be of great significance to discriminate between electrostatic and geometric (size) effects in this regard, and to compare the motion of radical cations with that of their neutral precursors as determined by NMR and neutron scattering methods.

In short, an exciting era is predicted.

Acknowledgement

We thank Dr Migg Roduner for sending us preprints of the work of his group which is 'in press' at the time of writing, so that we could include discussion of this in the present article, and for many enjoyable and stimulating discussions. In regard to our own studies of radicals in zeolites, we are grateful to the following companies: Laporte Inorganics, B.P. Ltd., Ceram Research Ltd.; and to the Science and Engineering Research Council of the United Kingdom and the Leverhulme Trust for support.

References

1. M. C. R. Symons: *Chem. Soc. Rev.* **13**, 393 (1984).
2. M. Shiotani: *Magn. Reson. Rev.* **12**, 333 (1987).
3. C. J. Rhodes: in *Specialist Periodical Reports on Electron Spin Resonance*, Vol. 13A. The Royal Society of Chemistry, Cambridge (1992).
4. K. Toriyama, K. Nunome, and M. Iwasaki: *J. Am. Chem. Soc.* **109**, 4496 (1987).
5. X.-Z. Qin and A. D. Trifunac: *J. Phys. Chem.* **94**, 4751 (1990).
6. M. V. Barnabas and A. D. Trifunac: *Chem. Phys. Lett.* **187**, 565 (1991).
7. M. V. Barnabas and A. D. Trifunac: *Chem. Phys. Lett.* **193**, 298 (1992).
8. X.-Z. Qin and A. D. Trifunac: *J. Phys. Chem.* **95**, 6466 (1991).
9. M. V. Barnabas and A. D. Trifunac: *J. Chem. Soc., Chem. Commun.* 813, (1993).
10. M. C. R. Symons and J. L. Wyatt: *J. Chem. Res.* (S), 362 (1989).
11. A. Hasegawa, M. Shiotani, and F. Williams: *Faraday Discuss. Chem. Soc.* **63**, 157 (1987).
12. C. J. Rhodes, C. Glidewell, and H. Agirbas: *J. Chem. Soc., Faraday Trans.* **87**, 3171 (1991).
13. K. B. Yoon and J. K. Kochi: *J. Chem. Soc., Chem. Commun.* 510, (1988).

14. K. Toriyama, K. Nunome, and M. Iwasaki: *J. Phys. Chem.* **90**, 6836 (1986).
15. K. Toriyama, K. Nunome, and M. Iwasaki: *J. Chem. Phys.* **77**, 5891 (1982).
16. K. Toriyama, K. Nunome, and M. Iwasaki: *J. Phys. Chem.* **84**, 2149 (1981).
17. M. Iwasaki, K. Toriyama, and K. Nunome: *Radiat. Phys. Chem.* **105**, 414 (1987).
18. M. Iwasaki, K. Toriyama, and K. Nunome: *J. Chem. Phys.* **79**, 2499 (1983).
19. M. Iwasaki, K. Toriyama, M. Fukaya, H. Muto and K. Nunome: *J. Phys. Chem.* **89**, 5278 (1985).
20. D. L. Griscom: *Phys. Rev. B* **40**, 4224 (1989).
21. R. V. Lloyd and R. V. Williams: *J. Phys. Chem.* **89**, 5379 (1985).
22. B. P. Roberts: *Tetrahedron Lett.* 5385 (1991).
23. I. D. Reid, C. J. Rhodes and E. Roduner: *Tetrahedron Lett.* 5617 (1992).
24. X.-Z. Qin and F. Williams: *J. Am. Chem. Soc.* **106**, 7640 (1984).
25. X.-Z. Qin and A. D. Trifunac: *J. Phys. Chem.* **94**, 3188 (1990).
26. K. Ushida, T. Shida, and K. Shimokoshi: *J. Phys. Chem.* **93**, 5388 (1989).
27. X.-Z. Qin, L. D. Snow, and F. Williams: *J. Phys. Chem.* **89**, 3602 (1985).
28. G. A. Olah, N. Hartz, G. Rasul, and G. K. Surya Prakash: *J. Am. Chem. Soc.* **115**, 6985 (1993).
29. K. Somekawa, K. Haddaway, P. Mariano and J. A. Tossel: *J. Am. Chem. Soc.* **106**, 3060 (1984).
30. C. J. Rhodes: *J. Am. Chem. Soc.* **110**, 4446 (1988).
31. H. D. Roth: *Acc. Chem. Res.* **20**, 43 (1987).
32. J. L. Gebicki, J. Gebicki, and J. Mayer: *Radiat. Phys. Chem.* **50**, 165 (1987).
33. M. I. Loktev and A. A. Slinkin: *Russ. Chem. Rev.* **45**, 807 (1976).
34. V. Ramamurthy, J. V. Caspar, and D. R. Corbin: *J. Am. Chem. Soc.* **113**, 594 (1991).
35. N. H. Sagert, R. M. L. Pouteau, M. G. Bailey, and F. P. Sargent: *Can. J. Chem.* **50**, 2041 (1972).
36. A. V. Kucherov, A. A. Slinkin, D. A. Kondratyev, T. N. Bondarenko, A. M. Rubinstein and Kh. M. Minachev: *J. Mol. Cat.* **37**, 107 (1986).
37. C. J. Rhodes: *J. Chem. Soc., Faraday Trans.* **87**, 3179 (1991).
38. S. Shih: *J. Catal.* 79, 390 (1983).
39. C. J. Rhodes: *Colloids and Surfaces A: Physicochemical and Engineering Aspects* **72**, 111 (1993).
40. A. V. Kucherov and A. A. Slinkin: *Kinet. Katal.* **24**, 947 (1983).
41. A. A. Slinkin and A. V. Kucherov: *Kinet. Katal.* **24**, 955 (1983).
42. P. L. Corio and S. Shih: *J. Catal.* **18**, 126 (1970).
43. C. J. Rhodes, I. D. Reid, and E. Roduner: *J. Chem. Soc., Chem. Commun.* 512 (1993).
44. Y. Kurita, T. Sonoda, and M. Sato: *J. Catal.* **19**, 82 (1970).
45. R. Erickson, M. Lindgren, A. Lund, and L. Sjoqvist: *Colloids and Surfaces A: Physicochemical and Engineering Aspects* **72**, 207 (1993).
46. I. R. Leith: *J. Chem. Soc., Chem. Commun.* 1282 (1972).
47. T. Ichikawa, M. Yamaguchi, and H. Yoshida: *J. Phys. Chem.* **91**, 6400 (1987).
48. E. Roduner, L.-M. Wu, R. Crockett, and C. J. Rhodes: *Catal. Lett.* **14**, 373 (1992).
49. R. Crockett and E. Roduner: to be published.
50. T. Shida, Y. Egawa, H. Kubodera, and H. Kato: *J. Chem. Phys.* **73**, 5963 (1980).
51. P. L. Corio and S. Shih: *J. Phys. Chem.* **75**, 3475 (1971).
52. S. Shih: *J. Phys. Chem.* **79**, 2201 (1975).
53. A. V. Kucherov and A. A. Slinkin: *Kinet. Katal.* **23**, 997 (1982).
54. A. Tabata and A. Lund: *Chem. Phys.* **75**, 379 (1983).
55. A. V. Veselov, V. I. Melekhov, O. A. Anisomov, Yu. N. Molin, K. Ushida, and T. Shida: *Chem. Phys. Lett.* **133**, 478 (1987).
56. R. Crockett and E. Roduner: *J. Chem. Soc., Perkin Trans. II*, 1503 (1993).
57. P. A. Jacobs and J. A. Martens: in H. Van Bekkum, E. M. Finigen, and J. C. Jansen (eds.), *Studies in Surface Science and Catalysis*, Vol. 58, Elsevier, Amsterdam (1991).
58. C. J. Rhodes and M. Standing: *J. Chem. Soc., Perkin Trans. II*, 1455 (1992).
59. R. Crockett and E. Roduner: *J. Chem. Soc., Perkin Trans. II*, (1994) in press.
60. K. Ohta, M. Shiotani, J. Sohma, A. Hasegawa, and M. C. R. Symons: *Chem. Phys. Lett.* **136**, 465 (1987).
61. G. E. Nelson and E. A. Williams: in R. W. Taft (ed.), *Progress in Physical Organic Chemistry*, Vol. 12, Wiley, New York (1976), p. 229.
62. J. Fujisawa, T. Takayasagi, S. Sato, and K. Shimokoshni: *Bull. Chem. Soc. Jpn.* **61**, 1527 (1988).
63. E. Butcher, C. J. Rhodes, M. Standing, R. S. Davidson, and R. Bowser: *J. Chem. Soc., Perkin Trans. II* 1469 (1992).

64. E. Roduner, R. Crockett, and L.-M. Wu: *J. Chem. Soc., Faraday Trans.* **89**, 2101 (1993).
65. J. O. Titilaze, S. C. Parker, F. S. Stone and C. R. A. Catlow: *J. Phys. Chem.* **95**, 4038 (1991).
66. A. V. Kucherov, A. A. Slinkin, K. M. Gitis, and G. V. Isagulants: *Catal. Lett.* **1**, 311 (1988).
67. F. R. Chen and J. J. Fripiat: *J. Phys. Chem.* **97**, 5796 (1993).
68. E. Roduner and R. Crockett: *J. Phys. Chem.* **97**, 11853 (1993).
69. L. C. Anderson, M. Xu, C. E. Mooney, M. P. Rosynek, and J. H. Lunsford: *J. Am. Chem. Soc.* **115**, 6322 (1993).
70. V. Amir-Ebrihimi, J. Grimshaw, E. A. McIlgorm, T. R. B. Mitchel, and J. J. Rooney: *J. Mol. Cat.* **27**, 337 (1984).
71. W. Matir and J. H. Lunsford: *J. Am. Chem. Soc.* **103**, 3728 (1981); D. J. Driscoll, W. Matir, J.-X. Wang and J. H. Lunsford: *J. Am. Chem. Soc.* **107**, 58 (1985); W. Matir and J. H. Lunsford: *J. Am. Chem. Soc.* **107**, 5062 (1985).
72. J. K. A. Clarke, R. Darcy, B. F. Hegarty, E. O'Donoghue, V. Amir-Ebrahimi, and J. J. Rooney: *J. Chem. Soc., Chem. Commun.* 425 (1986).
73. C. D. Chang, S. D. Hellring, and J. A. Pearson: *J. Catal.* **115**, 282 (1989).
74. H. Choukrous, D. Brunel, and A. Germain: *J. Chem. Soc., Chem. Commun.* 6 (1986).

Surface Trapped Electrons on Metal Vapour Modified Magnesium Oxide. Nature of the Surface Centres and Reactivity with Adsorbed Molecules

ELIO GIAMELLO and DAMIEN MURPHY
Dipartimento di Chimica Inorganica, Chimica Fisica e Chimica dei Materiali, Università di Torino, Via Pietro Giuria 9, 10125–Torino, Italy

(Received: 15 December 1993; accepted: 5 July 1994)

Abstract. The first part of the present paper reports a wide investigation on the interaction between metals of low ionisation energy (magnesium and alkali metals) and the surface of the ionic oxide MgO. The interaction basically consists in the ionisation of the metal and in the stabilisation of both the released electrons and the corresponding cations at the surface. Various types of paramagnetic electron centers have been identified including surface F centers, monoalkali cationic centers and ionic clusters. In the second part the reactivity of the electron rich MgO surface with simple inorganic molecules is described.

Key words. Magnesium oxide, zeolites, electron transfer, EPR.

1. Introduction

It is well known that the contact between atoms with low ionization energy (namely alkali and alkaline earth metals) and ionic solids (in particular metal oxides) causes the ionization of the metal with the stabilisation of both the released electrons and the positive ions in the solid matrix or at its surface [1–8]. There are several reasons underlying the attention devoted by various research groups to the addition of metal atoms to metal oxides:

(a) an intrinsic interest in the metal atom–oxide interaction and the relationship with the structural and physico chemical properties of the ionic surface;
(b) the possibility of obtaining a novel class of inorganic materials possessing particular physical properties. This is the case of the interaction between metal atoms and zeolites leading to the formation of novel 'inclusion' compounds with unusual electric and magnetic properties [9];
(c) the reactivity of the system resulting from metal addition to the oxide. These electron rich systems are able to transfer electrons towards a variety of molecules in contact with the surface easily forming anions and radical anions which may survive under suitable conditions and can therefore be investigated in the adsorbed state. The chemical behaviour of the metal doped oxides is in a sense analogous to that of the irradiated oxides which contain highly reactive surface electron donor centres and which have been investigated in the past [10–15].

Zeolites are generally considered ideal materials for the study of the metal vapour–ionic solid interaction because their aluminosilicate framework contains regular three dimensional arrays of cavities and channels of molecular dimensions. These cages act as templates for the formation of novel 'inclusion' compounds whose properties lie on the borderline of the nonmetal–metal transition [9] and the bulk of the past research has centred on these vapour modified zeolite systems. Although the interaction between alkali metals and nonporous oxides, such as the alkaline earth oxides [16] and aluminas [17], has been investigated in the past, the primary purpose of these studies was an examination of the catalytic properties of the vapour modified samples without giving a comprehensive description of the nature of the modified surface itself. The present study is instead devoted to understanding the structural features and the chemical reactivity of the ionic oxide MgO modified by the addition of low ionization energy metals. This interaction primarily results in the ionization of the metal atom with subsequent trapping of the released electrons sometimes at conventional and sometimes at unusual surface sites. The type of trapped electron centres formed depends both on the nature and on the amount of the added metal. The first part of the present study is therefore devoted to the description of the metal vapour modified surface. The ensuing chemical reactivity of the electron-rich MgO is illustrated in the second part by means of some selected examples of the interaction between electron-rich MgO and very simple inorganic molecules. The capability of electron transfer towards adsorbed molecules (with consequent formation of adsorbed anions) actually covers a wide range of inorganic and organic molecules which are in part under investigation. A detailed and systematic investigation of all the possible reactions between molecules and electron-rich MgO is in any case beyond the scope of the present work.

The leading technique in the investigation of both the structure of the electron-rich surface and its chemical reactivity is Electron Paramagnetic Resonance (EPR) since many of the surface centres originated by metal doping in addition to the many adsorbed anions generated by chemical reaction at the interface are paramagnetic. In both cases the EPR technique is the most suitable one for a detailed description at the molecular level of the structure of these paramagnetic centres.

Magnesium oxide has been adopted as the most suitable oxide and non-zeolitic system in this investigation for several reasons:

(a) activated MgO is a highly ionic material and for this reason is able to assist the ionisation of the metal atoms;
(b) the structure of MgO is very simple (rock salt type structure) and the surface morphology (illustrated in the next paragraph) well understood;
(c) the oxide can be prepared in high surface area form thereby allowing one to obtain appreciable amounts of surface products from the vapour-surface reaction.

Some important morphological aspects of high surface area MgO are presented below.

2. High Surface Area MgO: Morphological Aspects

Due to its simple crystal structure and high ionicity MgO constitutes an ideal system for the investigation of surface morphology. In the past it has been the object of many theoretical and experimental studies in this area whose illustration is not however among the scopes of the present paper. High area, thoroughly dehydrated MgO is obtained by the slow decomposition of other solids such as the corresponding hydroxide or carbonate followed by high temperature dehydration. The ex-hydroxide MgO employed in the present work is shown by Transmission Electron Microscopy [18] to be built up by very small, and despite the dimensions, quite regular cublets. The crystal faces exposed at the surface are thus the (100) type ones where each Mg^{2+} or O^{2-} ion is surrounded by five ions of opposite charge disposed according to the C_{4v} square pyramidal symmetry.

The main type of localised point defect available at the surface of MgO are isolated point defects or vacancies related to the absence of a single ion. Early theoretical studies on these point defects clarified that the surface vacancies were more stable than those in the bulk [19,20]. Of particular interest to the present paper are the surface anionic vacancies which can act as surface electron traps: the lack of one O^{2-} anion at the planar face produces a positive site constituted by an array of four Mg^{2+} ions in the same surface plane and by a fifth cation in the first subsurface layer. Similar vacancies also occur at the edges of the cubic microcrystals [21].

In addition to the presence of point defects the MgO surface also shows some morphological features larger in scale than the point defects such as steps, terraces, kinks and corners. As can be deduced from Figure 1, the presence of these morphological features increases the number of four- and three-coordinated surface ions. A relevant role for the morphological description of the surface of high area MgO has been played by UV-visible diffuse reflectance [22,23] and photoluminesence [24] spectroscopies as well as by spectroscopic characterisation in IR and EPR using suitable probe molecules [25]. The number of low coordination (four- and three-) sites in the high area samples of MgO has been evaluated at around 20–25% of the whole exposed sites [21,26,27].

3. Magnesium Oxide Preparation and Metal Addition

The starting $Mg(OH)_2$ polycrystalline powder is introduced in to the quartz bulb of a cell specially designed for MgO preparation, metal addition and EPR/UV-vis spectra recording. Simultaneously a small piece of the metal for the evaporation is introduced in to a side metal holder. A schematic drawing of the cell is shown in Figure 2. After connecting the cell to a vacuum line the hydroxide is decomposed at 523 K for 16 h to obtain high surface area MgO. The oxide is then slowly heated under vacuum up to 1073 K and kept at this temperature for 2 h to reach a controlled degree of dehydroxylation and finally cooled down at room temperature. The metal piece is then transferred from the side metal holder to the quartz bulb and the temperature slowly raised (under dynamic vacuum) to the level required for metal vapour generation, i.e. 850–870 K for Mg, 570–590 K for Li, 533–553 K for Na, 413–433 K for K and 363–393 K for Rb. A slightly modified version of the cell in Figure 2 was used to expose the MgO sample to the metal vapours with the oxide kept at a given temperature different from that required to generate the metal vapour.

The exposure time of the powder to the metal vapours is crucial and varied depending on the amount of metal required to be deposited on the sample. It was

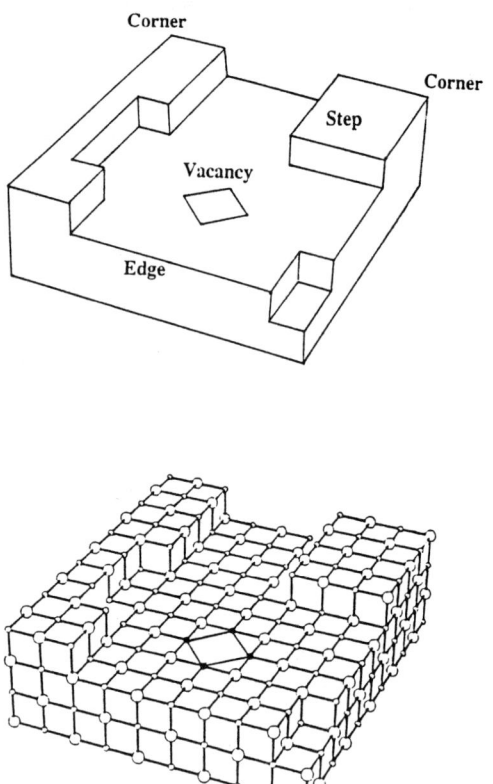

Fig. 1. Schematic view of the surface of an MgO microcrystal.

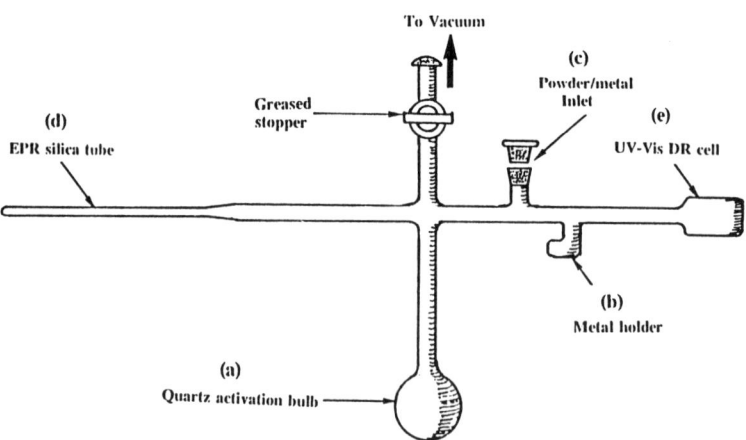

Fig. 2. The vacuum cell adopted for sample preparation and metal addition.

not possible to accurately control the amount of metal deposited on the sample. It was only possible to roughly control the experimental conditions according to the following procedures.

(a) *Low levels of added metal.* The powder and the metal were slowly heated under dynamic vacuum to a temperature about 10 K below that required for vapour production (for example about 560 K for Li). At the first appearance of a pressure increase in the vacuum line the cell was immediately removed from the furnace. Total exposure times of the powder to the vapours were therefore minimal, about 2–3 s. This procedure produced a very pale blue colored sample.

(b) *Medium levels of added metal.* The powder and metal were heated together under dynamic vacuum to the temperature range required for vapour production for each metal (temperatures given above). The powder was left in contact with the vapours for a length of time sufficient to generate a sample characterised by a rich blue colour. Typical exposure times were usually 3–10 s, varying in this time range as the amount or size of the metal pieces varied.

(c) *High levels of added metal.* The powder and sample were heated to a temperature approximately 20 K above that required for vapour production for example about 610 K for Li. The powder was left in contact with the vapours for a minimum of 10 s until the sample was coloured very deep blue or black.

4. Nature of the Trapped Electron Centres on M/MgO

Three main and distinct types of trapped electron centres have been identified and characterised by EPR spectroscopy on the vapour modified MgO surfaces (hereafter indicated as M/MgO where M = Li, Na, K, Rb or Mg). These are (4.1) surface colour centres (also called F_s centres), (4.2) 'monoalkali cation centres' and (4.3) alkali metal ionic clusters, respectively: a brief description of their features is given in the present section. The types of trapped electron centres formed depends on the nature and in particular on the amount of doping metal used. A representative example of the metal vapour oxide interaction can be seen from the EPR spectra of MgO exposed to different amounts of potassium metal vapours ranging from very low levels to very high levels of vapours (Figure 3). It can be clearly seen from this figure that a progressive transformation in the EPR spectra occurs since initially the spectrum (Figure 3a) is nearly symmetric and structureless whereas the spectra from Figures 3b–3d display groups of hyperfine lines (due to ^{39}K) of increasing complexity. This indicates that the paramagnetic centres originating at the surface by contact with the metal vapour tend to involve a progressively higher number of evaporated atoms with increasing the metal loading.

4.1. SURFACE F_S^+ CENTRES

An F centre in an alkaline earth oxide consists of one (F^+) or two (F) electrons trapped in an anionic vacancy surrounded by six cationic Mg^{2+} sites in octrahedral symmetry. When the O^{2-} vacancy, which can be indicated as F^{2+}, is located at the surface the corresponding colour centres are designated F_S^+ and F_S centres respectively. For F_S centres at the (100) face the electron trap consists of five

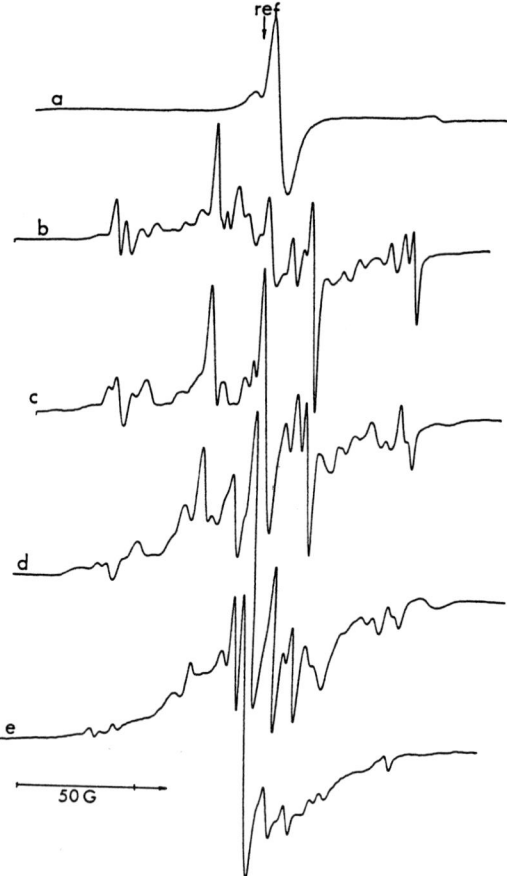

Fig. 3. EPR spectra of potassium vapour modified MgO. The amount of added K metal increases from (a) to (e).

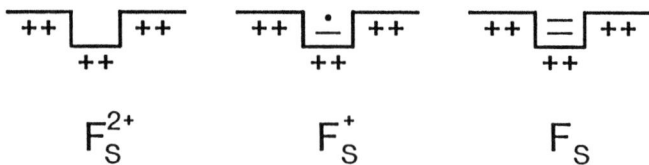

Fig. 4. Bidimensional schematic representation of a surface (a) F_S^{2+} anionic vacancy, (b) an F_S^+ colour centre and (c) an F_S colour centre in the alkaline earth oxide MgO.

Mg^{2+} cations in C_{4v} symmetry. For the sake of simplicity they can be schematically described according to the bidimensional scheme shown in Figure 4.

Surface F_S^+ and F_S centres can be generated on the surface of MgO and other alkaline earth oxides by a process known as 'radiative colouring' involving γ

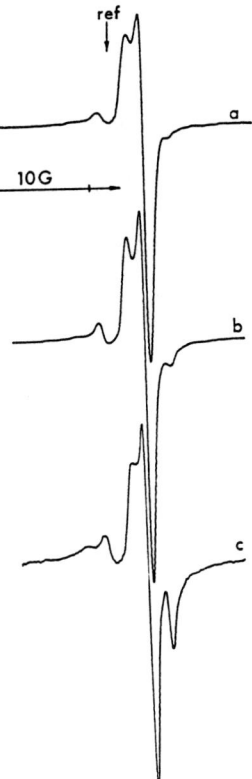

Fig. 5. EPR spectrum of γ irradiated MgO recorded at various powers. (a) 10 mW, (b) 1 mW, (c) 0.1 mW.

irradiation of the oxide or UV irradiation in a hydrogen atmosphere. Under these conditions the oxide turns to the blue-violet colour typical of these surface centres (also called colour centres) and exhibits a narrow EPR spectrum assignable to the one electron F_S^+ centres which are paramagnetic. The EPR spectra recorded in the case of the γ irradiated MgO at three different microwave powers are shown in Figure 5.

The spectrum obtained by addition of small amounts of Mg vapours to MgO is shown in Figure 6. The spectrum closely resembles those reported in Figure 5 and confirms the idea that the contact of alkaline earth metal vapours with MgO also produce surface colour centres [28]. This fact is further corroborated by the similarity between the UV-vis spectra of γ irradiated MgO and Mg vapour doped samples [29]. Surface F_S^+ centres have also been obtained on MgO by addition of alkali metals [28, 29]. Figure 7 compares the spectra obtained on various MgO samples (all derived from the same batch of the parent hydroxide) either by radiative colouring processes (Figures 7a and 7b) or by metal doping using Mg, Li, Na, K or Rb vapours respectively (Figures 7c–7g). With the exception of

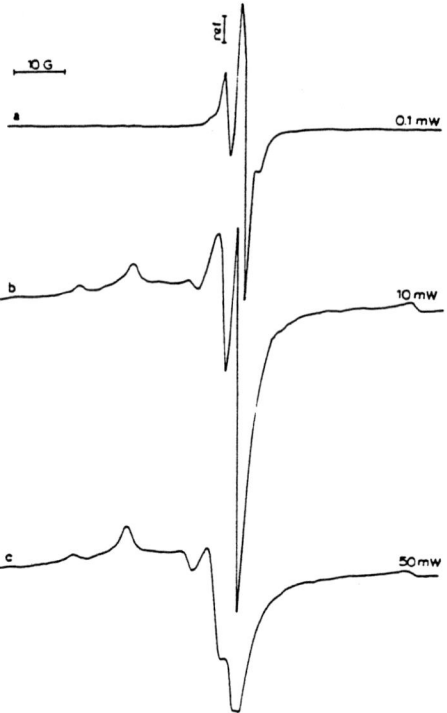

Fig. 6. EPR spectra at various powers of MgO treated with small amounts of Mg vapours.

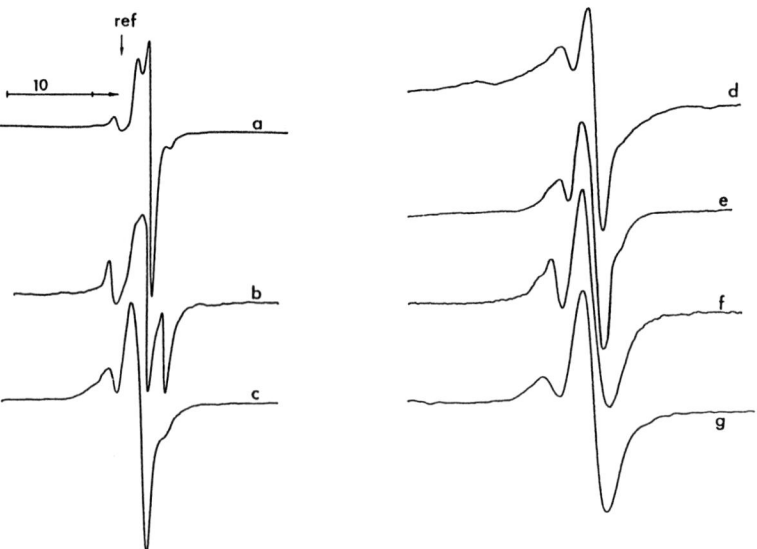

Fig. 7. EPR spectra of F_S^+ centres on MgO after (a) γ irradiation, (b) UV irradiation in hydrogen and after exposure to vapours of (c) Mg, (d) Li, (e) Na, (f) K and (g) Rb.

TABLE I
Spin Hamiltonian parameters for various F_S^+ centres on MgO.

F_S^+	$g_\perp$	$g_\parallel$	ΔH/gauss	A/gauss $^{25}Mg^{2+}$
γ irrad.	2.0014	2.0004	0.95 ± 0.05	–
UV/H$_2$ irr.	2.0011	1.9993	–	11.35
UV/D$_2$ irr.	2.0011	1.9993	–	10.2
Mg/MgO	2.0012	2.0004	1.5 ± 0.05	11.75
Li/MgO	2.0013	2.0003	1.95 ± 0.05	–
Na/MgO	2.0012	2.0004	2.5 ± 0.05	–
K/MgO	2.0013	2.0004	3.85 ± 0.05	–
Rb/MgO	2.0012	2.0004	3.4 ± 0.05	–

the line width all the spectra are quite similar as confirmed by the values of the g tensor component listed in Table I.

The EPR spectrum of the F centres is characterised by an axial g tensor with both components at values lower than the free spin resonance ($g_e = 2.0023$). This behaviour is typical of surface F_S^+ centres in MgO since bulk F^+ centres give rise to a symmetric resonance line at $g = 2.0023$ [29].

As reported before the accepted model of a surface F_S^+ centre is that of an electron occupying a surface anionic vacancy [30]. At such a site (when located at the (100) face) the F_S^{2+} surface vacancy has a square pyramidal C_{4v} type symmetry so that the unpaired electron in the F_S^+ centre has five neighbouring Mg^{2+} cations (as opposed to six in the bulk). In this environment the electron does not interact equivalently with all five cations and in fact it has been experimentally demonstrated that the electron interacts most strongly with the apical Mg^{2+} cation (i.e. the cation furthest from the surface) and less strongly but equivalently with the remaining four equatorial cations [30].

The observed shift from g_e of the g tensor components of the surface F_S^+ centres indicates that the preferential interaction with the apical Mg^{2+} cation involves a small degree of p orbital contribution from the cation itself to the electron wavefunction. Furthermore the small contribution from p states is affected by the axial symmetry of the potential well thus determining the axial nature of the g tensor with $g_\parallel > g_\perp$. The preferential interaction of the free electron with the apical Mg^{2+} cation is not only evident from the g values but also from the weak hyperfine structure arising from the 10.1% isotopic abundance of ^{25}Mg ($I = 5/2$ for ^{25}Mg). The hyperfine splitting (listed in Table I) was only observed in some spectra for the variously generated surface F_S^+ centres [31]. This fact also indicates that the unpaired electron must be localised to a limited extent in the s orbital of a $^{25}Mg^{2+}$ cation. The present hyperfine splitting of 11–12 gauss (Table I) compares well with the value of 11.2 gauss reported in the literature [30] and suggests about a 4.4% Mg^+ character.

The spectrum of the $F_S^+(H)$ centre (Figure 7b) is analogous to that of the F_S^+ centre except that the perpendicular component is split into a doublet with a hyperfine coupling constant of 2.05 gauss. This doublet arises from the weak interaction of the trapped electron with the proton ($I = 1/2$ for ^1H) of a nearby

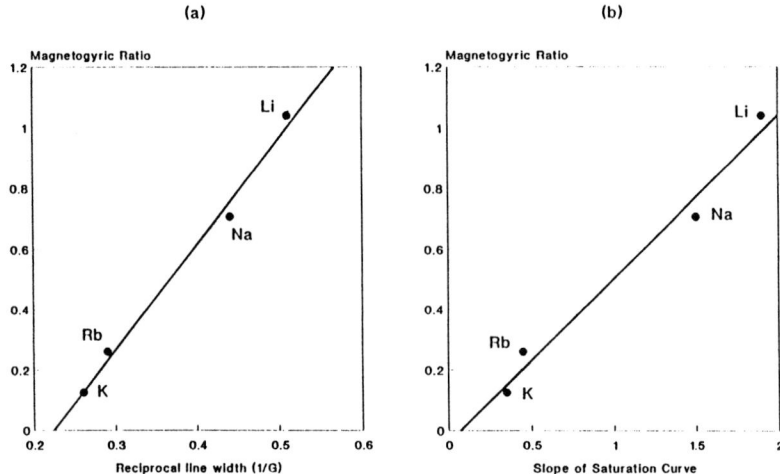

Fig. 8. (a) Reciprocal line widths of the F_S^+ centres as a function of the magnetogyric ratio (rad $G^{-1} \cdot 10^{-4}$) of the metal. (b) Slopes of the saturation curves for alkali metal generated F_S^+ centres versus the magnetogyric ratio of the metal.

surface hydroxly. This small hyperfine splitting indicates about 0.4% of the unpaired spin density at the s orbital of the hydroxylic hydrogen.

In principle, a weak hyperfine interaction between the F_S^+ centre on the metal doped surface and the parent cation should be expected. If the cations are stabilised at surface sites sufficiently remote from the trapped electrons then no hyperfine interaction will be observed. However the line widths (Figure 7, 8a and Table I) and saturation behaviours (Figure 8b) of the EPR signals were found to vary as a function of the metal used to dope the surface even though no hyperfine structure was resolved [29]. This indicates that the cations are stabilised at sites which are too far from the F_S^+ centres to produce a resolved hyperfine interaction but sufficiently close to influence the local magnetic environment. A close examination of the spectra generated on alkali metal doped MgO reveals a marked variation in the line width and saturation curves depending on the particular doping metal used (Li, Na, K or Rb). When the indivdual line widths (or reciprocal line widths as shown in Figure 8a) or slopes of the saturation curves (Figure 8b) are plotted against the magnetogyric ratio of the corresponding nuclei a clear relationship between these spectroscopic properties of the trapped electrons and the parent cations can be seen. This seems to indicate that an important role is played by the surface cations in the relaxation mechanism of the spin for the F_S^+ centres generated by metal doping.

4.2. MONOALKALI CATIONIC CENTRES

As the amount of doped metal is increased the EPR spectrum changes from that typical of F_S^+ centres to the signal ascribed to the presence of a new paramagnetic centre termed a 'monoalkali cationic centre'. This new surface centre was only

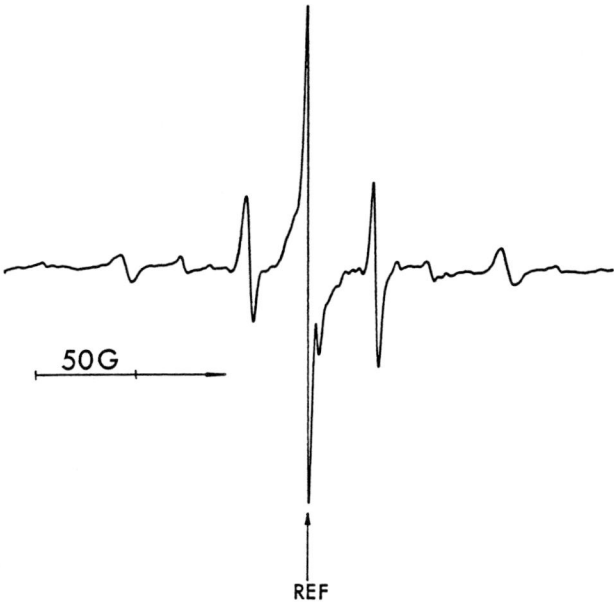

Fig. 9. Room temperature EPR spectrum of Na doped MgO (medium level of added metal).

observed on MgO exposed to Na, K and Rb. Exposure of MgO to similar levels of Li or Mg did not produce analogous spectra. The corresponding EPR spectra (for Na/MgO, K/MgO and Rb/MgO) are shown in Figures 9, 10 and 11 respectively.

The spectra are characterised by a multiplet structure of $2I + 1$ lines for each atom. A quartet structure was observed on the Na/MgO and K/MgO samples ($I = 3/2$ for ^{23}Na and ^{39}K). In the case of Rb/MgO both patterns of a sextet and a quartet are observed since Rb has two ($I \neq 0$ isotopes ($I = 5/2$ for ^{85}Rb, 72.15% and $I = 3/2$ for ^{87}Rb, 27.85%). The strong line close to the reference in the Na/MgO spectrum (Figure 9) is due to the presence of small Na metal particles [31]. In the Rb/MgO spectrum (Figure 11) a signal close to the reference has been attributed to surface F_S^+ centres while the most intense line in this spectrum at high field has been assigned to Rb small metal particles [31]. Although the presence of a four line quartet can be clearly seen in the spectrum of K/MgO (Figure 10) other minor signals are also present which arise from K trimeric ionic clusters (see below).

An electron in a pure s orbital of sodium, potassium or rubidium should produce a hyperfine structure arising from the Fermi contact interaction. Since each hyperfine transition can occur with equal probability all lines are expected to have the same intensity. This behaviour is not observed in Figures 9, 10 and 11. The reason for the increasing line widths and the decreasing line intensity towards the wings of the spectra have been found by means of computer simulations to be mainly due to small anisotropic contributions in both g and A tensors but also to surface heterogeneity effects which produce a variety of species with slightly different A

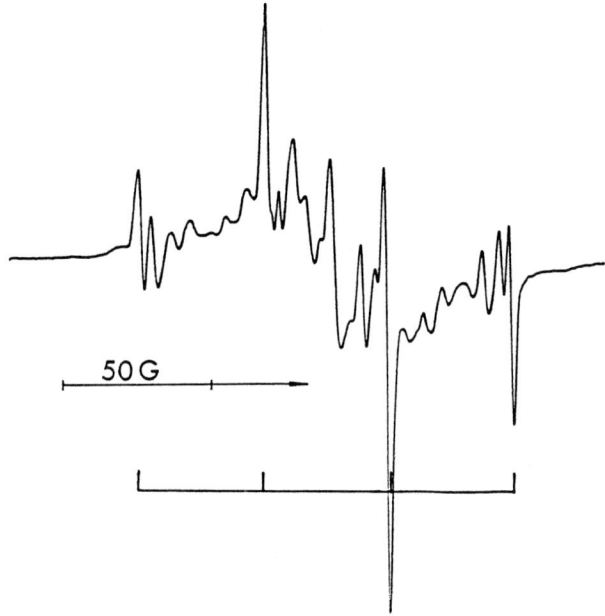

Fig. 10. Room temperature EPR spectrum of K doped MgO (medium level of added metal).

values [31]. The spin Hamiltonian parameters derived by simulation for each spectrum are listed in Table II. Assuming that the anisotropic coupling arises from the partial occupation of a Na, K or Rb based p orbital due to s–p hybridisation then $A_\parallel = a_{iso} + 2B$ and $A_\perp = a_{iso} - B$ for these systems. From the values of A_0 and B_0 for each atom and the spin Hamiltonian parameters listed in Table II, the total s and p spin density on Na, K and Rb was evaluated. Clearly anisotropy in the spectra is very small and the total spin density on the cation in the monoalkali cationic centre varies from Na/MgO (0.28) to K/MgO (0.54) to Rb/MgO (0.56). The reduced atomic character in the surface stabilised paramagnetic centres therefore arises from electron delocalisation away from the cation rather than a high p-like spin density.

Based on these results it appears that the 'monoalkali cationic centre' is characterised by the partial sharing of an electron between the parent atom and a suitable surface trap. This centre can be formally written as $M^{\delta+}(Trap)^{\delta-}$ and modelled as an association between the adsorbed alkali atom and a suitable surface site capable of trapping the electron. The unpaired electron in the centre is then effectively shared (for example in the case of sodium) between the partially ionised $Na^{\delta+}$ cation (26%) and the surface $(Trap)^{\delta-}$ (74%). However determining the structure and location of the above paramagnetic $Na^{\delta+}(Trap)^{\delta-}$ centre is not straightforward. On the basis of the present data it is not possible to exactly ascertain whether the traps involved in the formation of the novel paramagnetic centre are of the same type (F_S^{2+}) as those involved in trapping the released electrons in the early stages of the Na–MgO interaction producing the F_S^+ colour centres at the planar (100) face of MgO. In such a case the parent atom would

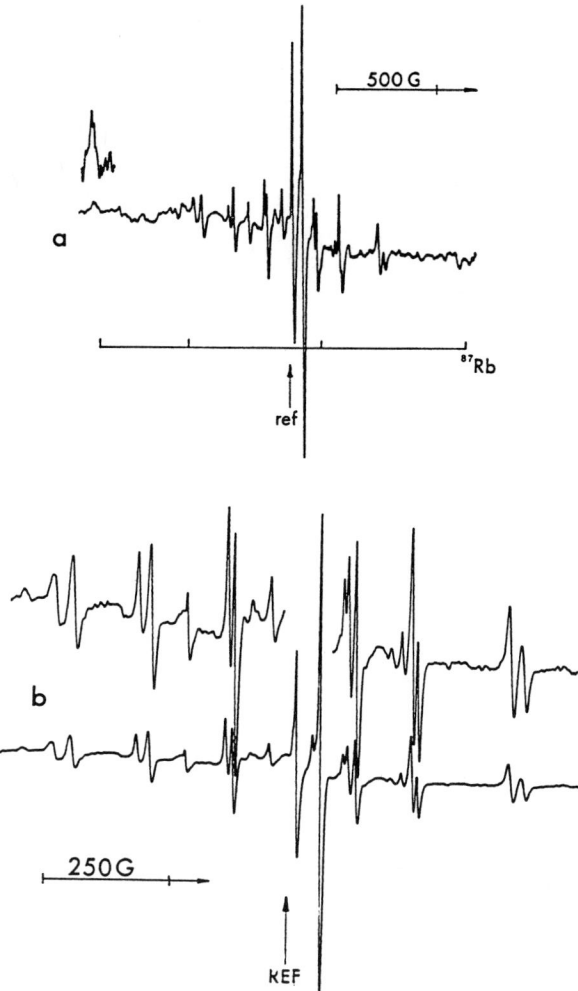

Fig. 11. Room temperature EPR spectrum of Rb doped MgO at two different scan ranges.

effectively 'cap' the F_S^{2+} vacancy on the (100) face thereby occupying the sixth coordination position on top of the square pyramidal trap. Based on the likely instability of the K and Rb atom at such a site, an alternative and more probable possibility is based on the localisation of the $Na^{\delta+}(Trap)^{\delta-}$ centre at a morphological defect such as a step where there is preferential stabilisation of the large monovalent dopants (i.e., K^+ and Rb^+) [32]. The original (Trap) prior to metal doping can be essentially viewed as a surface step with one O^{2-} anion removed as reported in Figure 12. Such a trap would be unlikely to directly trap an electron unless an alkali atom was stabilised nearby allowing the valence s electron to partially 'hop across' to the trap.

TABLE II

Spin Hamiltonian parameters derived by computer simulations for Na/MgO (Figure 9), K/MgO (Figure 10) and Rb/MgO (Figure 11). $C_S^2 = a_{iso}/A_0$ and $C_P^2 = B/B_0$.

Monoatomic Species	$g_\parallel$	$g_\perp$	$A_\parallel$/G	$A_\perp$/G	A_0/G	a_{iso}/G	B_0/G	B/G	C_S^2	C_P^2	Total spin density
Na/MgO	2.0018	2.0010	69	63.0	316	64.9	32.2	1.8	0.205	0.056	0.261
K/MgO	2.0012	2.0015	38.3	40.2	82.5	36.6	10.6	0.63	0.480	0.059	0.529
Rb/MgO	2.0015	1.9997	183	175	361	176.5	42.6	2.75	0.490	0.064	0.554

TABLE III

Spin Hamiltonian parameters for the trimeric Li, Na and K species on MgO.

Triatomic Species	$g_\parallel$	$g_\perp$	$A_\parallel$/G	$A_\perp$/G	A_0/G	a_{iso}/G	B_0/G	B/G	C_S^2	C_P^2	Total spin den. on M
Li	$g_{iso} = 2.0011$		–	–	143.3	8.26	–	–	0.173	0.0–0.10	0.173–0.273
Na	2.0010	2.0018	50.0	55.0	316.0	53.33	32.2	1.66	0.506	0.155	0.661
K	1.9979	1.9993	17.5	20.0	82.5	19.16	10.6	0.84	0.697	0.236	0.933

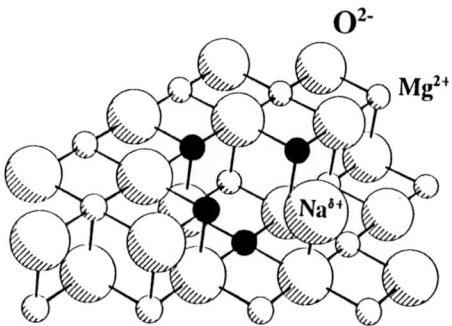

Fig. 12. Schematic drawing of a possible sodium monoalkali centre $Na^{\delta+}(Trap)^{\delta-}$.

4.3. ALKALI METAL IONIC CLUSTERS

When activated MgO is exposed to a higher amount of Li, Na and K metal vapours EPR spectra characterised by the presence of a 10 line multiplet are obtained. The corresponding spectra for these Li/MgO and K/MgO samples are shown in Figures 13 and 14.

Due to the presence of other interferring signals arising from F_S^+ centres and Li small metal particles, the 10 line multiplet on Li/MgO can be more clearly seen in the second and third derivative EPR spectra (Figure 13). The multiplet has the approximate intensity ratio of 1 : 3 : 6 : 10 : 12 : 12 : 10 : 6 : 3 : 1 which correctly fits the binomial distribution for an unpaired electron interacting with three equivalent nuclei of nuclear spin $I = 3/2$. The hyperfine separation in the spectrum was 8.26 gauss. Since ^7Li is the only nucleus present on the Li/MgO sample with $I = 3/2$ ($I = 3/2$ for ^7Li with a natural abundance of 92.58%) it can be assumed that the unpaired electron responsible for the 10 line multiplet interacts with three equivalent surface Li nuclei [33]. Similarly it can be shown that the 10 line multiplets on Na/MgO and K/MgO arise from a hyperfine interaction between an unpaired electron and three equivalent Na or K nuclei.

The spectra of Na/MgO and K/MgO exhibit an increase in line width towards the spectral wings, a situation reminiscent of the EPR spectra for Na, K and Rb 'monoalkali cationic centres' described in Section 4.2. Again based on computer simulations it appears that a combination of anisotropy and surface heterogeneity effects determine the final spectrum shape. It was found that the spectra of K/MgO and Na/MgO were characterised by slightly axial g and A tensors where $g_\parallel \approx g_\perp$ and $A_\parallel \approx A_\perp$. The total spin density over the three equivalent K nuclei, including the contributions from the p orbitals, was quite substantial (≈ 0.93) as reported in Table III. On the contrary a smaller extent of anisotropy and reduced atomic character was observed for the trimeric paramagnetic Na species. In this case the isotropic hyperfine coupling indicates a 0.506 spin density on the $3s$ orbitals of the Na atoms (Table III). The anisotropic contribution is however much smaller with a 0.155 spin density on the $3p$ valance orbitals of the three Na nuclei. The total spin density on the trimeric Na species is therefore 0.66 and less than the value observed for K/MgO.

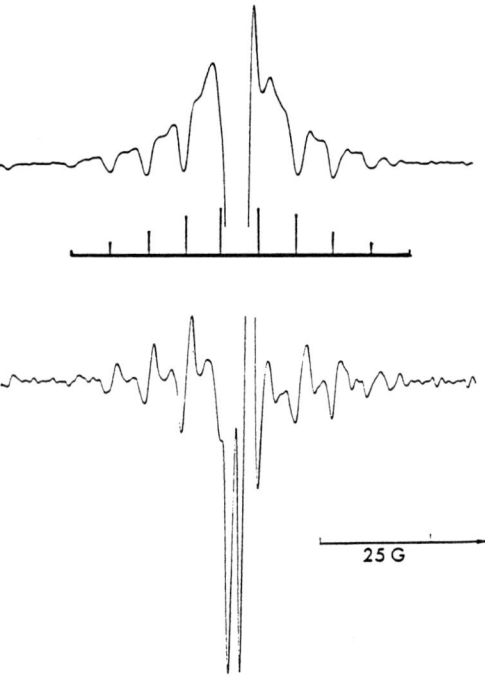

Fig. 13. (a) Second and (b) third derivative EPR spectrum of the trimeric ionic cluster on Li/MgO.

The percentage of s character in the trimeric Li species is only 17.3% over the three Li cations (Table III). The remaining 82.7% of the spin is not directly observed and this could in principle be accounted for either in terms of a substantial $s–p$ hybridisation contribution or alternatively due to extensive electron delocalisation away from the three surface nuclei. The calculated anisotropic hyperfine term due to an electron in a Li $2p$ orbital (B_0) is very small so that assuming the unpaired electron density unaccounted for in the Li $2s$ orbitals were distributed evenly between the three Li $2p$ orbitals (hence 28% each) the expected anisotropic hyperfine term would be less than 1 gauss and probably not detected in the present case where a_{iso} = 8.26 gauss [33]. It is unlikely however that the spin density in the p orbitals is greater than or equal to the spin density over the s orbitals in the trimeric species. An estimated remaining unpaired spin density of about 0.70 is then likely delocalised away from the three equivalent cations.

The assignment of the trimeric entities generated by alkali metal doping of MgO is not straightforward and should be improved by further work. Two types of well defined aggregates of three alkali metal atoms are paramagnetic namely the M_3 neutral cluster and the M_3^{2+} ionic cluster. Both clusters have been observed in the past: the former one in cryogenic Ar or hydrocarbon matrices for Li_3, Na_3 and K_3 [34–38] and the second in Y and X zeolites for Na_3^{2+} and K_3^{2+} [39–45]. The high values of the spin density over the metal atoms found in the present case (especially for Na and K) seem to better agree with the corresponding data

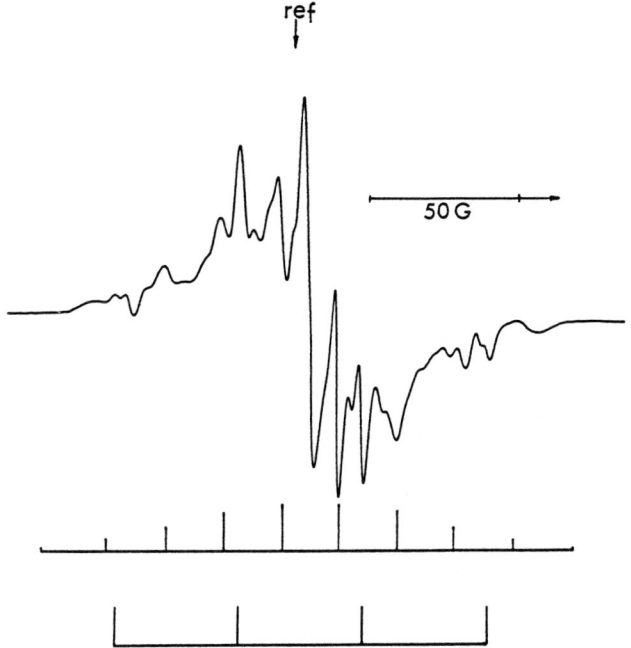

Fig. 14. Room temperature EPR spectra of K/MgO.

reported for the neutral Na_3 and K_3 clusters. The presence of such neutral molecules at the surface of ionic MgO seems however rather unlikely.

Possible candidates for the assignment of the trimeric paramagnetic entities observed on MgO are therefore the trimeric ionic clusters M_3^{2+} or alternatively ionised $M_3^{\delta+}$ clusters ($\delta < 1$). These latter clusters would result from a partial delocalisation of the unpaired electron from the neutral M_3 species into a suitable trap in analogy with what is observed for the $M^{\delta+}$ monoalkali cation centre described in Section 4.2. This complex assignment is currently under investigation. Nevertheless the findings reported in the previous paragraph represent the first detailed analysis of the complex interaction between low ionisation energy metals and an ionic surface. The paramagnetic centres reported in 4.2 and 4.3 have never before been observed at the surface of an oxide.

5. Chemical Reactivity of the M/MgO System

As discussed above a number of trapped electron centres are generated on the surface of MgO after exposure of the activated oxide to vapours of low ionisation energy metals. The presence of these excess trapped electrons on the surface significantly alters the chemical properties of the oxide. Indeed systems prepared by deposition of metal atoms on to polycrystalline MgO were studied in the past by Malinowski and coworkers [16, 46–48]. The systems prepared in this manner were classified as *superbasic catalysts* found to be active in some catalytic reactions

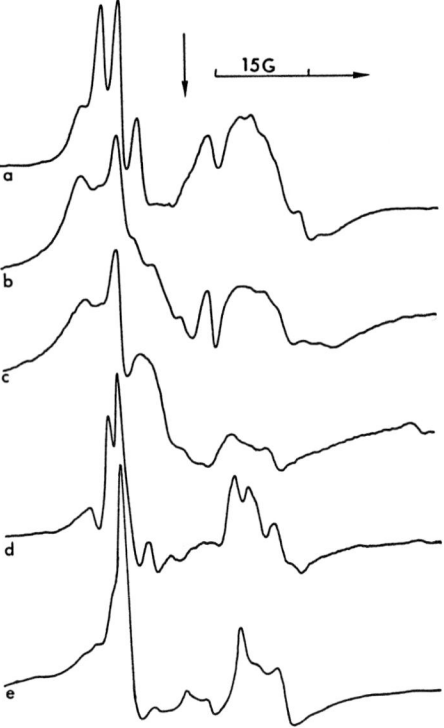

Fig. 15. Low temperature (77 K) EPR spectra of ^{13}CO adsorbed on different electron-rich MgO samples. (a, b, d, e) ^{13}CO on Mg/MgO and (c) ^{13}CO on Li/MgO.

such as isomerisation and dehydrogenation of hydrocarbons [16]. The reactivity of the metal vapour modified surface with simple probe molecules gives rise to an interesting surface redox chemistry which is only now being investigated. The metal doped MgO system is in fact a powerful electron donor system and reacts with a variety of molecules in contact with the surface. A first result of the reactivity (which can in some cases be very complex) is the formation of anions and of radical anions. The presence of the surface assists in the stabilisation of the newly formed radical anions on the surface itself thus allowing a detailed study of their structural properties. Some examples of this reactivity between surface trapped electrons and adsorbed molecules and the resulting structural properties of the radicals formed on the surface are presented below.

5.1. REACTIVITY WITH CARBON MONOXIDE

Exposure of the metal vapour modified surface to ^{13}CO at low temperature ($\approx$100 K) results in the immediate disappearance of the blue/violet colour (characteristic of the presence of surface F_S^+ colour centres) and the simultaneous formation of the EPR spectrum shown in Figure 15. The different spectra in Figure 15 were observed following adsorption of ^{13}CO on to different electron-rich MgO

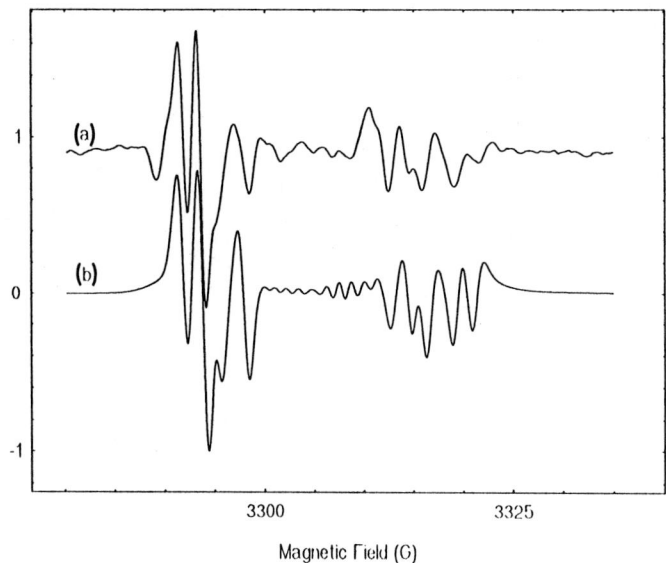

Fig. 16. Second derivative EPR spectrum of ^{13}CO adsorbed at $\approx$100 K on Mg/MgO. (a) Experimental and (b) computer simulation.

samples and are collected together in the figure to show the reproducibility of the experiments which, despite the complexity of the system under investigation, are quite satisfactory. Owing to the intrinsic complexity of these spectra the structure of the g tensor was better evaluated from the EPR spectrum obtained by the low temperature adsorption of ^{12}CO [49]. This analysis revealed an anisotropic g tensor with components at $g_1 \approx 2.0041$, $g_2 \approx 2.0014$ and $g_3 \approx 1.9817$. Additional information on the nature of the species responsible for the spectra in Figure 15 was obtained by employing ^{17}O-enriched CO (nuclear spin of ^{17}O is 5/2). The resulting EPR spectrum revealed a principal hyperfine pattern of six equally spaced lines indicating the interaction of the unpaired electron with one ^{17}O nucleus [49].

Computer simulations of the ^{13}CO low temperature spectra have allowed the interpretation of the experimental results. The second derivative experimental and computer simulated spectrum of ^{13}CO adsorbed on Mg/MgO at $\approx$100 K is shown in Figure 16. It was found that the main hyperfine coupling in the case of the ^{13}C and ^{17}O nuclei was observed along the g_2 direction: the spectra in Figures 15 and 16, however, are the result of the overlap between the signals of several very similar species slightly differing in the spin Hamiltonian parameters. The g and A tensor components of the four main species of the spectrum in Figure 16 (^{13}CO) are reported in Table IV. The total spin density of each CO radical species have been obtained on the basis of the observed ^{13}C and ^{17}O hyperfine splittings for ^{13}CO and C^{17}O (data not reported in Table IV for brevity).

The species responsible for the low temperature spectra has been assigned to the CO$^-$ radical [49]. Both the g and A tensors observed in the spectra are in agreement with this assignment. CO$^-$ is an 11 electron π diatomic radical with

TABLE IV

Spin Hamiltonian parameters derived by computer simulation of the spectrum of the ^{13}CO species on Mg/MgO at 100 K. (Fig. 16)

Species	Abundance	g_1	g_2	g_3	a_1 ^{13}C	a_2 ^{13}C	a_3 ^{13}C	C^2_{2s}	C^2_{2p}	Total spin density
I	0.18	2.0048	2.0012	1.9741	0.0	±29.6	0.0	0.009	0.298(C)	62.0%
II	0.25	2.0041	2.0014	1.9817	0.0	±25.3	0.0	0.008	0.265(C)	58.2%
III	0.29	2.0039	2.0018	1.9870	0.0	±22.0	0.0	0.007	0.226(C)	51.3%
IV	0.18	2.0034	2.0020	1.9917	0.0	±19.6	0.0	0.006	0.201(C)	46.0%

the unpaired electron located in one of the two π^*_g antibonding orbitals so that negative Δg shifts ($\Delta g_i = g_i - g_e$) are expected. The hyperfine coupling constants observed for the ^{13}C and ^{17}O nuclei also agree with the proposed assignment. For a π diatomic radical in the adsorbed state the electron is essentially confined to one of the π^*_g orbitals. The predicted experimental A tensor should have one component clearly higher than the other two as indeed observed. The values of the experimental hyperfine separation recorded for CO^- on Mg/MgO, however, are relatively small compared to those observed for other π diatomic radicals such as N_2^- [50], NO [51, 52] and $^{17}O_2^-$ [53]. Summation of the spin density over the two C and O atoms account for only a limited percentage of the unpaired spin ranging from 46% to 62% for the various adsorbed states of CO^- reported in Table IV. The various CO^- species observed are therefore related to the different degrees of electron transfer from the trapped electron centres on the surface to the CO molecule probably due to differences in the features of the surface site where the reaction occurs [49].

When ^{13}CO is adsorbed on the electron-rich surface at room temperature the blue/violet colour of the M/MgO sample immediately dissapears followed by the simultaneous appearance of a pink/orange colour and the formation of an EPR spectrum shown in Figure 17. The spectra shown in Figures 17b–d were observed immediately after ^{13}CO adsorption on different electron-rich MgO samples. Figure 17a was only observed after 30 days of ^{13}CO contact with bare thermally activated MgO. As reported elsewhere [49], the initial rate of the surface reaction leading to the spectra in Figure 17 was far lower for CO on the bare MgO sample compared to CO on the metal doped sample but after this initial effect the rates of reaction were similar and the final paramagnetic product of the interaction is the same in both cases as documented by the similarity of the spectra in Figure 17. It was concluded that the initial offset arose from CO rapidly reacting with the trapped electrons after which the excess CO continues to react with the MgO matrix similar to the rate of CO reaction on bare MgO [49].

Spectra similar to those reported in Figure 17 were indeed reported in the past and originally assigned to the $C_2O_2^-$ anion [54,55]. Later workers reassigned this spectrum to a more complex $C_6O_6^{3-}$ species [56,57]. However, recent work has given new evidence to suggest that the original assignment to a $C_2O_2^-$ radical may be correct [49]. Computer simulations of the room temperature ^{12}CO and ^{13}CO related radicals have shown that the spectra were characterised by anisotropic g

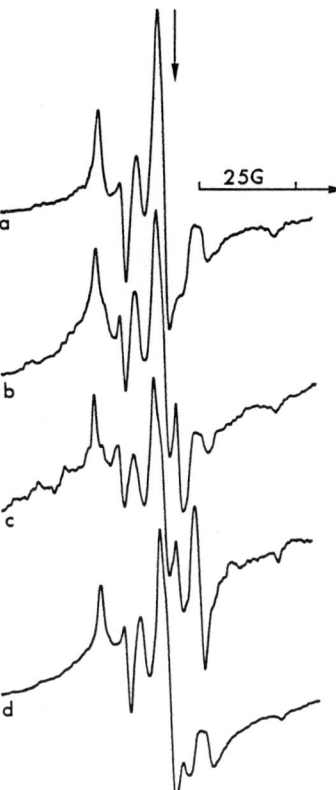

Fig. 17. Room temperature EPR spectra of ^{13}CO adsorbed at 300 K on (a) bare thermally activated MgO, (b) MgO previously UV irradiated in D_2, (c) Mg/MgO and (d) Na/MgO.

and A tensors, the latter indicating an interaction of the unpaired electron with two equivalent carbon atoms as expected for the $C_2O_2^-$ species.

In addition variable temperature studies in the range 77–298 K have shown that the spectra of the various CO$^-$ species observed at low temperature undergoes an irreversible evolution with increasing temperature, the final spectrum being the same as that observed following CO adsorption on Mg/MgO at 298 K. In the spectra recorded at variable temperature a mixing between the CO$^-$ features and the room temperature spectral features was observed even at low temperature (123–153 K). This is important since low temperature UV-vis diffuse reflectance spectroscopy has shown that the formation of large CO polymers (e.g. $C_6O_6^{3-}$) requires elevated temperatures and the actual low temperature polymerisation of CO on MgO is limited to entities containing no more than three carbon atoms [58]. The spectra reported in Figure 17 were therefore assigned to the $C_2O_2^-$ anion.

The low temperature reduction of CO at the surface of electron rich MgO as demonstrated by infrared work mainly leads to the formation of the acetylenediolate $C_2O_2^{2-}$ ion [59]. This species evolves upon increasing the temperature towards the formation of more complex oxocarbon anions ($C_6O_6^{2-}$, etc.). The results

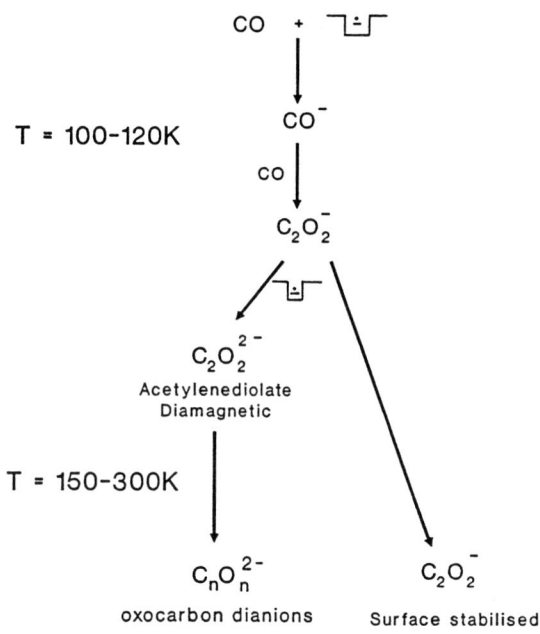

Fig. 18. Proposed reaction mechanism for CO interacting with surface electron-rich MgO.

reported here of the study of the reduction of CO at the electron rich surface of MgO are in close agreement with other experimental results and with theoretical studies of the same reduction process but in different media [60]. In particular in the present case both of the relevant intermediates hypothesised for CO reduction (i.e., CO^- and $C_2O_2^-$ formed by addition of CO to CO^-) are directly observed by EPR at the various heating steps of the temperature sensitive process. A fraction of the $C_2O_2^-$ ions survive unreacted at room temperature and do not evolve towards long chain carbon species. This fact can be explained by admitting that at particular surface sites the amount of electrons available are limited and the mobility of the ion on the ionic surface minimal. The reduction of CO on the electron rich MgO surface can therefore be summarised as shown in Figure 18.

5.2. REACTIVITY WITH OXYGEN

The blue/violet sample of Mg, Li or Rb doped MgO bearing surface trapped electrons exhibits EPR spectra characteristic of the various trapped electron centres on M/MgO. Upon contact with dioxygen a new asymmetric EPR signal is observed (Figure 19) which can be easily assigned to surface adsorbed superoxide O_2^- species.

The O_2^- species is paramagnetic when adsorbed at a cationic surface site (M^{x+}) and is easily detected by EPR spectroscopy. The electrostatic field at the surface removes the degeneracy between the two π^* oxygen orbitals. This factor gives rise to a spectrum of orthorhombic symmetry with three principal g_{xx}, g_{yy} and g_{zz}

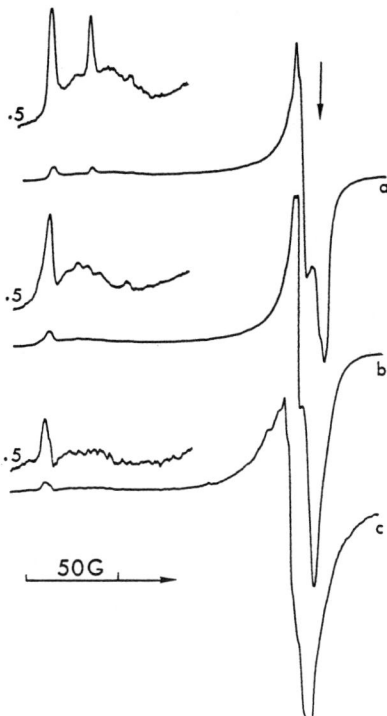

Fig. 19. O_2^- spectra recorded at 77 K on (a) Mg/MgO, (b) Li/MgO and (c) Rb/MgO.

values. The z direction is usually given as that of the O_2^- internuclear axis and the y direction as that perpendicular to the adsorption site. The extent of the energy splitting (Δ) between the two π^* orbitals depends on the electrostatic field felt by the superoxide species and this in turn determines the value of the g_{zz} component which is the most sensitive one to the electrostatic field [61]. The superoxide species may thus act as a probe of the local electrostatic field at the surface of the ionic system [25], allowing not only the identification of surface sites bearing a different charge (e.g. M^{2+}, M^{3+}, M^{4+} etc.) but also the distinction among cations with the same nominal charge which, because of different coordinative environments, exert different electrostatic fields. Previous work by Giamello et al. [62,63] has documented this in the case of O_2^- on MgO generated by a surface reaction between O_2 and a hydrogen containing molecule such as propene or toluene. A clear heterogeneity of adsorbed species was found with at least four distinct g_{zz} values, even though only one type of cation (Mg^{2+}) is present at the MgO surface [62,63].

The low field portion of Figure 19a indicates that five O_2^- species with different g_{zz} values are simultaneously present at the M/MgO surface. The three most intense and resolved components have g_{zz} values of 2.091, 2.077 and 2.064. Only the $g_{zz} = 2.091$ component is dominant in the spectrum of O_2^- on Li/MgO (Figure 19b) and Rb/MgO (Figure 19c). Two other less intense and broader components

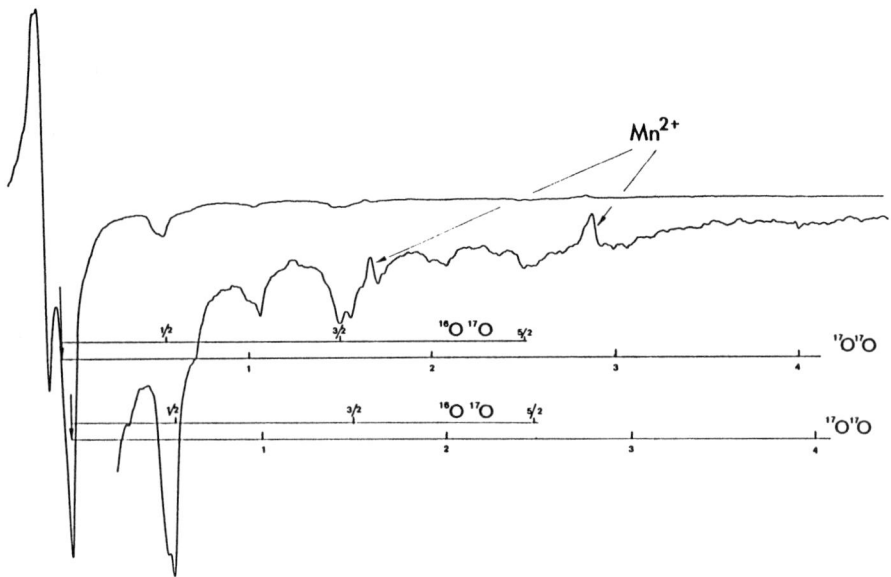

Fig. 20. Magnification of the high field portion of O_2^- on Mg/MgO obtained using a 37% enrichment of ^{17}O in the oxygen feed gas.

are present at $g_{zz} = 2.083$ and $g_{zz} = 2.072$. All five components are always present in the superoxide spectra on M/MgO although with variable relative intensities depending on the M/MgO sample, as seen in Figure 19. The $g_{zz} = 2.091$ component, however, is always the most intense and is more loosely bound to the surface than the other O_2^- species observed on MgO as revealed by the high g_{zz} value. On the other hand the second species with $g_{zz} = 2.064$ is tightly bound to the surface due to a strong interaction with the adsorption site.

Information on the structural features of adsorbed O_2^- may be gained by adsorption of ^{17}O-enriched oxygen on the vapour modified surface. The presence of a nonzero nuclear spin in the adsorbed superoxide allows one to study the corresponding hyperfine tensor and so to ascertain whether the spin density on the two oxygen nuclei is the same, i.e. whether the two species are adsorbed symmetrically ('side-on') or asymmetrically ('end-on') on the surface. The resulting $^{17}O_2^-$ spectrum on Mg/MgO is rather complex in the low field region due to the presence of the g_{zz} components from $(^{16}O^{16}O)^-$. Figure 20 therefore only shows the high field region of the $(^{17}O^{16}O)^-$ spectrum on Mg/MgO at 77 K. Two individual sextets are present, centred at different g_{xx} values but with the same hyperfine splitting (77 gauss). The fact that the two sextets are centred at different g_{xx} values indicate that two different O_2^- species are present, both with equivalent nuclei. The hyperfine coupling constant of 77 gauss indicates that the whole spin density on the two O_2^- species is about 1.0. In Figure 20 the vestiges of the low intensity 11 line structure related to the resonance of doubly labelled $(^{17}O^{17}O)^-$ species is also marked. The results show that the O_2^- species characterised by $g_{zz} = 2.091$ and

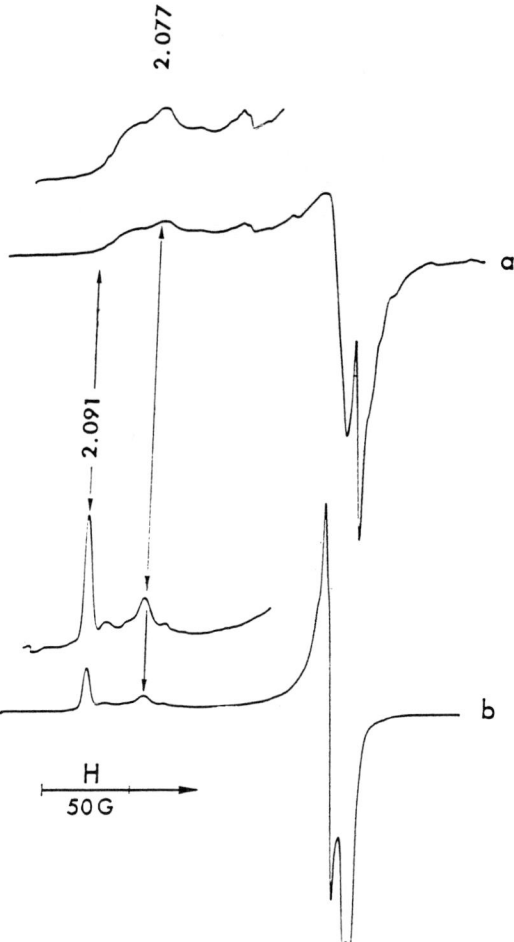

Fig. 21. Effects of temperature on the spectrum of O_2^- on Mg/MgO at (a) 298 K and (b) 77 K.

2.064 on Mg/MgO are symmetrically adsorbed on the surface as shown by the magnetic equivalence of the two oxygen nuclei [64].

Changes in the shape of the EPR spectrum recorded at different temperatures monitors the mobility of the paramagnetic species under study. A comparison of the spectra recorded for O_2^- on Mg/MgO at 298 K and 77 K is shown in Figure 21. The change in shape in passing from one temperature to the other is dramatic and indicates the onset of motional effects on varying the temperature. While the line at $g_{zz} = 2.077$ is still visible at 298 K, the line at $g_{zz} = 2.091$ (which was the most intense component at 77 K) is not observed at room temperature. This suggests that the species characterised by $g_{zz} = 2.091$ exhibits motional behaviour different from the other O_2^- species since its g_{zz} and g_{xx} components undergo averaging with increasing the temperature. The other species which are more tightly bound

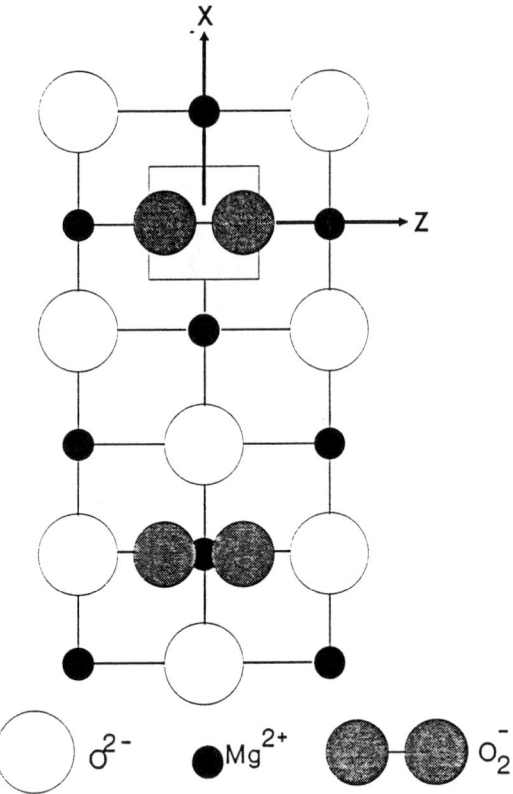

Fig. 22. Two O_2^- species on the flat (100) surface of MgO. Upper part O_2^- on an F_S^+ vacancy. Lower part O_2^- on a 5 coordinated Mg^{2+} cation.

to the surface (g_{zz} = 2.077 and 2.064) do not undergo motional averaging at room temperature and it is reasonable to consider them as immobile.

Based on the above results and the structural properties of the electron-rich M/MgO surface the adsorption sites of the two superoxide species peculiar to such samples (i.e. g_{zz} = 2.091 and 2.064) may be considered. Electron transfer from the F_S^+ centre to the adsorbed oxygen molecule generates the O_2^- species leaving an F_S^{2+} vacancy at the surface. This site should display a strong electrostatic potential towards the O_2^- anion higher than that of the four- and three-coordinated Mg^{2+} cations present at the edges and corners of the MgO microcrystals which are responsible for the species with g_{zz} = 2.083 and g_{zz} = 2.072 [25,62,63]. 'Side-on' adsorption of O_2^- at this site on the planar (100) face of MgO would adequately account for the species with g_{zz} = 2.064. The role of the five-coordinated ions on MgO can instead be invoked for the assignment of the g_{zz} = 2.091 component. The high g_{zz} values indicates a weak electrostatic interaction with the surface as expected for the five-coordinated Mg^{2+} ion exposed at the cubic planar face of the crystal. The lower crystal interaction at this site is also confirmed by the motional effects observed with increasing temperature. Both superoxide species

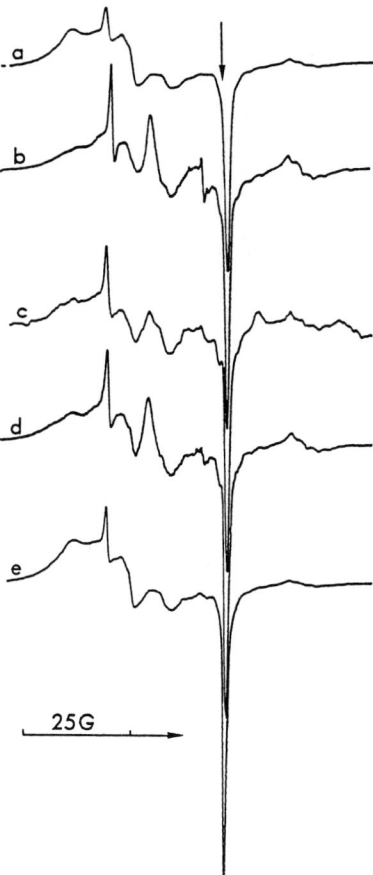

Fig. 23. EPR spectra at 77 K of O$^-$ on different Mg/MgO samples.

lie parallel to the surface exhibiting magnetic and structural equivalence of the oxygen atoms [64] (Figure 22).

5.3. REACTIVITY WITH NITROUS OXIDE

When Mg/MgO and K/MgO are exposed to nitrous oxide at low temperature ($\approx$100 K) the EPR spectrum typical of the trapped electron species is immediately replaced by the spectra shown in Figure 23 (for N$_2$O adsorbed on Mg/MgO). These signals have been assigned to O$^-_{(sur)}$ (the subscript (sur) refers to a surface species) ions [31]. The observed g values were $g_\parallel = 2.0016$ and $g_\perp = 2.0368$ for Mg/MgO and $g_\parallel = 2.0016$ and $g_\perp = 2.045$ for K/MgO.

The spectrum of O$^-_{(sur)}$ of K/MgO is typical of the ion in axial symmetry on the MgO surface. The symmetry of the O$^-_{(sur)}$ ion on Mg/MgO is, however, less certain. According to theory the g components are calculated to be for first order and axial symmetry as $g_\parallel = g_{zz} \approx g_e$ and $g_\perp = g_{xx} = g_{yy} = g_e + (2\lambda/\Delta E)$ [65, 66]. λ

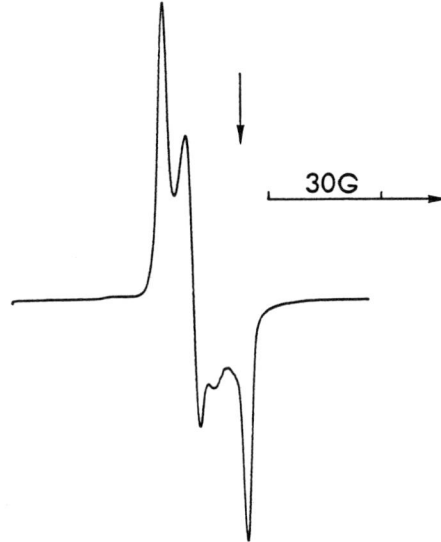

Fig. 24. Room temperature EPR spectrum of O_3^- formed by exposure of oxygen to a sample of Mg/MgO containing $O^-_{(sur)}$ ions.

is the spin orbit coupling constant of oxygen and ΔE the energy difference between the p_z level and the two degenerate p_x and p_y levels. ΔE depends on the crystal field at the adsorption site so that variations in ΔE will give rise to a large distribution in the g values which will broaden the $g_\perp$ line but not the $g_\parallel$ line [66]. The distribution of $g_\perp$ in the $O^-_{(sur)}$ spectrum on Mg/MgO therefore probably reflects the consequences of surface heterogeneity on the axial signal rather than arising from a species in orthorhombic symmetry.

When the samples of Mg/MgO and K/MgO containing $O^-_{(sur)}$ were exposed to oxygen at room temperature the signal due to O^- ions was immediately destroyed and replaced by the new spectrum which is shown in Figure 24. This spectrum has been assigned to the surface ozonide O_3^- ion [31]. The g values for this radical were $g_1 = 2.0017$, $g_2 = 2.011$ and $g_3 = 2.0159$. The stability of this ion on the modified surface is however unusual since it decomposes to give $O^-_{(sur)}$ in the absence of hydrogen and O_2^- in the presence of hydrogen.

The ozonide ion is a 19 electron radical isoelectronic with AB_2 type radicals, such as SO_2^- and NO_2^{2-}. In these ions the energy levels are well separated and because they are not significantly perturbed by the surface crystal field the g tensor has been used to fingerprint the species [53]. The g values observed in the present case are quite similar to the values reported in the literature for O_3^- stabilised on MgO [14,15].

When Mg/MgO containing O^- ions was exposed to excess CO_2 at room temperature the spectrum of the O^- ion was immediately replaced by the spectrum shown in Figure 25a. The new signal was characterised by a g tensor with $g_1 = 2.006$, $g_2 = 2.0081$ and $g_3 \approx 2.028$ and with all three values greater than g_e. The CO_3^- ion has 23 electrons and therefore is isoelectronic with NO_3. Both radicals have been observed by EPR in single crystals and the reported g values for CO_3^- were $g_1 =$

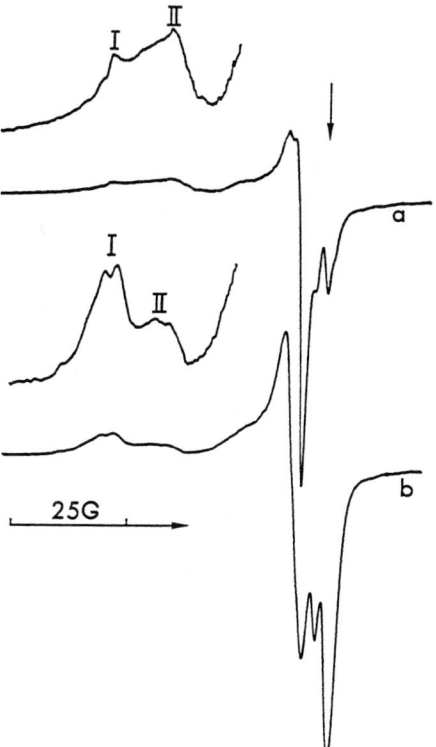

Fig. 25. EPR spectrum at 77 K of CO_3^- on Mg/MgO (a) after CO_2 addition at 300 K to O^- on Mg/MgO and (b) after annealing at 373 K for 1 hour.

2.0066, $g_2 = 2.0086$ and $g_3 = 2.0184$ [68]. The deviation from axial symmetry is expected if the planar symmetric radical is distorted. This occurs on the surface of the ionic oxide producing unequal O—C—O angles and giving rise to the asymmetric g tensor [69].

When the sample containing CO_3^- ions is annealed at 373 K for 1 h the spectrum in Figure 25a is replaced by the spectrum in 25b. This shows that the $g_3 \approx 2.028$ component is in fact composed of at least two components at $g_3 = 2.024$ and $g_3 = 2.032$. This indicates that there are at least two surface stabilised environments for the CO_3^- ion on Mg/MgO. This situation is not surprising and similar to the heterogeneity of adsorbed states observed for O_2^- [64], O_s^- [31] and CO^- [49] on the same system. The thermal stabilities of these two CO_3^- species are different. As seen in Figure 25b species I is more stable at higher temperature compared to species II although at room temperature species II is more abundant (Figure 25a).

6. Conclusions

The interaction between vapours of metals with low ionization energies (Mg, Li, Na, K and Rb) and the surface of an oxide generates an electron-rich M/MgO surface. The structural properties and chemical reactivity of this metal vapour modified sample are consequently altered. Three different types of trapped paramagnetic electron centres have been identified on these M/MgO systems which are: (a) F_S^+ colour centres; (b) a paramagnetic centre called a 'monoalkali cationic centre'; and (c) ionic alkali metal clusters. The type of centre formed depends on the amount and type of metal used. These studies have demonstrated that the surface of the modified oxide contains an excess of trapped electrons and may be described as surface electron-rich MgO. The results are promising since they offer a new insight into the types of trapped electron centres which can be generated on the MgO surface by doping with an excess of metal vapours.

The chemical reactivity of MgO modified with metal vapours is markedly altered and has very strong reducing properties compared to bare undoped MgO. The electron transfer process from the electron-rich surface to adsorbed molecules initiates an interesting surface redox chemistry where small radical anions are generated and stabilised on the surface. A number of small inorganic anions including $O^-_{(sur)}$, O_2^-, O_3^-, CO^-, $C_2O_2^-$ and CO_3^- have been formed and identified on these M/MgO samples.

References

1. H. Weber: *Z. Phys.* **130**, 392 (1951).
2. W. C. Ward and E. B. Hensely: *Bull. Am. Phys. Soc.* **10**, 307 (1965).
3. J. C. Kemp, W. M. Ziniker, and E. B. Hensely: *Phys. Lett. A* **25**, 43 (1967).
4. W. C. Dash: *Phys. Rev.* **92**, 68 (1953).
5. W. Carson, D. F. Holcomb, and H. Ruchardt: *J. Phys. Chem. Solids.* **12**, 66 (1959).
6. B. Henderson and J. E. Wertz: *Adv. Phys.* **17**, 749 (1968).
7. J. E. Wertz, J. W. Orton, and P. Auzins: *J. Appl. Phys. (Suppl.)* **33**, 322 (1962).
8. R. L. Sproull, R. A. Bever, and G. Libowitz: *Phys. Rev.* **92**, 77 (1953).
9. P. A. Anderson and P. P. Edwards: *J. Am. Chem. Soc.* **114**, 10608 (1992).
10. P. H. Kasai: *J. Chem. Soc.* **43**, 3322 (1965).
11. P. H. Kasai and R. J. Bishop, Jr.: *J. Phys. Chem.* **77**, 2308 (1973).
12. A. J. Tench and P. Holroyd: *J. Chem. Soc., Chem. Commun.* 471 (1968).
13. J. H. Lunsford and J. P. Jayne: *J. Chem. Phys.* **44**, 1487 (1966).
14. A. J. Tench, T. Lawson, and J. F. J. Kibblewhite: *J. Chem. Soc. Faraday Trans. II* **68**, 1169 (1972).
15. N. B. Wong and J. H. Lunsford: *J. Chem. Phys.* **55**, 3007 (1971).
16. J. Kijenski and S. Malinowski: *J. Chem. Soc. Faraday Trans. I* **74**, 230 (1978).
17. L. R. M. Martens, P. J. Grobet, W. J. M. Vermeiren, and P. A. Jacobs: in Y. Murakami, A. Iijima, J. W. Ward (eds.), *New Developments in Zeolite Science and Technology*, Kodansha, Tokyo, p. 935 (1986).
18. S. Coluccia, M. Baricco, L. Marchese, G. Martra, and A. Zecchina: *Spectrochimica Acta* **49A**, 1289 (1993).
19. C. R. A. Catlow, I. D. Faux, and M. J. Norgett: *J. Chem. C (Solid State Phys.)* **9**, 419 (1976).
20. W. C. Mackrodt and R. F. Stewart: *J. Chem. C (Solid State Phys.)* **12**, 431 (1979).
21. S. Coluccia, S. Lavagnino, and L. Marchese: *Mater. Chem. Phys.* **18**, 445 (1988).
22. A. Zecchina, M. G. Lofthouse, and F. S. Stone: *J. Chem. Soc. Faraday Trans. I* **71**, 1476 (1975).
23. S. Coluccia and L. Marchese: *Proc. Int. Sym. Acid Base Cat.* 207 (1988).
24. S. Coluccia: in M. Che and G. C. Bond (eds.), *Adsorption and Catalysis on Oxide Surfaces,*, Elsevier Science Publishers, Amsterdam, p. 58 (1985).
25. M. Che and E. Giamello: in J. L. G. Fierro (ed.), *Spectroscopic Characterisation of Heterogeneous Catalysts, Part B*, Elsevier Science Publishers, Amsterdam, p. 57, B265 (1987).
26. A. Zecchina, G. Spoto, and S. Coluccia: *J. Mol. Catal.* **14**, 351 (1982).
27. P. Ugliengo, G. Borzani, and D. Viterbo: *J. Appl. Crystallogr.* **21**, 75 (1988).

28. E. Giamello, A. Ferrero, S. Coluccia, and A. Zecchina: *J. Phys. Chem.* **95**, 9385 (1991).
29. E. Giamello and D. Murphy: *J. Phys. Chem.* **98**, 7329 (1994).
30. A. J. Tench and R. L. Nelson: *J. Colloid and Interface Sci.* **26**, 364 (1968).
31. D. Murphy: *Ph.D. Thesis, Trinity College Dublin* (1993).
32. E. A. Colburn: *Surface Science Reports* **15**, 281 (1992).
33. D. Murphy, E. Giamello, and A. Zecchina: *J. Phys. Chem.* **97**, 1739 (1993).
34. J. A. Howard, R. Sutcliffe, and B. Mile: *Chem. Phys. Lett.* **114**, 84 (1984).
35. J. A. Howard and B. Mile: in *Electron Spin Resonance Volume 11B*, Specialist Periodical Report, p. 136 (1989).
36. J. A. Howard, C. A. Hampson, M. Histed, H. Morris, and B. Mile: in *Physics and Chemistry of Small Clusters*, Eds., P. Jena, B. K. Rao and S. N. Khanna, Plenum, 421 (1987).
37. D. M. Lindsay and G. A. Thompson: *J. Chem. Phys.* **77**, 1114 (1982).
38. G. A. Thompson and D. M. Lindsay: *J. Chem. Phys.* **74**, 959 (1981).
39. B. Xu and L. Kevan: *J. Chem. Soc., Faraday Trans I* **87**(17) 3843 (1991).
40. B. Xu and L. Kevan: *J. Phys. Chem.* **96**, 2642 (1992).
41. M. R. Anderson, P. P. Edwards, J. Klinowski, J. M. Thomas, D. C. Johnson and C. J. Page: *J. Solid State Chem.* **508**, 165 (1984).
42. P. A. Anderson, R. J. Singer, and P. P. Edwards: *J. Chem. Soc., Chem. Commun.* 914 (1991).
43. B. Xu and L. Kevan: *J. Chem. Soc., Faraday Trans I* **87**(19) 3157 (1991).
44. P. P. Edwards, M. R. Harrison, J. Klinowski, S. Ramdas, J. H. Thomas, and D. C. Johnson: *J. Chem. Soc., Chem. Commun.* 982 (1984).
45. P. A. Anderson, D. Barr, and P. P. Edwards: *Angew. Chem.* **103**, 1511 (1991).
46. S. Malinowski and J. Kijenski: *Catalysis* 130 (1981).
47. J. Kijenski, and S. Malinowski: *React. Kinet. Catal. Lett.* **18**, 317 (1981).
48. J. Kijenski, M. Marczenski, and S. Malinowski: *React. Kinet. Catal. Lett.* **7**, 151 (1977).
49. E. Giamello, D. Murphy, L. Marchese, G. Martra, and A. Zecchina: *J. Chem. Soc., Faraday Trans.* **89**, 3715 (1993).
50. M. C. R. Symons: *J. Chem. Soc.* 570 (1963).
51. J. H. Lunsford: *J. Chem. Phys.* **46**, 4374 (1967).
52. M. Primet, M. Che, C. Naccache, M. V. Mathieu, B. Imelik: *J. Chem. Phys.* **67**, 1629 (1970).
53. M. Che and A. J. Tench: *Adv. Catal.* **32**, 1 (1983).
54. D. Cordischi, V. Indovina, and M. I. Occhiuzzi: *J. Chem. Soc., Faraday Trans. I* **76**, 1147 (1980).
55. R. M. Morris, R. A. Kaba, J. G. Groshens, K. J. Klabunde, R. T. Blatsiberger, N. I. Woolsey, and V. I. Stenberg: *J. Am. Chem. Soc.* **102**, 3419 (1980).
56. R. M. Morris and K. J. Klabunde: *J. Am. Chem. Soc.* **105**, 2633 (1983).
57. D. Cordischi and V. Indovina: in M. Che and G. C. Bond (eds.), *Adsorption and Catalysis on Oxide Surfaces*,, Elsevier Science Publications, Amsterdam, p. 209 (1985).
58. M. Bailes and F. S. Stone: *Mat. Chem. Phys.* **29**, 489 (1991).
59. L. Marchese, S. Coluccia, G. Martra, E. Giamello, and A. Zecchina: *Mat. Chem. Phys.* **29**, 437 (1991).
60. S. Olivella, M. A. Pericas, F. Serratosa, and A. Messegener: *J. Mol. Struct (Theohem)* **105**, 91 (1983).
61. W. Kanzig and M. H. Cohen: *Phys. Rev. Lett.* **3**, 509 (1959).
62. E. Giamello, P. Ugliengo, and E. Garrone: *J. Chem. Soc., Faraday Trans. I* **85**, 1373 (1989).
63. E. Giamello, P. Ugliengo, E. Garrone, M. Che and A. J. Tench: *J. Chem. Soc., Faraday Trans. I* **85**, 3987 (1989).
64. E. Giamello, D. Murphy, E. Garrone, and A. Zecchina: *Spectrochimica Acta* **49A**, 1323 (1993).
65. J. R. Brailsford, J. R. Morton, and L. E. Vannotti: *J. Chem. Phys.* **49**, 2237 (1968).
66. J. R. Brailsford and J. R. Morton: *J. Chem. Phys.* **51**, 4794 (1969).
67. M. Che and A. J. Tench: *Adv. Catal.* **31**, 777 (1982).
68. G. W. Chanty, A. Horsfield, J. R. Morton, and D. H. Whiffen: *Mol. Phys.* 589 (1962).
69. P. W. Atkins and M. C. R. Symons: in *The Structure of Inorganic Radicals*, Elsevier Science Publications, Amsterdam (1967).

Radicals on Surfaces Formed by Ionizing Radiation

MASARU SHIOTANI
Department of Applied Chemistry, Hiroshima University, 724 Higashi-Hiroshima, Japan

and

MIKAEL LINDGREN
Department of Physics and Measurement Technology, Chemical Physics, University of Linköping S-581 83 Linköping, Sweden

(Received: 15 December 1993; accepted: 5 July 1994)

Abstract. An overview is presented of radicals generated on porous metal oxide surfaces such as zeolites whose main source of generation has been ionizing radiation. Attention is primarily paid to ESR studies on structures and reactions of organic neutral and ionic radicals. A short introduction is also given to paramagnetic metal ions and clusters formed in zeolites and other related materials.

Key words. ESR, radical, radical ions, ionizing radiation, paramagnetic metal ions, metal clusters, metal oxides, zeolites, alkane, alkenes, alkynes, benzene, fluorinated benzenes, sodium clusters, silver clusters, silica clathrates.

1. Introduction

The importance of the accumulation of chemical knowledge in studies of free radicals in surface environments have been introduced and treated at other places in this volume. Here we shall review studies of radicals and radical ions where the main source of generation has been ionizing radiation, such as X or γ-rays. Attention will primarily be on organic molecular radicals and radical ions. but a short introduction to paramagnetic metal ions and clusters is also given, since the topics are connected in view of potential use in catalytic applications.

The study of organic radical cations is closely related to the methods of their generation and stabilization. In the past decade, a number of organic radical cations have been extensively studied by means of ESR and optical absorption spectroscopy combined with the low-temperature halocarbon matrix method and ionizing radiation [1]. For smaller organic radical cations such as methane the inert-gas matrices, especially neon, have been applied [2]. The use of porous metal oxides such as zeolites in combination with radiation is another way of generating and stabilizing organic radical cations and a variety thereof have been subjected to ESR studies.

Zeolites possess several advantages over halocarbon and inert-gas matrices in the generation and stabilization of radical cations. (a) In common with other matrices, the zeolites, when exposed to ionizing radiation such as *X*- or γ-rays, are capable of oxidizing a wide range of adsorbed and/or solvated molecules. The ejected electron is captured by atoms constituting the zeolite framework, such as oxygen atoms, or, among exchanged cations, since the electron back-transfer reaction apparently is prevented. Also, the ionization potential of the zeolite[1] can be larger than that of adsorbed molecules so as to readily transfer the hole from the zeolite matrix to the molecules. (b) The limited temperature range is a serious problem of the inert gas matrix for temperature-dependent studies of radical cations. The study of temperature dependent ESR lineshapes can generally provide important information about inter- and/or intramolecular dynamics as well as thermally activated chemical reactions. Halocarbon matrices have the most widespread use due to a considerable temperature range available up to *ca.* 200 K. Radicals and radical cations in zeolites can generally be studied up to room temperature, the stability of the adorbed radicals depending on size and mobility of molecules adsorbed, the molecule–matrix interactions, zeolite channel and cavity dimensions, etc. (c) The reactivity of radical cations can be controlled by environmental factors and may thus be matrix dependent. Zeolites can provide a variety of experimental conditions in addition to temperature, such as polarity and reaction volume. Thus, zeolites can be regarded as versatile microreactors allowing a wider range of reactions to take place and also with greater experimental control of the physical and chemical conditions.

Ionic clusters and small metal particles can be stabilized in surface environments such as zeolites, aluminosilica clathrates, clays, etc. They have been studied for fundamental reasons, for example when trying to understand the detailed mechanisms behind the formation of electronically conducting metals, as reviewed recently by Edwards and coworkers [3]. Also the studies on catalytic electron transfer systems involving metal particles (atoms), metal ions and metal ion clusters were initiated by Kasai, Rabo and coworkers and an extensive number of papers have been published. For a comprehensive description of charge and electron transfer reactions in zeolites involving adsorbed organic molecules it is referred to a recent review by Yoon [4]. Owing to space limitations we shall here restrict our attention to characterization by ESR, and related methods, of the small metal particles and ionic clusters stabilized in surface environments.

2. Organic Radicals and Radical Ions

2.1. METHANE AND ETHANE

A limited number of studies have been reported for irradiated saturated hydrocarbons adsorbed on metal oxide surfaces. We start by presenting the cases of most simple alkanes, methane and ethane, the results date back to the 1970s.

Methane adsorbed on zeolite 4A was irradiated by γ-rays at 77 K [5,6]. ESR spectra of three different kinds of methyl radicals were observed depending on heat pretreatment temperature of the zeolite. The first one was described as a 'normal' methyl radical with an isotopic hfsc[2] of 2.35 mT due to three equivalent hydrogens, as expected from comparisons with the results obtained involving the methyl radical generated in homogeneous systems. The two remaining methyl radicals were designated 'abnormal' methyl radicals, type I and type II. Type I radical was found with a hfsc to one extraneous hydrogen with an anisotropic[3] hfsc, $A_\parallel = 1.89$ and $A_\perp = 0.84$ mT, in addition to those of the three equivalent hydrogens. The ESR spectrum of the abnormal type II methyl radical showed large hfsc to one of the three hydrogens, $a = 3.63$ mT, the hfs of the other two

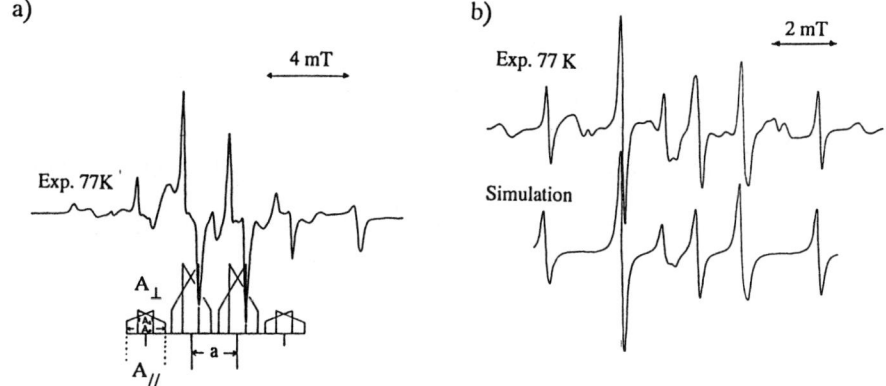

Fig. 1. ESR spectra of the abnormal methyl radicals trapped on activated zeolite 4A. (a) Type I; $A_\perp$ and $A_\parallel$ for the additional hfsc indicated in the stick plot; a denotes the isotropic hfsc due to three equivalent hydrogens. (b) Type II; the simulation is made using hfs constants as given in the main text. (Redrawn from [5].)

being smaller in magnitude, $a = 2.27\,\text{mT}$. ESR spectra of the two abnormal types of methyl radicals are shown in Figure 1. Using CD_4 as adsorbate it could be established that the hydrogen atom liberated from irradiated CH_4 (or CD_4) is trapped on the zeolite surface and converted into H^+. The extra hf splittings similar to type II have been observed for irradiated ethane adsorbed on zeolite 4A [7]. The anisotropic ^{13}C-hfs constant of the abnormal methyl radical, type II, $A_\parallel = 7.58$ and $A_\perp = 1.055\,\text{mT}$, indicates that this methyl radical interacts with the surface through a $2p$ unpaired-spin orbital and rotates about a C_3 symmetry axis.

The formation of a methane radical cation as the primary step of the reaction may be possible in these cases, however, this radical cation has never been stabilized and detected in a zeolite. Upon formation, it probably dissociates instantly into a methyl radical and a proton due to either a unimolecular or ion–molecular reaction. Alkane radical cations with a longer chain have been generated and stabilized in a similar irradiated zeolite system [8] as will be discussed in a following section. Knight has observed the methane radical cation in a neon matrix at 4 K [9].

Abnormal methyl radicals have been observed not only in zeolite system, but also with other metal oxides such as silica gel and Vycor glass. UV irradiation of methane adsorbed on silica gels gives rise to two different methyl radicals, with 2.07 and 2.12 mT hfs constants at 77 K [10]. Both hfs constants are slightly smaller than that of the normal methyl radical. It has been suggested that these abnormal methyl radicals originate from a stabilization at oxygen atom of siloxane ($\equiv$Si—OH) on silica surfaces. When methane adsorbed on porous Vycor glass was UV irradiated at 77 K, an abnormal methyl radical could be detected with an extraordinary small isotropic hfsc of 1.9 mT [11]. In subsequent ESR studies the ^{13}C-hf splitting of the abnormal methyl radical was found to be smaller than that of the normal methyl radical and interpreted in terms of a spin-delocalization onto the oxygen atom of the surface siloxane bridge, keeping the structure at the radical center planar [12]. Through similar studies Fujita and coworkers have demonstrated that vanadium oxide and other transition metal oxides dispersed on the surface of porous Vycor glass are good photosensitizers for the formation of alkyl radicals from the corresponding alkyl hydrocarbons, by irradiation with light of wavelength longer than 350 nm at 77 K. The Vycor glass is superior to others because of its ability to transmit light of $\lambda > 220$ nm efficiently. For example, a report on the photochemical decomposition of ethane and methylamines adsorbed on the surface of porous Vvcor glass can be found in Reference [13].

2.2. LINEAR AND BRANCHED ALKANES

The radical cations of n-$C_6H_{14}^+$ and n-$C_8H_{18}^+$ were generated from adsorbed parent molecules in ZSM-5 by X-irradiation at 4 K [8]. A planar extended structure was suggested for both radical cations, which is in contrast with halocarbon matrix isolation studies. In such matrices the extended conformation of linear alkane cations, $C_nH_{(2n+2)}$, $n > 3$, are usually observed together with more folded conformational structures [14]. In ZSM-5 the folding of the linear chain is restricted owing to the narrow tubular shape of the zeolite channels in which the alkanes become trapped. With the increase of alkane concentration neutral 1-alkyl radicals were formed upon irradiation at 4.2 K. It was suggested that the parent molecules are becoming adsorbed as a dimer, from which the 1-alkyl radical is preferentially formed. It was taken as a proof that n-alkane radical cations undergo prompt deprotonation via an ion–molecule reaction when a neighboring alkane molecule exists as a proton acceptor, and that the driving force behind the site selectivity (1-alkyl) was consistent with the higher unpaired electron density at the terminal C—H bond of the cation. However, this might also have been an accidental coincidence, in fact, numerous studies of alkyl-substituted cyclohexane radical cations found no such relationship between the locations of high hydrogen spin density in the cation and the site of the alkyl radicals formed therefrom [15]. An elimination process followed by an ion–molecule reaction has also been studied for hexamethylethane (HME) and trimethylbutane (TMB) radical cations generated in ZSM-5 zeolite by γ-radiolysis [16]. For example, the tetramethylethylene radical cation intermediate formed by CH_3 elimination from TMB^+ was observed in the temperature range 120–200 K. The deprotonation product, a $(CH_3)_3C$—$C^{\cdot}(CH_3)_2$ radical, was observed at temperatures above 200 K.

The variation in microstructure of the zeolite framework can sometimes have large impact on the stabilization of radical cations. Small variations in hfs constants are of course not unexpected, but more dramatical changes may also occur. Barnabas and Trifunac have detected different electronic ground state structures for *cis*- and *trans*-decalin radical cations generated by ionizing radiation in various zeolites [17]. The two lowest electronic states of the decaline radical cations differ in energy by *ca.* 10 kcal/mol. The calculated hf splittings for these states of *cis*- and *trans*-decalin$^+$ have been reported by Lund and coworkers: for *cis*-decalin$^+$, the lowest state is $^2A_1$ (4.65 mT) and the high-energy state is $^2A_2$ (3.21 mT); for *trans*-decalin$^+$, the lowest state is ^2A_g (5.06 mT) and the high-energy state is ^2B_g (2.98 mT) [18]. Both the $^2A_1$ and $^2A_2$ states of *cis*-decalin were observed at different temperatures in silicates (S-115), but in ZSM-34 and Na-Ω-5 one species was observed preferentially: $^2A_1$ (ZSM-34) and $^2A_2$ (Na-Ω-5). For *trans*-decalin$^+$ the ^2A_g was observed in ZSM-34, and the ^2B_g state was observed in Na-Ω-5. In silicate, only the ^2A_g state was observed at low temperature. Note that ZSM-34 and Na-Ω-5 are the same zeolites with the same type of cage, but with different cage sizes. A similar situation has been found for certain methylcyclohexane radical cations in halocarbon matrices, particularly for the 1,3-dimethyl-substituted derivatives. Here, different matrices have been shown to stabilize different ground states, and in some cases different states were observed simultaneously in the same matrix [19]. The energy difference of the electronic states in question can be expected smaller in these cases since a near degeneracy of SOMO^4 is expected in view of the doubly degenerate HOMO of the unsubstituted cyclohexane molecule.

It would be very interesting to learn from theoretical investigations what determines the particular choice of electronic state in these cases. The 'stiffness' as well as size confinement would have an impact on the vibrations and thereby average bond lengths within a molecule, whereas the effective polarizability at the site of the cation could alter its charge distribution. Both effects could alter an electronic ground state to a situation different from the gas phase case.

2.3. BENZENE AND RELATED RADICAL CATIONS

Benzene and some other aromatics can be spontaneously oxidized by introducing them to activated porous metal-oxides such as ZSM zeolite, H-mordenite and aluminosilicates, and their radical cations have been subjected to ESR studies [20]. They include benzene, aniline, and trimethylbenzene, and have been treated at other places in this volume. In other cases, the radical cations can be generated by photoillumination [21] or ionizing radiation, which will be focused on here.

It has been known for a long time that ionic radicals of aromatic compounds formed by ionizing radiation in the adsorbed states at 77 K remain stable. However, in earlier studies it was difficult to ambiguously attribute the experimental ESR spectra to the radical cations since the difference in the ESR hf parameters between the positive and negative radical ions is sometimes small. For the benzene radical ions, the following approximate ESR hf parameters were expected: $g = 2.0024$ and $a(6H) = 0.444$ mT [22] for the cation and $g = 2.0028$, $a(6H) = 0.375$ mT for the anion [23], based on the results in the solution studies. By an examination of the ESR spectra of irradiated benzenes adsorbed on silica gel and silica alumina, Edlund *et al.* were first to unambiguously attribute their ESR spectra with $g = 2.0022$ and $a(6H) = 0.44$ mT to the benzene radical cation [24]. In subsequent studies, the selective formation of aromatic molecular cations was confirmed by Komatsu *et al.* for toluene in the irradiated adsorbed state on silica gel and Vycor glass [25]. The degeneracy of the π-electron HOMO/LUMO of the benzene ring can be lifted by methyl substitution. Adding one electron-repelling methyl substituent will form the toluene molecule. With the addition of an electron the degenerate e_{2u} LUMO originating from the unsubstituted symmetric benzene is split to give a Ψ_a^- SOMO, as shown below, with large spin density at the C2, C3, C5 and C6 carbons. In a similar manner the positive charge distribution in the ring dictates a Ψ_s^+ SOMO as shown below, originating from the e_{1g} orbital, with high spin density on C1 (methyl) and C4. Using this and similar guidelines the selective formation of cations in a series of aromatics was shown to occur for the irradiated adsorbed states.

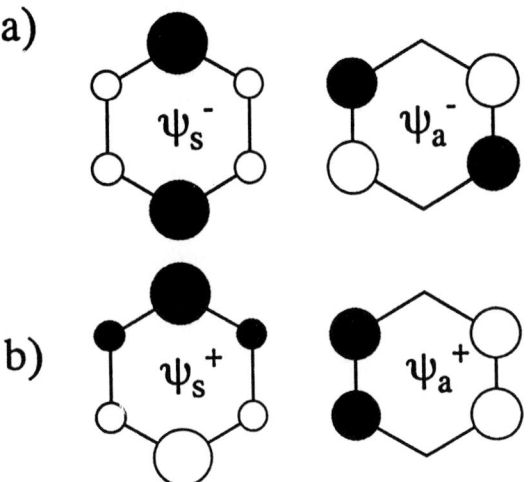

When HY zeolite was used as adsorbent, an enhanced resolution was achieved in the ESR spectra of the benzene radical cation, and changes of the temperature-dependent line shape were studied up to 240 K [25]. At 203 K the anisotropic hf and g value parameters governing the ESR spectra become axially symmetric with $A_\perp = 0.405$ mT, $A_\parallel = 0.500$ mT, and $g_\perp - g_\parallel = 1.8 \cdot 10^{-4}$. The axial symmetry suggests that rapid rotation takes place about the molecular six-fold axis. Upon lowering the temperature of the sample, the associated changes in the ESR spectra could not be reproduced either with axial or rhombic symmetry of the g value and hf tensors, indicating that the rotation frequency then became comparable to the hf splitting anisotropy.

Here, we also note that ESR evidence for a static distortion of the benzene radical cation ($C_6H_6^+$) has been suggested by Iwasaki *et al.* [26]. The orbital degeneracy of e_{1g} in the neutral D_{6h} benzene is removed at 4 K in a $CClF_3$ matrix and the unpaired electron was shown to occupy the $b_{2g}(\Psi_s^+)$ orbital with D_{2h} symmetry. Upon warming the sample, six hydrogens become equivalent by dynamic averaging, as observed for benzene on metal oxide surfaces, however, at a considerably higher temperature (*ca.* 90 K) than in the adsorbed case. Recently ENDOR and ESE methods have been applied to reinvestigate, at lower temperatures, the benzene radical cation formed on HY zeolite and silica gel by X-ray irradiation [27]. The electronic structure and particularly the dynamics were found to be different from the halocarbon matrix ($CFCl_3$) stabilized species. The monomer cation undergoes a pseudo-rotation even at 3.5 K with an ESR signal due to six averaged hydrogens. The time domain deuterium ESE ($C_6D_6^+$) was found to be consistent with axial symmetry of the hfs constants: $T_\perp = 0.8$ MHz, $a = 2.17$ MHz in consistency with the ENDOR result obtained at higher temperatures. The pseudo rotation was shown to slow down by the exchange of one hydrogen with deuterium, to yield a broad hf quintet at 3.5 K. It was not possible to ascertain, however, whether this corresponds to the rigid state. Although this is consistent with a Ψ_a^+ SOMO, as shown above, stabilized by a static Jahn–Teller distortion of the degenerate HOMO, a Ψ_s^+ SOMO dynamically exchanging between different sites at certain rates could also give a similar ESR lineshape. Further studies seem necessary to understand the relations between an eventual rigid ground state structure and its dynamical states upon stabilization in heterogeneous systems.[5]

Based on the studies of irradiated adsorbed metal oxide systems such as zeolite and silica gels, the selective generation of organic radical cations can be summarized by the following reactions. First the adsorbent (A) is ionized by irradiation:

$$A \xrightarrow{\gamma,\ X\text{-ray}} A^+ + e^-$$

In the case of higher ionization potential (IP_1) of A than that of the adsorbed molecule (M), the hole can be transferred to M to yield the cation:

$$A^+ + M \rightarrow A + M^+$$

This is essentially the mechanism which is well accepted for the formation of molecular cation radicals in irradiated halocarbon matrix at low temperature [1].

The ionization potentials (IP_1) of fluorinated benzenes are known to increase with number of fluorine atoms: benzene (9.24 eV). m-difluorobenzene (9.32 eV), 1.3.5-trifluorobenzene (9.62 eV), 1,2,4,5-tetrafluorobenzene (9.36 eV), pentafluorobenzene (9.64 eV), and hexafluorobenzene (9.90 eV) [28]. We have recently investigated the radiation-induced oxidation of a series of fluorinated benzenes adsorbed on ZSM-5 zeolite and silica gel (SiO_2) and confirmed the selective formation of the radical cation up to pentafluorobenzene. The ESR spectra observed for the 1,2,4,5-tetrafluorobenzene radical cation on ZSM-5 and SiO_2 are shown in Figure 2, together with the spectrum observed in the perfluorocyclohexane (cC_6F_{12}) matrix for comparison. As indicated by the stick plots, the spectra are easily explained in terms of axial symmetric hf and g-tensors. The observation of four equivalent fluorine hf splittings with $A_\parallel = 10.2$ mT and $A_\perp = 0$ mT can be assigned to the radical cation with a planar π-type structure with Ψ_a^+ SOMO (in the benzene carbon framework). It is noted that the ESR spectra of the 1,2,4,5-tetrafluorobenzene radical anion is completely different.[6] The full series of results of these studies will appear elsewhere.

2.4. BENZENE DIMERIC CATIONS AND CYCLOHEXADIENYL RADICALS

In ESR spectra of the irradiated benzene/zeolite and benzene/SiO_2 systems, additional hf lines with $a = 0.21$ mT evolve for samples in which the concentration of benzene is increased (see Figure 3). The hf splitting is ca. 50% of that of the monomeric benzene radical cation, and the ESR spectrum has been attributed to a dimeric form. In a recent ENDOR study it was possible to distinguish several forms of the dimer. One form was observed in SiO_2 regardless of temperature, and had a hfs constant of 6.5 MHz. When lowering the temperature below ca. 110 K with moderate concentration of benzene in the HY sample, the first form gradually transforms to a second form [27]. The latter one is characterized by its ENDOR spectrum with hfs constants of 4.8 MHz and 8.9 MHz.

On both wings of the ESR spectrum of the benzene cation radicals in the zeolite and SiO_2 systems, weaker hf lines attributable to the cyclohexadienyl radical

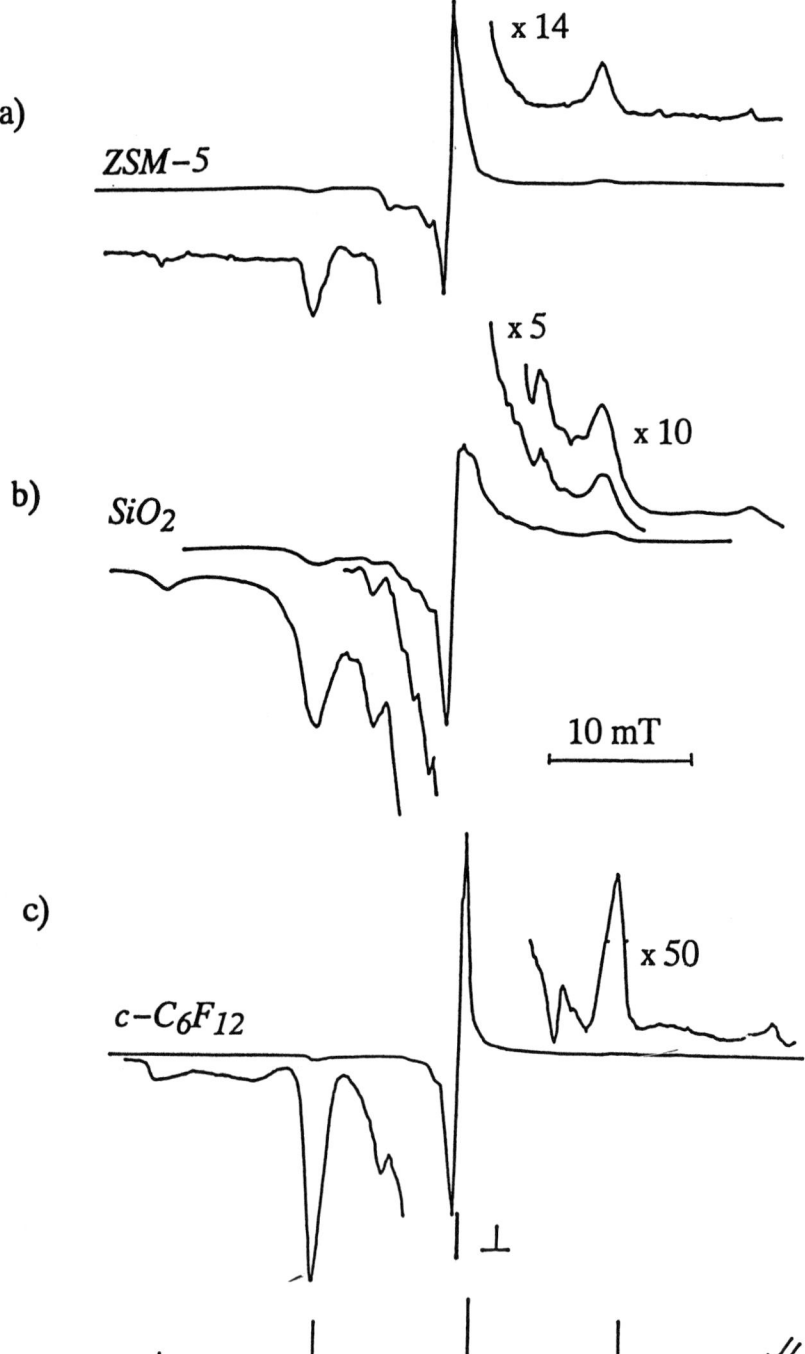

Fig. 2. The ESR spectra of γ-irradiated 1,2,4,5-tetrafluorobenzene adsorbed on activated (a) ZSM-5 zeolite, (b) silica gel (SiO_2), and in (c) perfluorocyclohexane (cC_6F_{12}). All spectra recorded at 77 K (unpublished results).

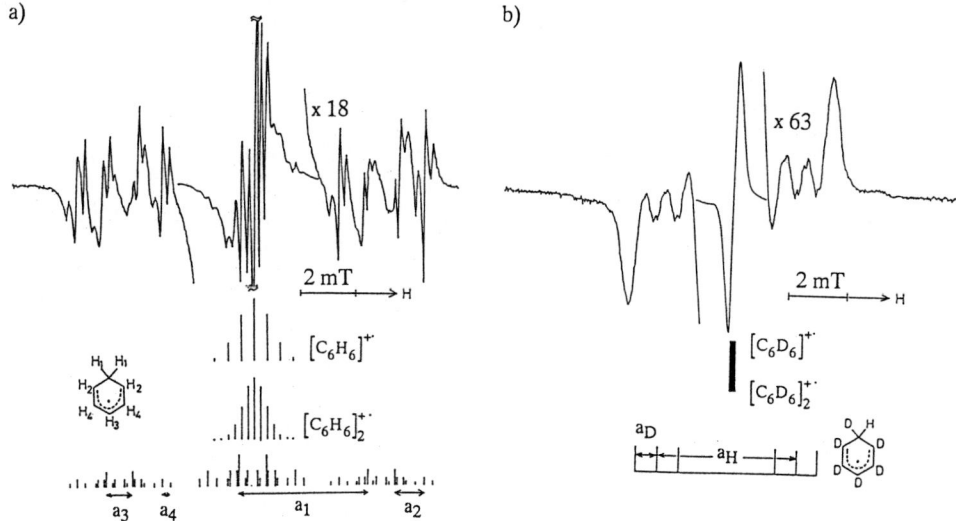

Fig. 3. The ESR spectra of γ-irradiated (a) benzene and (b) perdeuterated benzene adsorbed on activated ZSM-5 zeolite, recorded at 100 K (unpublished results). The ESR spectra of the cyclohexadienyl radical together with monomeric and dimeric benzene cation radicals are indicated as stick plots and insets.

($C_6H_7^·$) are generally observed. In the case of benzene/ZSM-5 system, giving a hf structure with splittings: $a_1(2H) = 4.84$: $a_2(2H) = 0.94$: $a_3(2H) = 1.33$: $a_4(1H) = 0.26$ mT. When C_6D_6 is adsorbed instead of C_6H_6, the $C_6D_6H^·$ radical is formed. The two cases are shown in Figure 3. These results strongly suggest that a Brønstead acid site (H^+ in scheme below) on the surface can be a precursor of the hydrogen atom which reacts with benzene to form the cyclohexadienyl radical. A similar reaction to form the hexadienyl radical has been observed for irradiated acidic ice containing a small amount of benzene. The suggested reaction mechanism is summarized as follows:

$$(ZSM\text{-}5) \xrightarrow{\gamma\text{-ray}} (ZSM\text{-}5)^+ + e^-$$

$$H^+ + e^- \rightarrow H^·$$

$$H^· + C_6H_6 \rightarrow C_6H_7^·$$

The proton site acts essentially as an electron acceptor to generate a neutral radical. Thus, the mechanism can be described as the reduction route of the ionization process. As we will see in following sections, similar reactions take place in irradiated and by other means reduced zeolites and similar metal oxide systems exchanged with alkaline cations.

Brønsted acids can be strong enough to oxidize some alkenes and alkynes to form the associated radical cations which can be detected by ESR. However, we did not succeed to observe the spontaneous formation of benzene radical cation on ZSM-5 with [Si]/[Al] = 60, although Vedrine et al. have reported that benzene is spontaneously oxidized upon introducing it to activated ZSM-5 with [Si]/[Al] = 30 [29].

2.5. ALKENES AND ALKYNES

Spontaneous formation of alkene and alkyne radical cations on activated zeolites and related metal oxides occurs in a vast number of cases, and have been studied extensively since the pioneering works of Turkevich [30] and Hirschler [31] in the middle of the 1960s. Studies of inter- and/or intramolecular rearrangements of radical cations of, for example, butenes [32,33], tetramethylene (TME) [34], cycloalkenes [35–37], some dienes [38], and some alkynes [32,36] on ZSM-5 and H-mordenite, and have been reviewed at other places in this volume. Only a few studies have been reported for irradiated alkenes and alkynes in the adsorbed state. These include the reactions of quadricyclane and norbornadiene radical cations on ZSM-5 and silicate [34,39], TME$^+$ on zeolites [40], and Jahn–Teller distortions of the tetramethylallene radical cation on Na-Y zeolite [41].

Here we briefly introduce the results on irradiated acetylene and propylene on ZSM zeolites. These demonstrate another potential use of ESR combined with ionizing radiation in studies of molecular rearrangements of molecules on solid catalysts such as zeolites.

Exposure of activated H-ZSM-5 zeolites (or ZSM-11) to acetylene at room temperature gives rise to an ESR singlet without any resolved hf structure, centred at $g = 2.002$ [42]. Associated with the appearance of ESR signal is a color: the sample changes from being white to red-brown or black. When the sample thereafter was irradiated by γ-rays, the same hf singlet signal was observed in addition to signals attributed to monomeric and dimeric benzene cation radicals, i. e., spectra very similar to the ones shown in Figure 3, discussed in a previous section. This provides indirect ESR evidence of benzene formation through trimerization of acetylenes within the internal void spaces of the zeolites. The detailed origin of the hf singlet is not known but it may be attributable to one, or more probably, several, conjugated radical species.

Three independent groups have recently reported the acid-catalytic reaction of acetylenes in ZSM-5 zeolite and concluded the formation of simple aromatics and oligomer/polymer chains [43–45], in accordance with our ESR observation. For example, Pereira et al. [43], have characterized the reactions products by means of temperature-programmed desorption (TPD), thermogravimetric analysis (TGA), transmission IR, reflectance UV-visible, and ^{13}C-NMR. The TPD-TGA results showed that the products decomposed above ca. 550 K, similar with polyacetylene, with 70% of the hydrocarbons in the zeolite desorbing as simple aromatics and olefins. From the changes observed in IR spectra it was concluded the reactions where occurring at Brønsted sites in the zeolite. Oligomerization and/or polymerization of unsaturated hydrocarbons and related compounds have been known to proceed easily within the channels of zeolites. For example, we have observed a propagating radical of polytetrafluoroethylene in the tetrafluoroethylene/ZSM-5 system where the initiator of the reaction was γ-ray irradiation and the detection method ESR [46].

We will close this section by pointing out one more example of potential usefulness of ESR in combination with ionizing radiation. Activated H-ZSM-5 exposed to propene gas gives an ESR signal but the hf-pattern is not clearly resolved. After irradiating the same sample with γ-rays one readily obtains a well-resolved 13-line pattern with binomial intensity distribution and a hfsc of 1.75 mT [42]. This is the well known spectrum of the TME cation radical which has been observed in several other cases (see above). Ichikawa and coworkers have studied the generation of TME molecules as a byproduct of acid-catalyzed reactions of butene on H-mordenites [33]. Thus, it is reason to conclude similar reactions of propene to yield TME on ZSM-5. The detailed mechanism, however, is not known.

3. Ionic and Atomic Metal Clusters

3.1. IONIC SODIUM CLUSTERS IN ZEOLITES

The first observation of cationic Na clusters in zeolites was made by Kasai [47]. Linde Type Y zeolite with sodium in the cationic sites was subjected to γ- and X-irradiation. Following irradiation in vacuum, the typical 13-line pattern with hyperfine interaction due to four Na nuclei was observed with $a = 3.23$ mT at $g = 1.999$. Associated with the ESR signal was a distinct pink color and the paramagnetic center was explained in terms of the void space at the 0.5;0.5;0.5 lattice point, coordinated with four tetragonal Na atoms. In a following report, using a sodium vapour to reduce cationic sites in Linde Type Y and X zeolites, Kasai and coworkers detected the same Na_4^{3+} cluster as the one formed from irradiation in Type Y zeolite [48]. In Linde Type X zeolite another cluster was observed, characterized by a 19-line hf pattern with $a = ca.$ 2.2 mT at $g = 1.999$, suggesting the interaction between an unpaired electron and six sodium atoms to form a Na_6^{5+} cluster. No attempt was made to find a plausible coordination site for this paramagnetic center.

Following these pioneering works, the presence of $Na_n^{(n-1)+}$ clusters, and related alkali cationic clusters, has been observed in several systems, some of them are discussed more in following sections. As an illustrative example of the radiation chemistry of γ-irradiated metal ion-exchanged zeolite surfaces we show in Figure 4 the simultaneous presence of the Na_4^{3+} cluster and the NH_2 radical formed by radiation of a system consisting of NH_3 adsorbed on zeolite 4A.[7]

The sodium cluster, Na_4^{3+}, and clusters formed from potassium in K-exchanged zeolites, were both shown to spontaneously give away its unpaired electron to adsorbed oxygen, thus forming the O_2^--radical [49]. The ESR spectrum of the latter, having a characteristic line shape due to a nearly axial g-tensor,[8] was formed during the concomitant disappearance of the signal due to the ionic clusters. This and parallel reports also gave several other examples of electron transfer reactions between adsorbed (inorganic) molecules and the monovalent or divalent cations in the zeolite framework [50]. For example: Cu^{2+} (or Ni^{2+}) exchanged Y zeolites were shown to give Cu^+ (or Ni^+) and NO^+ upon adsorption of NO. The reduced cationic Cu^+ (or Ni^+) site could thereafter be reoxidized into Cu^{2+} (or Ni^{2+}) by exposure to NO_2, the latter being converted to an NO_2^- species. Although the reactions considered are endothermic in the gas or liquid phases, they pro-

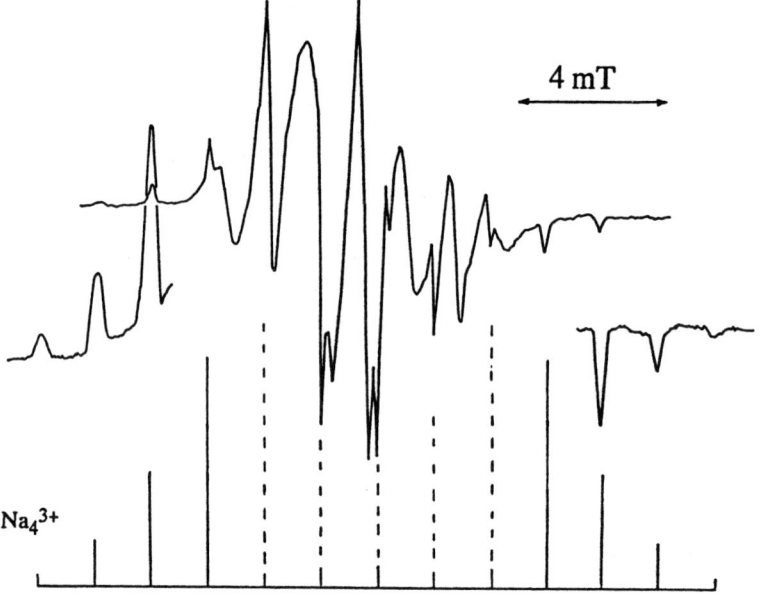

Fig. 4. The characteristic 13-line ESR hyperfine pattern of the Na_4^{3+} cluster together with the ESR signal of the $\cdot NH_2$ radical formed by γ-irradiation of NH_3 adsorbed on zeolite 4A. The 13 lines of the Na_4^{3+} cluster is indicated with a stick plot. (Figure redrawn from [5].)

ceeded readily within the zeolite framework. In these initial work it was concluded that the origin of this charge transfer reactions within the zeolites is to gain crystalline energy by rearranging the adsorbed phase in the void open spaces into cationic–anionic form.

Here we have given a brief introduction to the pioneering work on electron transfer reactions by Kasai and coworkers in the sixties and early seventies. We will now turn to more recent studies on structure and reactions of the ionic clusters, as characterized by ESR and related methods.

3.2. CLUSTERS FORMED FROM ALKALI METAL VAPOUR

By exposing alkaline cationic exchanged zeolites to various alkali metal vapours a vast number of cluster structures and electron transfer reactions can be obtained and studied. The preparation method, initiated by Kasai, Rabo et al., was further developed by Edwards and coworkers, who started by treating Na^+, K^+ and Rb^+ exchanged Y-type zeolites with the atomic vapours [51]. It was concluded that, upon low exposure, the same kind of Na and K cationic clusters which had previously been observed, were formed in the Na^+ and K^+ exchanged zeolites, irrespective of which kind of atomic vapour was used, the reactions being summarized as follows:

$$4\text{Na}^+ + \text{M}^0 \rightarrow \text{Na}_4^{3+} + \text{M}^+$$
$$4\text{K}^+ + \text{M}^0 \rightarrow \text{K}_4^{3+} + \text{M}^+$$

Since the same kind of complex was formed in Na$^+$-A and Na$^+$-sodalite [52], it was proposed that the Na$_4^{3+}$ and K$_4^{3+}$ clusters were initially formed in the sodalite cages. Clusters ascribable to Rb$_4^{3+}$ could not be detected.

A vast number of Type Na-Y zeolites exchanged by other monovalent (Li$^+$, K$^+$, and Cs$^+$), divalent (Mg^{2+}, Ca^{2+}, Sr^{2+}, and Ba^{2+}) and trivalent (La^{3+}), were treated with either sodium or potassium vapour [53]. In all cases except Cs-Y and Ba-Y the ESR spectra associated with M$_4^{3+}$ (M = Na or K) were detected, together with contributions from neutral metal clusters and particles, M$_x^0$, resulting in notable variations of the ESR parameters. From FT-ESEEM spectra it was possible to estimate distances in the range of 5 Å between the center of cationic clusters and sodium atoms in the neighbouring cage.

Edwards and coworkers later described the formation of new kind of clusters, for example, potassium vapour within zeolites K-X and K-A gave rise to the ionic cluster K$_3^{2+}$, characterized by the ESR parameters: $a = 1.28$ mT (10-lines) at $g = 1.9992$ [54]. The simultaneous presence of Na$_5^{4+}$ and Na$_6^{5+}$ clusters in Na-X were shown to give ESR spectra with the apparent number of lines varying between 16 and 19 with $a = 2.5$ mT at $g = 2.0022$ [55]. The ESR characterizations of sodium clusters all assumed that the participating Na nuclei contribute with equivalent hf splitting constant to give the ESR lineshape, as would be expected if the unpaired electron spin density at each site is identical. Recent X-ray diffraction studies of Na-X, reduced by Rb metal vapour, concluded a full occupancy of Na$_5^{4+}$ clusters in the available sodalite cages; however, with one of the sodium atoms being in a crystallographic unique position at the center of the T_d sodalite cage, only the remaining four Na atoms being equivalent by virtue of symmetry [56].

More detailed investigations of the relations between structure and spin density distribution, of metal ion clusters seem necessary in order to clarify particularly the origins of the vast number of reported ESR spectra. Parallel X-ray diffraction studies seem particularly useful for such comparisons. In addition to structural information, the positions of all products originating from the reduction reaction can be detenmined, and thus, indirectly clarifying many details of the mechanisms of electron transfer. The results of such detailed X-ray experiments have been reported in studies of reactions of Na$_{12}$-A with Rb vapour [57], formation of hexasilver clusters in various Ca^{2+} and Ag$^+$ exchanged zeolite A [58], and finally, structure of Ag$_{5.6}$K$_{6.4}$-A and its reactions with Cs vapour [59].

3.3. CLUSTERS FORMED BY IMPREGNATION METHODS

An alternative method for preparation of ionic clusters in *zeolites* was introduced by Martens and coworkers [60]. Typically, dried activated zeolite powder was added to an NaN$_3$/methanol slurry, just enough to fill the zeolite pore volume. By carefully controlled heating, the sodium azide decomposes and leaves sodium atoms to reduce cations in the zeolite, forming clusters in the void cages. In this manner it is possible to have a detailed control of the amount of sodium. Both Na$_4^{3+}$ (in zeolite Y) and Na$_6^{5+}$ (in zeolite X) were identified by their appearances as characteristic 13-line and 19-line ESR-hyperfine patterns. The method also enabled detailed studies of acid and base catalysed reactions. The ionic sodium clusters were shown to possess typical basic catalytic properties at low reaction temperatures as concluded from isomerization and hydrogenation reactions of alkenes and alkynes.

Encouraged by the development of the new impregnation method to expose the zeolite channels to metal atoms. Kevan and coworkers undertook systematic ESR and ESEEM studies of alkali metal cation exchanged X and Y zeolites, to obtain further details of the metal particles [61] and sodium and potassium ionic clusters [62]. Some alkaline earth cation-exchanged X zeolites were studied by combining ESR, far-infrared and diffuse reflectance spectroscopy [63]. The series of five alkali metal (M_2) vapours formed by alkali metal azide decomposition were reacted with the series of alkali metal (M_1) cation exchanged X-zeolites to form alkali metal particles. Following the earlier work of Edwards, the alkali metal particles were distinguished from their apparent g-value: Li = 2.0034; Na = 2.0016; K = 1.9997; Rb = 1.9929; Cs = 1.9686.[9] In M_2/Na-X and M_2/K-X (except for Cs/K-X and some uncertain cases) metal particles originating from M_2 were formed. For Rb^+ and Cs^+ exchanged X zeolites, Rb_n and Cs_n metal particles were found to be formed, regardless of the choice of M_2 vapour, so that the cluster atoms must originate from the exchanged zeolite. These results were discussed following the original works of Kasai, Rabo, and coworkers, in terms of electron-transfer energetics, electron transfer via the zeolite framework, ion and atom sizes relative to ring sizes in the zeolite framework, and ion/atom migration.

The Na_4^{3+}, Na_6^{5+}, K_4^{3+} and, K_3^{2+} ionic clusters in X and Y zeolites were studied by combined ESR and ESEEM. NaN_3 exposed Na-exchanged zeolites gave: Na-X (19 hf-lines; a = 2.59 mT, g = 2.0013, assigned to Na_6^{5+}), Na-Y (13 hf-lines; a = 3.25 mT, g = 2.0012, assigned to Na_4^{3+}), similarly; KN_3 exposed K-exchanged zeolites gave: K-X (10 hf-lines; a = 1.66 mT, g = 1.999, assigned to K_3^{2+}), K-Y (13 hf-lines; a = 1.28 mT, g = 1.999, assigned to K_4^{3+}). The ESR parameters of the ionic cluster are essentially the same as those reported by Kasai et al. and Edwards et al., suggesting similar structures.

Owing to the different relaxation rates of metal ion clusters and the metal particles, their ESR signals could be separated by the routinely used field-sweep two-pulse ESE. This gave a hf pattern indicating that the electron spin in the K_3^{2+} cluster is not distributed equally over all the nuclei, and a distorted structure was suggested. ^{27}Al nuclear modulation was detected for all the cluster species. Semiquantitative simulation of the ^{27}Al nuclear modulation showed that the average number of aluminum nuclei interacting with the electron spin in these ionic clusters is around 9; the isotropic hf coupling is ca. 0.8–1.1 MHz and the average distance between the unpaired electron and the aluminum nuclei is within 0.30–0.36 nm.

Synthesis of ionic clusters in zeolites was advanced further as reported by Yoon and Kochi [64]. They showed a very feasible way of reducing the cationic framework of a zeolite by the addition of, for example, an hexane solution of *n*-butyllithium, or the exposure of Na-Y/etheral slurrys containing phenyllithium to mild UV light (λ = 350 nm). The advantages with the methods are that they proceed easily at room temperature, and that no heat treatment is necessary, only precaution against oxygen (atmosphere) exposure is needed.[10]

3.4. SPECIAL METAL CLUSTERS FORMED BY IRRADIATION AT LOW TEMPERATURE

By performing γ- or X-ray irradiation of materials at low temperature, charged molecular radical ions or neutral radicals, as well as inorganic defect centers in inorganic materials, can he conveniently generated and studied. At low temperature one detects usually the very primary, and most unstable, species, which convert into more stable products as the temperature is gradually increased. Adopting this technique also rare species of metal clusters have been generated. It is in any case remarkable that although the method was used in the very initial stages of studies of metal ion clusters, studies of a more systematic character seem to have started considerably later. Here we will review the studies of radiation induced sodium clusters. Silver clusters will be discussed in a subsequent section.

The irradiated Na forms of zeolites at room temperature generally give sodium clusters as discussed in foregoing sections. However, by irradiation at 77 K one detected the Na_2^+ (seven-line spectrum) cluster in Na-A [65] and a family of related Na_2^+ clusters in partially Li^+-exchanged Type X and Y zeolites [66]. The ESR spectra of Na_2^+ clusters detected at 77 K have large hf splitting constants, typically 8.5 mT in $K_{76}Na_4X$ and 10 mT in Li_7Na_5A, in addition to overlapping signals due to free electron/hole and hydrogen atoms. Following irradiation at higher temperature, 196 K, one detected a completely different 10-line ESR signals attributed to Na_3^{2+} clusters. This cluster was also reported to form irreversibly from the 77 K Na_2^+ cluster by warming. Similar Na_3^{2+} clusters seem to have been generated in Li^+ exchanged X and Y zeolites by the reduction induced by solvated electrons [67].

An elegant way of simultaneously elucidating structural information and time-resolved kinetics is to combine pulse radiolysis and ESR. The former method is employing a pulse of electrons with high energy to irradiate a sample, and gives the kinetics in terms of decay rates of reactive products at a particular temperature, usually identified and detected by optical absorption data. From the known decay rates at various temperatures it is possible to calculate the energy of activation associated with evolving, or decaying products. The same products, if being free radical species, are easily identified in parallel ESR experiments of samples which has been irradiated by means of X- or γ-rays at lower temperatures.

Trapped electrons are usually absorbing from ca. 500 nm towards the IR, and have been studied in a large number of systems. It is considered to be the primary reducing species during radiolysis of alcoholic glasses, ammonia, DNA/water, etc., to name a few organic systems of fundamental interest.

Using the combined pulse radiolysis/ESR method the sodium clusters were studied in Zeolite systems [68]. It is interesting to note the identification of the Na_4^{3+} cluster as an electron trapped by a diamagnetic Na_4^{4+} cluster in Na-X, Na-Y and Na-Sodalite zeolites with the aborption maximum (λ_{max}) at ca. 500–560 nm. Similarly, only the Na_2^+ cluster was explained as a trapped electron observed in Na-A zeolite with $\lambda_{max} = 750$, and the Na_3^{2+} cluster was observed in Na-A and

Na-sodalite zeolite with $\lambda_{max} = 680$–700 nm. All kinetic data are in consistence with the decay of associated seven-line, 10-line and 13-line ESR spectra at various temperatures, as had been reported previously in a separate communication originating from the same group [66]. The quenching reaction with oxygen was then used to determine the position of the ion clusters; the sites inside the sodalite cage not being equally accessible to oxygen exposure.

3.5. SILVER CLUSTERS AND COMPLEXES IN ZEOLITE SYSTEMS

Silver atoms and combinations of silver atoms and ions seem to possess an outstanding ability to form various charged and neutral clusters. A complexity arises since these clusters apparently are stable under a wide range of conditions. For example: Ag^0, Ag_2^+ and Ag_3^{2+} have been reported in γ-irradiated systems of $AgNO_3$ and/or $AgCl_4$ diluted in benzene or toluene [69]. Ag_3^0, and Ag_5^0 metal clusters are formed by condensation of silver atoms together with benzene or N_2 on single crystals or on cold surfaces at 4.2 K [70]. In surface environments a similar variety of clusters can be stabilized. Such studies followed the pioneering work by Uytterhoven and Jacobs who dealt with the catalytic activity in Ag-exchanged zeolites in the end of the seventies [71]. The clusters or complexes can be formed by redox reactions induced by adsorbed hydrogen or oxygen in Ag-exchanged zeolites, or conveniently by irradiation of the same, or related. materials. We will start by shortly introducing the basic autoreduction of Ag-cations in zeolite systems and related redox couples. The more recent ESR work undertaken to investigate the detailed structure and reactions of Ag-clusters and cluster–adsorbate complexes, will thereafter be reviewed in more detail.

3.5.1. *Autoreduction of Ag Cations in Zeolites*

Vacuum thermal hydration of silver zeolite A causes reduction of silver ions. The mechanisms seem to have been clarified in pioneering work by Uytterhoven and Jacobs [71], although many details concerning the paramagnetic silver clusters have been debated thereafter. The reactions involve the autoreduction by zeolite water at low temperature:

$$2(Ag^+ZO^-) + H_2O \xrightarrow{300\text{--}525\ K} \frac{1}{2}O_2 + 2Ag^0 + 2ZOH$$

and lattice oxygen at somewhat elevated temperatures:

$$2(Ag^+ZO^-) \xrightarrow{400\text{--}650\ K} \frac{1}{2}O_2 + 2Ag^0 + ZO^- + Z^+$$

Here ZO^- denotes the zeolite lattice, ZOH the lattice hydroxyl, and, Z^+ a Lewis acid site.[11]

Reduction by hydrogen was shown to follow similar schemes with the stoichiometric relation between Ag and hydrogen as shown below:

$$Ag^+ - Ag^0 - Ag^+ + H_2 \xrightarrow{100\text{--}140\ °C} Ag^0 - Ag^0 - Ag^0 + 2H^+,$$

and the reversible reaction induced by oxygen treatment, summarized in a similar manner, as follows:

$$Ag^0 - Ag^0 - Ag^0 + 2H^+ + \frac{1}{2}O_2 \xrightarrow{330\,°C} Ag^+ - Ag^0 - Ag^+ + H_2O,$$

Through detailed studies of the far-infrared spectra associated with these reactions Ozin et al. were able to identify the particular silver clusters involved in the electron transfers represented above, and also clarify a reversibility of process using hydrogen and oxygen as mediators [72]. Parallel work at that time, combining optical absorption, mass spectroscopy, magnetic susceptibility, and ESR, concluded that the predominant charged silver clusters formed under the mild conditions as introduced above, are not paramagnetic [73], opposing the initial suggestions [74]. From this standpoint we now turn to the ESR studies of silver and silver ionic clusters, which still seem to attract considerable attention.

3.5.2. Identification of Paramagnetic Ag Clusters in Zeolites by ESR

As indicated in the foregoing section there exists a very large number of studies, and in several cases diverging conclusions, concerning the silver clusters stabilized in the zeolite systems. The first silver cluster known to be characterized by ESR is Ag_6^{n+} formed from H_2 reduction of Ag_{12}-A zeolite [75]. The cluster is characterized by an isotropic hf septet with hfsc 7.2 mT,[12] and seemed to be the same cluster as had been identified earlier by X-ray diffraction [76]. The X-ray work predicted the existence of eight additional Ag ions in each corner of a cube coordinating the octahedral Ag_6 cluster. This was beautifully confirmed in work by Morton and Preston who detected superhyperfine structure (ca. 0.55 mT) to this second shell in γ-irradiated samples of Ag-4A [77]. Similar clusters with six identical silver atoms in the first shell was also found by H_2 reduction of Ag_6-M^+A zeolite (M^+ = Na, K, Cs, Ca) [78], and by γ-irradiation of Ag_1-NaA [79]. Six-membered silver clusters appear to be the most stable as they are generally detected in most systems at elevated temperatures, regardless of generation method.

Clusters smaller than the hexa-silver cluster have been observed in some heterogeneous systems, and in those cases the formation has been initiated by γ-irradiation. For example, Ag^0 atoms have been observed, formed by irradiation in hydrated Ag-NaA at 4.2 K characterized by an ESR doublet of ca. 53 mT. At 77 K a hf doublet of ca. 71 mT was assigned to the same species [80]. A silver cluster assigned to Ag_4^{3+} was observed by warming a sample from 77 to 273 K [81].

In dehydrated Ag_{12}-A zeolite Preston and Morton (see above) detected in addition to the $Ag_6^+ \cdot 8Ag^+$ cluster both Ag_2^+ and Ag_3^{2+} clusters during the course of annealing reactions. Perhaps the most systematic study of irradiated silver loaded zeolites was carried out by Michalik and Kevan who studied dehydrated Ag_r-NaX, with r = 0.25, 1.0 and 5.0. In γ-irradiated samples they observed an Ag_3^0 cluster at 77 K for the two lowest ratios of Ag loading. In Ag_1-NaX the Ag_3^0 cluster was shown to rearrange into an Ag_6^+ cluster [79].[13] Only the Ag_6^+ was observed for highly loaded samples, irrespective of generation method. Various silver clusters were recently reinvestigated by Michalik and coworkers in detail, using Ag_1-NaA zeolites with various water contents [82]. Thus, the mechanisms behind the irradiation-induced formation of paramagnetic Ag clusters seem to have been clarified. It was concluded that in hydrated samples, silver clustering is determined by the mobility of Ag^0. During the dehydration, which usually takes place at high temperature (typically 373–600 K), the migration of silver ions and smaller clusters can be accountable for the agglomeration. This leads to the Ag_3^{2+} and Ag^+ clusters in the hydrated zeolites, and Ag_6^{n+} in dehydrated ones, notably, the latters always being diamagnetic, and becoming paramagnetic first upon chemical reduction or irradiation. The conclusions seem to be in agreement with the formation of organic Ag-complexes by adsorbing various molecular species such as ammonia and methanol [83].

3.6. METAL CLUSTERS IN SILICA CLATHRATES

Clathrate structures similar to the analogue gas and liquid hydrates are formed spontaneously upon thermal degradation of alkalisilicides and alkaligermanides (X—Si and X—Ge: X = Na, K, Rb, Cs) [84]. The interesting thing with these compounds is that the alkali metal atoms are trapped in cage structures, where typically 20–28 Si (or Ge) atoms forming the cage are arranged in a manner similar to C_{60} and other fullerenes. The fullerenes have been shown to possess superconductivity at rather high temperatures when doped by alkali atoms [85]. Its structural similarities with C_{60} makes more thorough investigations interesting, particularly since the Si-Na clathrates are known to undergo an insulator-to-metal transition upon gradually increasing the sodium content [86]. The search for high temperature superconductivity in these materials has as yet resulted in only negative results, however, the subject seems interesting enough to promote new research aiming on a further development of these materials [86].

The possibility to control the structure and alkali content make these materials also interesting for studies of cluster atoms and their electron transfer reactions with adsorbed organic molecules.

Acknowledgements

This work has been supported by the subsidy for Science Research of the Ministry of Education in Japan; Grant No. 04640473, and the Swedish Science Research Council. M. L. thanks JSPS (Japan Society for the Promotion of Science) for a postdoctoral scholarship.

Notes

1. At least locally at the site of the adsorbed molecule. The framework of cations and anions within the zeolite creates locally large electric fields which can drastically 'shift' the ionization potentials and electron affinities of an adsorbed molecule.
2. Abbreviation, hfsc: hyperfine splitting constant, and shortened versions will be used throughout.
3. Notation upon discussions of ESR spectra and radical structures: a and g denote isotropic values of hyperfine interaction and line position (g value). T_i, A_i and g_i ($i = 1, 2, 3$ or x, y, z) are the

principal values of the dipolar, hyperfine ($a + T_i$) and g value tensors. With axial symmetry index 1 = 2, designated as ⊥ (perpendicular); index 3 is here designated ∥ (parallel).
4. SOMO: Singly Occupied Molecular Orbital, HOMO: Highest Occupied Molecular Orbital, LUMO: Lowest Unoccupied Molecular Orbital.
5. The unpaired electron of perfluorobenzene radical cation ($C_6F_6^+$) in cC_6F_{12} was recently shown to occupy the b_{1g} orbital with D_{2h} symmetry. [A. Hasegawa, M. Shiotani, Y. Hama: *J. Phys. Chem.*, 1994, **98**, 1834].
6. We have stabilized the anion of the 1,2,4,5-tetrafluorobenzene by irradiation of a 2-MTHF glass doped with the solute at low temperature. This is a well known method to study molecular anion radicals. See A. Hasegawa in [1].
7. From the resolved hf splittings due to Na^+ the NH_2 radical was shown to interact with exchanged cations in the zeolite surface [5]. See Y. Suzumo, M. Shiotani, and J. Sohma: *Chem. Phys. Lett.* **44**, 177 (1976).
8. The g-value parameters for the Na_4^{4+}–O_2^- complex was in this case: $g_1 = 2.0016$; $g_2 = 2.0066$; $g_3 = 2.113$. Similar values were obtained from other M–O_2^- complexes (M = alkali metal or alkali metal cluster ions).
9. g-Values for metal particles formed from M_2 vapour in cationic M_1–X zeolite where $M_2 = M_1$.
10. As has been known for a long time, Kasai *et al.* found that air exposure leads to the formation of superoxide ions; water gives an unknown diamagnetic product.
11. We have used the notation of Ozin *et al.* to summarize these reactions [72].
12. Essentially the same parameters in all six-membered silver clusters. Small differences are observed depending on zeolite, temperature and generation method.
13. An alternative interpretation of the associated ESR spectrum was later given by Morton *et al.* (*J. Phys. Chem.* **91**, 2117–20 (1987)), who assigned the signal to an Ag_5^{4+} cluster. If this is more correct, it means that also the mechanisms suggested by Michalik and Kevan concerning the Ag_3^0 species formed by hydrogen reduction also must be changed accordingly.

References

1. A. Lund and M. Shiotani (Eds.): *Radical Ionic Systems. Properties in Condensed Phases*, Kluwer Academic Publishers, Dordrecht (1991).
2. L. B. Knight, Jr.: 'Generation and study of inorganic cations in rare gas matrices by electron spin resonance', in A. Lund and M. Shiotani (Eds.), *Radical Ionic Systems. Properties in Condensed Phases*, Kluwer Academic Publishers, Dordrecht (1991).
3. P. P. Edwards, L. L. Woodall, P. A. Anderson, A. R. Armstrong, and M. Slaski: *Chem. Soc. Rev.* 305 (1993).
4. K. B. Yoon: *Chem. Rev.* **93**, 321 (1993).
5. J. Sohma and M. Shiotani: *Magnetic Resonance in Colloidal and Interface Science*: ACS Symposium Series, **34**, 141 (1976).
6. M. Shiotani, F. Yuasa, and J. Sohma: *J. Phys. Chem.* **79**, 2669 (1975).
7. S. Kudo, A. Hasegawa, T. Komatsu, M. Shiotani, and J. Sohma: *Chem. Lett.* 705 (1973).
8. K. Toriyama, K. Nunome, and M. Iwasaki: *J. Am. Chem. Soc.* **109**, 4496 (1987).
9. L. B. Knight, Jr., J. Steadman, D. Feller, and E. R. Davidson: *J. Am. Chem. Soc.* **106**, 3700 (1984).
10. S. Kubota, M. Iwaizumi, and T. Isobe: *Bull. Chem. Soc. Jpn.* **44**, 2684 (1971).
11. T. Katsu, M. Yanagida, and Y. Fujita: *J. Phys. Chem.* **75**, 4064 (1971).
12. Y. Fujita, T. Katsu, M. Sato., and K. Takahashi: *J. Chem. Phys.* **61**, 4307 (1974).
13. K. Hatano, N. Shimamoto, T. Katsu, and T. Fujita: *Bull. Chem. Soc. Jpn.* **47**, 4 (1974).
14. (a) K. Toriyama, K. Nunome, and M. Iwasaki: *J. Chem. Phys.* **77**, 5891 (1982). (b) G. Dolivo and A. Lund: *J. Phys. Chem.* **89**, 3977 (1985). (c) M. Lindgren, A. Lund, and G. Dolivo: *Chem. Phys.* **99**, 103 (1985).
15. M. Shiotani, M. Lindgren, F. Takahashi, and T. Ichikawa: *Chem. Phys. Lett.* **170**, 201 (1990).
16. M. V. Barnabas, D. W. Werst, and A. D. Trifunac: *Chem. Phys. Lett.* **204**, 435 (1993).
17. M. V. Barnabas, D. W. Werst, and A. D. Trifunac: *Chem. Phys. Lett.* **187**, 565 (1991).
18. V. I. Melekhov, O. A. Anisimov, L. Sjöqvist, and A. Lund: *Chem. Phys. Lett.* **174**, 95 (1991).

19. M. Lindgren, M. Matsumoto, and M. Shiotani: *J. Chem. Soc., Perkin Trans. II*, 1397 (1992).
20. (a) V. A. Bol'shov and A. M. Volodin: *React. Kinet. Catal. Lett.* **43**, 87 (1991). (b) C. J. Rhodes: *J. Chem. Soc., Chem. Commun.* **13**, 900 (1991). (c) Ch. Gullyev, L. A. Surina, S. A. Surin, and G. D. Chukin: *Neftekhimiya* **30**(4), 492 (1990) (Russ).
21. V. A. Bol'shov and A. M. Volodin: *React. Kinet. Catal. Lett.* **46**, 337 (1992).
22. M. K. Carter and G. Vincow: *J. Chem. Phys.* **47**, 292 (1967).
23. T. R. Tuttle, and S. I. Weissman: *J. Am. Chem. Soc.* **80**, 3542 (1958).
24. O. Edlund, P.-O. Kinell, A. Lund, and A. Shimizu: *J. Phys. Chem.* **46**, 3679 (1967).
25. T. Komatsu, A. Lund, and P.-O. Kinell: *J. Phys. Chem.* **76**, 1721 (1972).
26. M. Iwasaki, K. Toriyama, and K. Nunome: *J. Chem. Soc., Chem. Commun.* 320 (1983).
27. R. Erickson, M. Lindgren, A. Lund, and L. Sjöqvist: *Colloids and Surfaces* **A:72**, 207 (1993).
28. (a) D. G. Streets and G. P. Ceaser: *J. Phys. Chem.* **66**, 257 (1977). (b) J. D. Clark and D. C. Frost: *J. Am. Chem. Soc.* **89**, 244 (1967).
29. J. C. Vedrine *et al.*: *J. Catalysis* **59**, 248 (1979).
30. D. N. Stamieres and J. Turkevich: *J. Am. Chem. Soc.* **86**, 749 (1964).
31. A. E. Hirchler, W. C. Neikam, D. S. Barmby, and R. L. James: *J. Catal.* **4**, 628 (1965).
32. S. Shih: *J. Catal.* **79**, 390 (1983).
33. T. Ichikawa, H. Yamaguchi, and H. Yoshida: *J. Phys. Chem.* **91**, 6400 (1987).
34. M. V. Barnabas and A. D. Trifunac: *J. Chem. Soc., Chem. Commun.* 813 (1993).
35. P. L. Corio and S. Shih: *J. Phys. Chem.* **75**, 3475 (1971).
36. C. J. Rhodes: *J. Chem. Soc., Faraday Trans.* **87**, 3179 (1991).
37. R. Crockett and E. Roduner: *J. Chem. Soc., Perkin Trans II* (1993).
38. C. J. Rhodes and M. Standing: *J. Chem. Soc., Perkin Trans II*, 1455 (1992).
39. M. V. Barnabas, D. W. Werst, and A. D. Trifunac: *Chem. Phys. Lett.* **206**, 21 (1993).
40. M. V. Barnabas and A. D. Trifunac: *Chem. Phys. Lett.* **193**, 298 (1992).
41. X. Z. Qin and A. D. Trifunac: *J. Phys. Chem.* **95**, 6466 (1991).
42. M. Shiotani and K. Ohta: unpublished results.
43. C. Periera, G. T. Kokotalio, and R. J. Gorte: *J. Phys. Chem.* **95**, 705 (1991).
44. S. D. Cox and G. D. Stucky: *J. Phys. Chem.* **95**, 710 (1991).
45. S. Bordiga, G. Ricchiardi, G. Spoto, D. Scarano, L. Carnrelli, A. Zecchina, and C. O. Arean: *J. Chem. Soc., Faraday Trans.* **89**, 1843 (1993).
46. M. Shiotani and K. Yamaji: unpublished results.
47. P. H. Kasai: *J. Chem. Phys.* **43**, 3322 (1965).
48. J. A. Rabo, C. L. Angell, P. H. Kasai, and V. Schomaker: *J. Chem. Soc., Disc. Faraday Soc.* **41**, 328 (1966).
49. P. H. Kasai and R. J. Bishop, Jr.: *J. Phys. Chem.* **77**, 2308 (1973).
50. P. H. Kasai and R. J. Bishop, Jr.: *J. Am. Chem. Soc.* **94**, 5560 (1972).
51. P. P. Edwards, M. R. Harrison, J. Klinowski, S. Ramdas, J. M. Thomas, D. C. Johnson, and C. J. Page: *J. Chem. Soc., Chem. Commun.* 982 (1984).
52. M. R. Harrison: *J. Solid. State Chem.* **54**, 330 (1984).
53. R. E. H. Breuer, E. de Boer, and G. Geismar: *Zeolites* **9**, 336 (1989).
54. P. A. Anderson, R. J. Singer, and P. P. Edwards: *J. Chem. Soc., Chem. Commun.* 914 (1991).
55. P. A. Anderson and P. P. Edwards: *J. Chem. Soc., Chem. Commun.* 915 (1991).
56. Y. Kim, Y. W. Han, and K. Seff: *J. Phys. Chem.* **97**, 12663 (1993).
57. S. H. Song, U. S. Kim, Y. Kim, and K. Seff: *J. Phys. Chem.* **96**, 10937 (1992).
58. S. H. Song, Y. Kim, and K. Seff: *J. Phys. Chem.* **95**, 9919 (1991).
59. M. S. Jeong, Y. Kim, and K. Seff: *J. Phys. Chem.* **97**, 10139 (1993).
60. L. R. M. Martens, P. J. Grobet, and P. A. Jacobs: *Nature* **315**, 568 (1985).
61. B. Xu and L. Kevan: *J. Phys. Chem.* **96**, 2642 (1992).
62. B. Xu, X. Chen, and L. Kevan: *J. Chem. Soc., Faraday Trans.* **87**, 3157 (1991).
63. B. Xu and L. Kevan: *J. Phys. Chem.* **96**, 3647 (1992).
64. K. B. Yoon and J. K. Kochi: *J. Chem. Soc., Chem. Commun.* 510 (1988).
65. G. A. Kuranova: *High Energy Chem.* **91**, 25 (1992).
66. X. Liu and J. K. Thomas: *Chem. Phys. Lett.* **192**, 555 (1992).
67. P. A. Andersson, D. Barr, and P. P. Edwards: *Angew. Chem. Intern. Ed. Engl.* **30**, 1551 (1991).
68. K.-K. Iu, X. Liu, and J. K. Thomas: *J. Phys. Chem.* **97**, 8165 (1993).

69. (a) L. Shields: *Trans. Faraday. Soc.* **62**, 1042 (1966). (b) C. E. Forbes and M. C. R. Symons: *Mol. Phys.* **27**, 467 (1974). A. D. Stevens and M. C. R. Symons: *Chem. Phys. Lett.* **109**, 514 (1984).
70. (a) J. A. Howard, K. F. Preston, and B. Mile: *J. Am. Chem. Soc.* **103**, 6226 (1981). (b) J. A. Howard, R. Sutcliffe, and B. Mile: *J. Phys. Chem.* **87**, 2268 (1983). K. Kernisant, G. A. Thompson, and D. M. Lindsay: *J. Chem. Phys.* **82**, 4739 (1985).
71. (a) H. K. Beyer, P. A. Jacobs, and J. B. Uytterhoeven: *J. Chem. Soc., Faraday Trans.* **72**, 674 (1976). (b) P. A. Jacobs, J. P. Linart, H. Nijs, J. B. Uytterhoeven, and H. K. Beyer: *J. Chem. Soc., Faraday Trans.* **73**, 1745 (1977). (c) P. A. Jacobs, J. B. Uytterhoeven, and H. K. Beyer: *J. Chem. Soc., Faraday Trans. I* **73**, 1755 (1977). (d) P. A. Jacobs and J. B. Uytterhoeven: *J. Chem. Soc., Faraday Trans. I* **75**, 56 (1979). (e) L. R. Gellens, W. J. Mortier, R. A. Schoonheydt, and J. B. Uytterhoeven: *J. Phys. Chem.* **85**, 2783 (1981). (f) L. R. Gellens, W. J. Mortier, and J. B. Uytterhoeven: *Zeolites* **1**, 11 (1981).
72. (a) G. A. Ozin, M. D. Baker, and J. Godber: *J. Phys. Chem.* **88**, 4902 (1984). (b) M. D. Baker. G. A. Ozin, and J. Godber: *J. Phys. Chem.* **89**, 305 (1985).
73. J. Texter, R. Kellermann, and T. Gonsiorowski: *J. Phys. Chem.* **90**, 2118–24 (1986).
74. L. R. Gellens, W. J. Mortier, R. A. Schoonheydt, and J. B. Uytterhoeven: *J. Phys. Chem.* **85**, 2783 (1981).
75. D. Hemmerschmidt and R. Haul: *Ber. Bunsen-Ges. Phys. Chem.* **84**, 902 (1980).
76. Y. Kim and K. Seff: *J Am. Chem. Soc.* **100**, 6989 (1978).
77. (a) J. R. Morton and K. F. Preston: *J. Magn. Reson.* **68**, 121 (1986). (b) J. R. Morton and K. F. Preston: *Zeolites* **7**, 2 (1987). (c) J. R. Morton, K. F. Preston, A. Sayari, and J. S. Tse: *J. Phys. Chem.* **91**, 2117 (1987).
78. (a) P. Grobet and R. A. Schoonheydt: *Surf. Sci.* **156**, 893 (1985). (b) R. A. Schoonheydt and H. Leeman: *J. Phys. Chem.* **93**, 2048 (1989).
79. J. Michalik and L. Kevan: *J. Am. Chem. Soc.* **108**, 4247 (1986).
80. M. Narayana, A. S. W. Li, and L. Kevan: *J. Phys. Chem.* **85**, 132 (1981).
81. M. Narayana and L. Kevan: *J. Chem. Phys.* **76**, 3999 (1982).
82. (a) A. van der Pol, E. J. Reijerse, E. de Boer, T. Wasowicz, and J. Michalik: *Mol. Phys.* **75**, 37 (1992). (b) T. Wasowicz and J. Michalik: *Radiat. Phys. Chem.* **37**, 427 (1991).
83. (a) J. Michalik, A. van der Pol, E. J. Reijerse, T. Wasowicz, and E. de Boer: *Appl. Magn. Reson.* **3**, 19 (1992). (b) J. Michalik, T. Wasowicz, J. Sadlo, and A. van der Pol: *Colloids and Surfaces A* **72**, 81 (1993).
84. (a) C. Cros and J.-C. Benejat: *Bull. Soc. Chim. France* **5**, 1739 (1972). (b) C. Cros, M. Pouchard, and P. Hagenmuller: *J. Sol. State. Chem.* **2**, 570 (1970).
85. A. F. Hebard, M. J. Rosseinsky, R. C. Haddon, D. W. Murphy, S. H. Glarum, T. T. Palstra, A. P. Ramirez, and A. R. Kortan: *Nature* **350**, 600 (1991).
86. S. B. Roy, K. E. Sim, and A. D. Caplin: *Philisophical Magazine B* **65**, 1445 (1992).

Photostimulated Formation of Radicals on Oxide Surfaces

ALEXANDER M. VOLODIN, VADIM A. BOLSHOV and
TATIANA A. KONOVALOVA
Boreskov Institute of Catalysis, Prosp. Akad. Lavrentieva 5, Novosibirsk, 630090, Russia.

(Received: 15 December 1993; accepted: 5 July 1994)

Abstract. We present results of *in situ* EPR investigations of the mechanism of photostimulated processes resulting in radical and ion-radical particle formation on the surfaces of oxide dielectrics (magnesium, calcium, aluminum oxides, zeolites). Three types of reactions are discussed:
1. Formation of oxygen anion-radicals on MgO and CaO surfaces.
2. Formation of benzene cation-radicals on ZSM-5 zeolites.
3. Formation of radical particles from aromatic nitrocompounds adsorbed on alumina.

On the basis of investigation of the spectral relationships and the properties of surface active centre, it is concluded that light is absorbed by coordinatively unsaturated surface sites in the first system, whereas in the other processes, electron donor–acceptor (EDA) complexes between adsorbed molecules and surface active sites are supposed to be key intermediates. These EDA complexes are shown to incorporate donor solvent molecules as well. In this case the energetic characteristics of the photoprocesses are substantially determined by the ionization potential of solvent molecules.

Mechanisms of photo- and thermostimulated processes are compared and possible similarities are discussed for all the reactions studied.

Key words. Radicals, ion-radicals, surfaces, magnesium oxide, calcium oxide, aluminum oxide, zeolites, benzene, nitrobenzene.

1. Introduction

Photoinduced reactions proceeding at the gas–solid interface can proceed both in the area of band-to-band absorption of the adsorbents and at photon energies which are substantially smaller than their band gap [1–18]. In the latter case there is particular interest in processes in which light is absorbed either by surface defects of the solid or by compounds adsorbed or impregnated on the surface. Here, these processes will be called photoprocesses in the field of the adsorbent surface absorption.

It should be noted that not all photostimulated reactions induced by photon energies smaller than the band gap are connected with the surface absorption of light: an admixture centre participating in the absorption can also be situated in the near-surface layer and in the bulk. It is its accessibility to gas phase molecules that seems to us to be the sole criterion of the surface position of the centre. This means that the centre can be created, destroyed or modified by adsorption processes under mild conditions, when one may neglect the defect diffusion into the bulk.

Optical methods are widely used to obtain information about mechanisms of photostimulated processes on oxide surfaces. But neither diffuse reflectance spectroscopy nor luminescence studies unambiguously reveal the nature of the absorption centre and reaction intermediates. This ambiguity is clearly manifested in the interpretation of observed absorption bands for magnesium oxide. In several papers [4–8] these bands have been connected with the existence of surface sites of low coordination $(Mg^{2+}-O^{2-})_{LC}$, whereas other authors [9–11] attribute them to surface and near-surface F^+ centres.

The present paper presents a mechanistic analysis of the elementary stages of the photoprocesses in the field of oxide surface absorption, resulting in charge separation and formation of radical and ion-radical particles stabilized on the surface. Investigation of the nature of electron and hole centres resulting from the charge separation not only enables us to understand their further chemical transformation pathways, but also to obtain information about the nature and properties of primary surface structures responsible for the absorption of light.

Three types of photostimulated reactions resulting in paramagnetic particle formation on the surface of oxide dielectrics will be discussed:

1. Formation of oxygen anion-radicals on MgO and CaO surfaces.
2. Formation of benzene cation-radicals on ZSM-5 zeolites.
3. Formation of radical particles from aromatic nitrocompounds adsorbed on *alumina*.

These processes are substantially different in terms of the nature of the light absorbing centres. In the first case light is absorbed by surface structural defects of alkaline earth oxides (coordinatively unsaturated sites $(Me^{2+}-O^{2-})_{LC}$) [12–14]. In the two other cases electron donor–acceptor (EDA) complexes emerging after adsorption of gas molecules on the surface active sites are responsible for the absorption [15, 16]. In such a case the energetic characteristics of the photoprocesses depend to a great degree on the properties of the adsorbed molecules.

It should be noted that a large number of mechanistic investigations of photo-induced processes on the surface of oxide dielectrics have already been conducted; the majority of the studies used luminescent methods or the diffuse reflectance technique. In spite of the fact that radical intermediates are often detected in the photostimulated processes, there have been scarcely any systematic investigations using EPR. One can suppose that such studies would give new information on the structure of the intermediates and the nature of light absorbing centres as well as on the mechanism of photostimulated processes on the surface.

1.1. EXPERIMENTAL SECTION

Experimental results were obtained on an *in situ* EPR installation based on ERS-221 spectrometer. Figure 1 shows the cavity block of the installation.

A quartz ampoule containing a catalyst and connected to a high-vacuum apparatus is placed in the spectrometer resonator. The catalyst temperature (90–835 K) is controlled by a special thermostat during the course of pretreatment and experiments. The vacuum apparatus allows prolonged pretreatment and gas adsorption, both under dynamic vacuum (10^{-7} torr) and under controlled pressure (up to 800 torr). The presence of a monochromator mounted in the lighting device allows the study of spectral characteristics of photostimulated processes in the range of

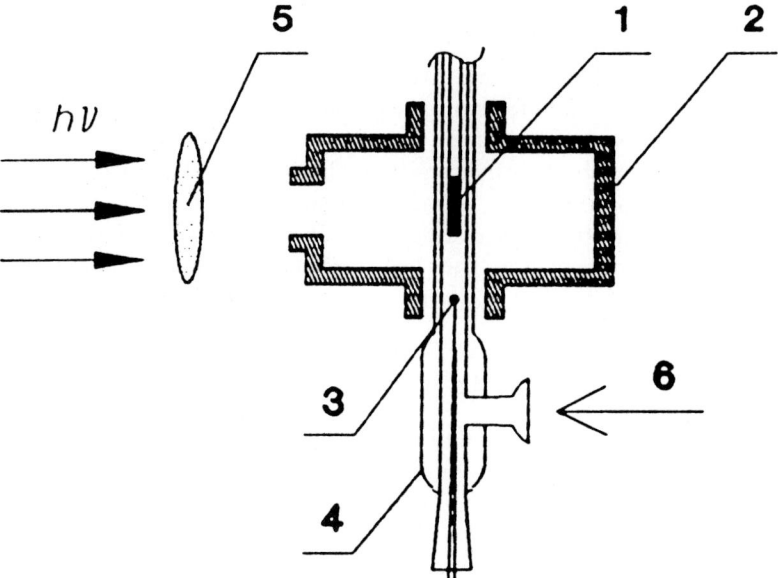

Fig. 1. Cavity block of the experimental installation. 1– ampoule with a catalyst; 2– cavity of the ERS-221 spectrometer; 3– control thermocouple; 4–quartz Dewar tube; 5– quartz lens; 6– stream of nitrogen vapour from the thermostat.

quantum energies 1–5 eV. This installation permits ESR spectra to be recorded and other information to be obtained on the state of the adsorption layer after any treatment step, as well as during the course of a catalytic process. Sample treatment, conducted to clean the surface of organic pollutants, water and CO_2, consisted in alternating heating in oxygen (0.1–10 Torr) and vacuum (10^{-7}–10^{-6} Torr) at 700–800 K. After each experiment the sample was regenerated as described above. The regeneration procedure and characteristics of the experimental complex gave results that were reproducible to within ±10% of the radical particle concentration.

Detailed descriptions of the sample pretreatment and experimental method are given elsewhere [12–16].

2. Oxygen Anion-Radical Formation on MgO and CaO Surfaces

2.A. MgO

Photoinduced processes for magnesium oxide, as for many other oxides, can be induced with photon energies substantially smaller than the band gap. For titanium [17], zinc [18] and tin [19] oxides these processes have been shown to proceed via electron transfer from the $2p$ orbitals of an O^{2-} surface anion to a metal cation, followed by charge separation and stabilization of the electron and hole centres:

$$(Me^{n+}-O^{2-}) \stackrel{h\nu}{\rightleftarrows} (Me^{(n-1)+}-O^{-})^* \tag{1}$$

$$(Me^{(n-1)+}-O^-)^* \to [e]_{st} + O_{st}^- \qquad (2)$$

The charge separation and stabilization can be caused by oxide surface potential fields [18] or by the chemical interaction of electron or hole centres with adsorbed molecules moving the equilibrium in reaction (2) to the right [13, 17, 19].

For the magnesium oxide surface, stage 1 can be reliably considered to proceed under illumination with photon energies smaller than the band gap, the absorption being caused by coordinatively unsaturated surface sites [4–8]. Bands at around 230 nm and 274 nm are easily distinguished in the photoluminescence excitation and diffuse reflectance spectra. They are attributed by the authors to the excitation of surface pairs $(Mg^{2+}-O^{2-})_{LC}$ with four- and three-coordinated anions O_{LC}^{2-}, respectively:

$$(Mg^{2+}-O^{2-})_{LC} \rightleftarrows (Mg^{2+}-O^{2-})_{LC}^* \qquad (3)$$

So, under certain conditions one can expect stabilized electron and hole centres to form an MgO surface under illumination in the field of surface absorption due to the excitation of $(Mg^{2+}-O^{2-})_{LC}$ pairs (stage 1) with further charge separation (stage 2). Electron centres may be revealed as reduced metal cations, F^+ centres, etc., or as products of their interaction with adsorbed molecules (O_2^- in the case of O_2, CO_2^- in the case of CO_2, etc.). Hole centres may be revealed either as O_{st}^- anion-radicals or as products of their interaction with adsorbed molecules ($[O^- \cdot O_2]$ in the case of O_2, CO_2^- in the case of CO, etc.). Parallel with the hole centres O_{st}^-, one may expect O_{ads}^- anion-radicals to result from the adsorption of an oxygen atom on an electron centre [13]:

$$[e] + O \rightleftarrows O_{ads}^- \qquad (4)$$

Generally, O_{st}^- and O_{ads}^- anion-radicals may differ both in radiospectroscopic characteristics and in reactivity. Their destruction mechanism should be substantially different. A hole center O_{st}^- should be destroyed by a temperature rise due to a recombination process, regenerating O^{2-} anion, whereas O_{ads}^- anion-radical destruction should be accompanied by oxygen desorption (reverse reaction (4)) followed by electron recombination.

It should be noted that the efficiency of the photoformation of stabilized centres is very low in vacuum [13, 20]. That is why preliminary photoreduction of the MgO surface in hydrogen followed by the formation of stabilized electron centres (F^+ – centres) with their further interaction with N_2O is usually applied to obtain a relatively high concentration of the anion-radicals [21–23]. The reaction follows the mechanism:

$$[e]_{st} + N_2O \to O_{ads}^- + N_2 \uparrow \qquad (5)$$

O_{ads}^- electron centres are assumed to form on the surface.

High concentrations of oxygen anion-radicals on the MgO surface were also shown to result from UV illumination in the presence of molecular oxygen without preliminary surface photoreduction [24]. The reaction mechanism proposed by the authors involves adsorption of an oxygen atom emerging from the oxygen photodissociation on an electron centre, i.e. the formation of O_{ads}^- electron centres.

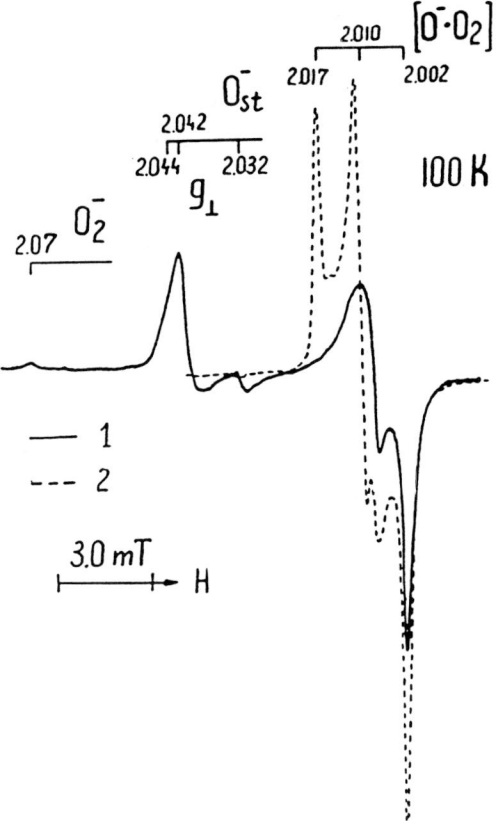

Fig. 2. (1) EPR spectrum obtained after illumination of MgO at 100 K and $P_{O_2} = 0.03$ Torr ($h\nu = 4.43$ eV) with subsequent evacuation at 323 K; (2) After additional exposure in O_2 (10^{-3} Torr, 1 min) at 100 K and subsequent evacuation. Temperature of recording EPR spectra $T = 100$ K.

The fate of the hole centres necessarily formed under UV irradiation is not discussed in the paper.

We have shown earlier [17–19] that enhancement of the charge separation efficiency followed by the stabilization of the electron and hole centres on the surface, according to the schemes (1) and (2), can be observed upon illumination of oxide semiconductors (TiO_2, ZnO and SnO_2) in the presence of acceptor oxygen molecules.

Here we show the results of the mechanistic investigation of the formation and destruction of the electron and hole centres on the MgO surface under illumination and analyze the results obtained on the basis of the approach developed in References [13, 14, 17–19].

Figure 2 presents typical EPR spectra appearing after the illumination of MgO in the presence of oxygen. The hole centres are revealed (in the presence of oxygen in the gas phase) as $[O^-_{st} \cdot O_2]$ complexes ($g_1 = 2.017$; $g_2 = 2.010$; $g_3 =$

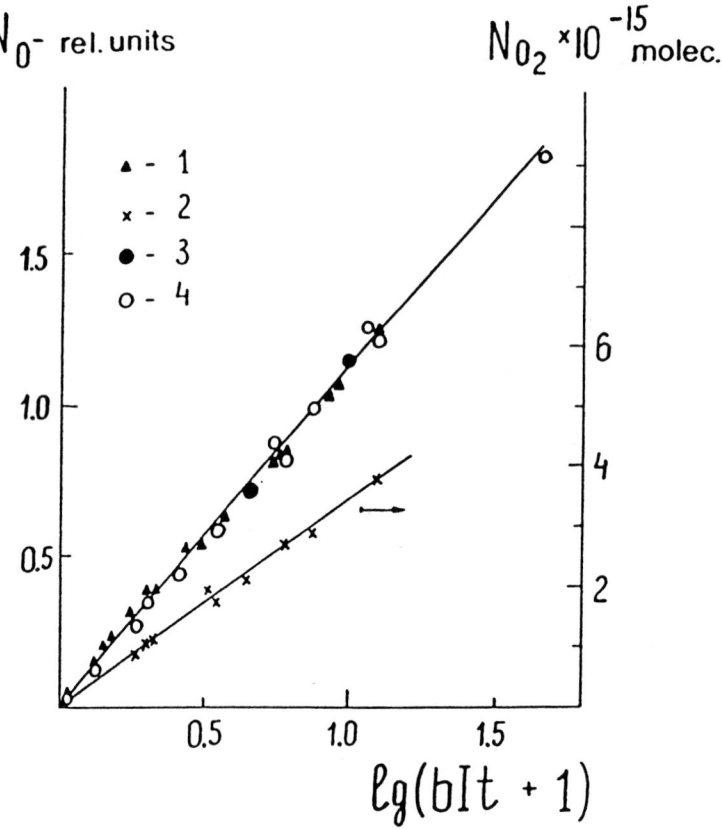

Fig. 3. Kinetics of: oxygen photoadsorption at 298 K and gas phase pressure 0.04 Torr (1); accumulation of strongly photosorbed oxygen not removable by evacuation at 298 K (2); accumulation of $(O_{st}^-)_{3C}$ anion-radicals under illumination in the presence of 10 Torr O_2 (3) and 0.04 Torr O_2 (4). ($h\nu = 4.43$ eV, $I = 2.5 \cdot 10^{14}$ photon/s, $bI = 8 \cdot 10^{-4}$ s^{-1}).

2.002). Upon evacuation at 298 K they are reversibly transformed into O_{st}^- anion-radicals ($g_\perp^1 = 2.042$; $g_\perp^2 = 2.033$; $g_\perp^3 = 2.044$) according to the reaction:

$$O_{st}^- + O_2 \rightleftarrows [O_{st}^- \cdot O_2], \tag{6}$$

The electron centres manifest themselves in the form of O_2^- anion-radicals (Figure 2).

Kinetic dependencies of the O_{st}^- accumulation under irradiation in the field of the surface absorption of MgO are shown (Figure 3) to lie satisfactorily on a line in the coordinates of the equation suggested by Solonitzyn [25]:

$$N(t) = A \cdot \log(bIt + 1) \tag{7}$$

the kinetic curves of the oxygen photoadsorption lie on the same linear coordinates with the same b values (Figure 3).

The efficiency of electron and hole centre formation ceases to increase at oxygen

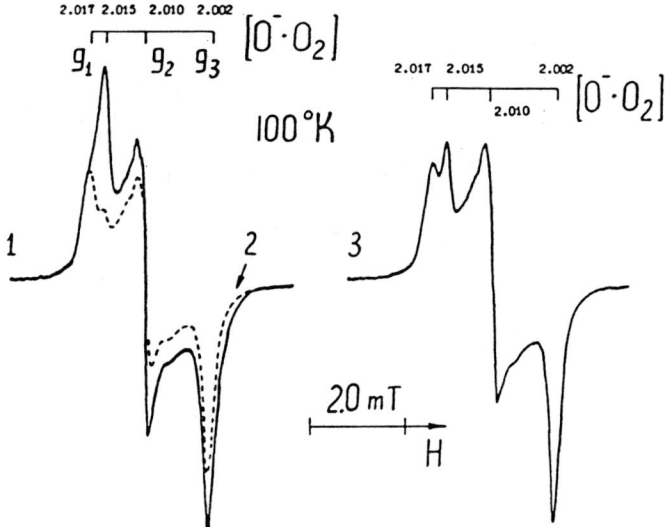

Fig. 4. (1) EPR spectra obtained under illumination of MgO at 100 K and $P_{O_2} = 0.03$ Torr ($h\nu = 4.43$ eV) with subsequent evacuation at 100 K. (2) After additional evacuation at 200 K. (3) After exposure in O_2 at 100 K. Temperature of recording EPR spectra: $T = 100$ K.

pressures as low as 10^{-3} Torr. Further increase of the oxygen gas phase pressure does not result in the enhancement of the number of the stabilized hole (O_{st}^-) and electron (O_2^-) centres, as can be seen from Figure 3. Thus, on increasing the oxygen pressure above 10^{-3} Torr, the collision frequency between the oxygen molecules and the adsorption centres does not limit the charge separation process. This means that the rate of the reaction:

$$(Mg^+-O^-)^* + O_2 \to O_2^- + O_{st}^- \tag{8}$$

is higher than the rate of the recombination reaction (3). In such a case, the minimum lifetime of the excitation before the recombination, using the Eley–Rideal mechanism for the reaction (8), can be estimated to be $\approx 10^{-3}$ s if reaction (8) proceeds with the probability ≈ 1 upon collision of an oxygen molecule with an adsorption centre. The experimental value of the excitation lifetime for MgO is much greater than the analogous characteristics for TiO_2 and SnO_2 [17–19] and is close to the one for the excitation of a surface complex with participation of low coordinated sites on the MgO surface, according to References [8, 26].

The EPR spectra in Figure 2 and the kinetic relationships in Figure 3 assume oxygen anion-radicals to be stable at room temperature (290–300 K). If the illumination of the initial MgO samples in O_2 is carried out at 100 K with subsequent evacuation and spectrum recording at the same temperature, the formation of a new type of complex (hereafter designated as $[X \cdot O_2]$) with $g_1 = 2.015$, $g_2 = 2.010$ and $g_3 = 2.002$ is observed (Figure 4, spectrum 1).

Heating and brief evacuation at 200 K are accompanied by the practically complete destruction of $[X \cdot O_2]$ complexes (Figure 4, spectrum 2). In this case, however, about half the centres are preserved and subsequent adsorption of oxygen partly restores their number (Figure 4, spectrum 3). But our attempt to detect X centers in EPR spectra gave no results [12]. Thus $[X \cdot O_2]$ complexes with $g_1 = 2.015$ can be treated as a probe for the occurrence of their stabilization centres, which are not observed in EPR spectra. It is most likely that the X centres are hole centers O^-_{st}, whose EPR spectra cannot be observed under our experimental conditions.

The established X centres on MgO possess some properties that are typical of the hole centres O^-_{st}:

1. They are formed under illumination and destroyed with increasing temperature together with electron centres (F-centres, O_2^-).
2. They react with H_2 and CH_4 molecules even at 90 K.
3. They interact with molecular oxygen in the reaction:

$$X + O_2 \rightleftarrows [X \cdot O_2]$$

to produce complex $[X \cdot O_2]$, observed by EPR.
4. They interact with CO in the reaction:

$$X + CO \rightarrow CO_2^-$$

to produce anion-radical CO_2^-, observed by EPR.

Partial destruction of the X centres with increasing temperature (Figure 4, spectra 1 and 3) is due to a recombination process. Their initial concentration can be regenerated by repeated illumination at 100 K in O_2.

The existence of absorption by two types of surface centers with peak maxima at $\lambda = 274$ and 230 nm is detected in both luminescence excitation and diffuse reflectance spectra of powdered MgO. According to References [4–8], this absorption is due to the excitation of charge transfer complexes for low-coordinated surface anion O^{2-}_{LC} (Equation 3). In this case the bands at $\lambda = 274$ and 230 nm are associated with absorption by three- (O^{2-}_{3C}) and four-coordinated (O^{2-}_{4C}) anions, respectively.

Thus, the formation of hole centers (anion-radicals O^-_{st} and X centres) stabilized on MgO under illumination in the region of MgO surface absorption can be suggested to follow the same mechanism (Equation 3) for complexes $(Mg^{2+}-O^{2-})_{3C}$ and $(Mg^{2+}-O^{2-})_{4C}$. For example, the most effective formation of centers is observed at $h\nu = 5.0$ eV, close to the spectral characteristic for the formation of X and F centres, and correlates with the absorption band of O^{2-}_{4C}. The most effective formation of O^-_{st} centres is observed at $h\nu = 4.43$ eV and correlates with the absorption band of O^{2-}_{3C}.

In accordance with these results, we can suggest the following mechanism for the formation of the surface anion-radicals O^-_{st} on MgO (giving typical EPR spectra) under illumination:

$$(Mg^{2+}-O^{2-})_{3C} \stackrel{h\nu_1}{\rightleftarrows} (Mg^+-O^-)^*_{3C} \rightarrow [e]_{st} + (O^-_{st})_{3C} \qquad (9)$$

A similar mechanism can be suggested for the formation of hole centers X (not detected by EPR):

$$(Mg^{2+}-O^{2-})_{4C} \overset{h\nu_2}{\rightleftarrows} (Mg^+-O^-)^*_{4C} \to [e]_{st} + (O^-_{4C})_{st} \tag{10}$$

Here, $(O^-_{st})_{3C}$ and $(O^-_{st})_{4C}$ correspond to O^-_{st} and X centres, respectively. Since, in the above schemes, not only the initial states $((Mg^{2+}-O^{2-})_{3C}$ and $(Mg^{2+}-O^{2-})_{4C})$ but also the reaction products $((O^-_{st})_{3C}$ and $(O^-_{st})_{4C}$, respectively) are different, charge separation on MgO surface can be suggested to be due to the electron transfer, the hole centre being localized at a site of the corresponding surface defect.

The existence of two types of surface stabilized hole centres on MgO with different spectral characteristics for their formation and essentially different regions of their thermal stability, indicates that the products of their interactions can be different not only with O_2 molecules but also with the other reactants. In all cases one must take into account that not only the observed hole centers but also those that are not detected in EPR spectra can take part in the reactions. In several cases (high quantum energies, low temperatures), they can play a decisive role.

2.B. CaO

Oxygen adsorption on the initial CaO sample does not lead to appearance of a noticeable amount of radical products. Irradiation of this sample at 300 K (λ = 303 nm) in the presence of gaseous oxygen leads to the formation of oxygen radicals stabilized on the surface. Their EPR spectrum is represented in Figure 5 (spectrum a). In this case manometric measurements show oxygen photoadsorption. Evacuation of the oxygen at 300 K only slightly affects the number of paramagnetic centres (Figure 5, spectrum a), indicating a sufficiently high bond strength between the observed forms of photo-adsorbed oxygen and CaO surface.

The analysis of the formation conditions, thermal stability and kinetic relationships for the photostimulated formation of oxygen radicals has permitted us to discriminate reliably between three types of essentially different complexes with molecular oxygen giving the EPR spectra illustrated in Figure 5. These are the anion-radicals O_2^- with EPR spectral parameters $g_x = 2.003$ and $g_y = 2.007$ (g_z cannot be observed under our conditions due to its low intensity). These anion-radicals can be obtained in practically pure form after evacuation of the sample at 360 K, giving spectrum a (Figure 5). Along with the formation of O_2^-, photoadsorption of oxygen at 300 K is accompanied by the production of molecular oxygen complexes with hole centres (hereafter designated as X) $[X \cdot O_2]$ (Figure 5) with $g_1 = 2.002$, $g_2 = 2.009$, $g_3^1 = 2.015$ and $g_3^2 = 2.016$ [27, 28]. The X centres themselves are not detected by EPR method under our conditions ($T = 90\text{--}350$ K). Evacuation at 350 K leads to oxygen desorption from these complexes due to a shift of the reaction equilibrium to the right:

$$[X \cdot O_2] \rightleftarrows X + O_2$$

and hence to the disappearance of the signal from $[X \cdot O_2]$ complexes from the EPR spectra. At this temperature the X centres themselves are sufficiently stable,

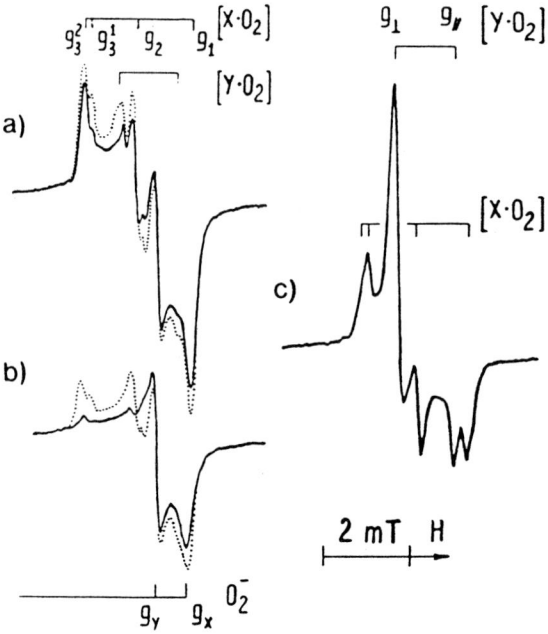

Fig. 5. (a) ESR spectrum obtained after irradiation of CaO at 300 K and P_{O_2} = 0.1 Torr (λ = 303 nm) with further 20 min. evacuation at 300 K (solid lines) and after additional exposure in O_2 (0.03 Torr, 30 s) at 140 K (dots).
(b) After 30 min evacuation of sample at 353 K giving spectrum a (solid line) and after additional exposure in O_2 (0.03 Torr, 30 s) at 140 K (dots).
(c) Spectrum obtained after irradiation of initial CaO sample at 140 K and P_{O_2} = 0.03 Torr (λ = 303 nm). Recording temperature of all ESR spectra was 140 K.

and further exposure of the sample in O_2 regenerates about a half of their number (Figure 5, spectrum b).

Photoadsorption of O_2 at lower temperatures leads to the formation of a new type of complex with molecular oxygen, hereafter designated as [Y · O_2] (Figure 5, spectrum c) with EPR parameters: $g_\parallel$ = 2.004 and $g_\perp$ = 2.012. The enthalpy of formation of this complex is considerably lower than for [X · O_2], hence its decomposition under evacuation starts already at 200 K. At this temperature Y centres are stable and further adsorption of O_2 regenerates the initial spectrum once again. The Y centres themselves, as well as the hole centres X, are not observed in the EPR spectra at 90–200 K. The Y centres are much less thermally stable than the X centres, and even after heating at 300 K their concentration is insignificant (Figure 5, spectrum a). Thermal decomposition of all the photo-induced centres with increasing temperature takes place due to recombination processes and is accompanied by regeneration of the initial state. Repeated irradiation at the relevant temperatures once again leads to the formation of all the above paramagnetic centres. Note that complexes with molecular oxygen with EPR parameters close to the above values are formed during O_2 adsorption on

FORMATION OF RADICALS ON OXIDE SURFACES

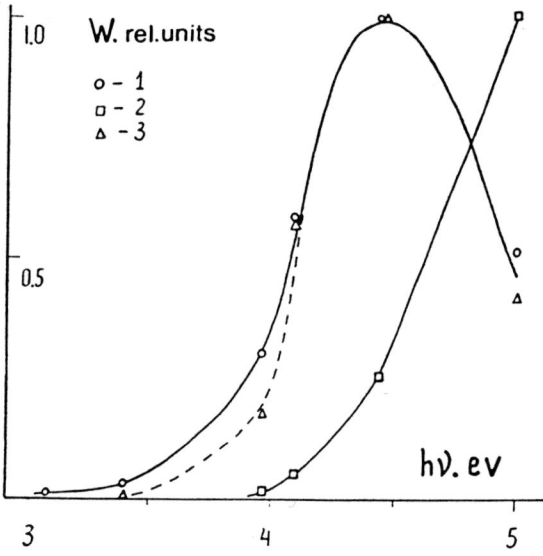

Fig. 6. Spectral relationships of the oxygen photoadsorption initial rates at 298 K and O_2 pressure 0.03 Torr over CaO (1) and MgO (2) and of the $[Y \cdot O_2]$ complex photoformation over CaO at 200 K and O_2 pressure 0.03 Torr. Initial rate $W = 1.0$ corresponds to the O_2 photoadsorption quantum yield equal to 2% for MgO and 0.5% for CaO.

the thermally activated surface of CaO and are known from the literature [29–30].

It is important to emphasize that not all the types of surface electron and hole centres produced under irradiation of the CaO sample in vacuum are detected by EPR. Information about their generation and number can be obtained from the formation of their complexes with molecular oxygen, giving the above typical EPR spectra.

The spectral dependences of the initial rate of the manometrically recorded O_2 photoadsorption and of photoinduced formation of the oxygen radical forms are presented in Figure 6. They are moved to substantially smaller photon energies in comparison with the case of MgO (Figure 6, curve 2). This is connected with the smaller band gap value for CaO and consequently with smaller photon energies necessary for the realization of the photoprocesses with participation of the low coordinated surface ions.

Studies of the chemical properties of the X and Y centres show that they are highly reactive, reacting with CO, H_2 and CH_4 molecules even at 100–150 K, whereas their complexes with molecular oxygen do not react with these molecules at these temperatures.

The anion-radicals CO_2^- with typical EPR spectra result from the reaction of the X and Y centres with CO molecules. It should be noted that the conditions for formation and decomposition, thermal stability, and EPR spectra for complexes $[X \cdot O_2]$ and $[Y \cdot O_2]$ correlate with the corresponding characteristics for surface complexes $[O_{3C}^- \cdot O_2]$ and $[O_{4C}^- \cdot O_2]$ on MgO.

TABLE I
Parameters of EPR spectra for different types of O⁻ anion-radicals and [O⁻ · O₂] complexes on the MgO and CaO surfaces.

	O_{3C}^-		$[O_{3C}^- \cdot O_2]$			O_{4C}^-	$[O_{4C}^- \cdot O_2]$		
	$g_\perp$	$g_\parallel$	g_1	g_2	g_3		g_1	g_2	g_3
MgO	2.033–2.044	2.002	2.017	2.01	2.002	n.o.	2.015	2.01	2.002
CaO		n.o.	2.015–2.016	2.009	2.002	n.o.	$g_\perp = 2.012$	$g_\parallel = 2.004$	

n.o. = anion-radicals not observable EPR.

Table I presents EPR parameters of these complexes and centres of their stabilization (O_{3C}^- & O_{4C}^-) on the surface of magnesium and calcium oxides.

Thus, the formation of adsorbed oxygen radicals on CaO under irradiation in the range of surface absorption, can be suggested to follow the mechanism proposed previously for MgO with the participation of low-coordinated surface sites:

$$(Ca^{2+}-O^{2-})_{3C} \underset{}{\overset{h\nu_1}{\rightleftarrows}} (Ca^+-O^-)^*_{3C} \rightarrow [e]_{st} + (O_{st}^-)_{3C} \tag{11}$$

$$(Ca^{2+}-O^{2-})_{4C} \underset{}{\overset{h\nu_2}{\rightleftarrows}} (Ca^+-O^-)^*_{4C} \rightarrow [e]_{st} + (O_{4C}^-)_{st} \tag{12}$$

In the presence of molecular oxygen, electron centres manifest themselves as anion-radicals O_2^-, hole centres, O_{3C}^-, as complexes $[O_{3C}^- \cdot O_2]$ (or $[X \cdot O_2]$); and hole centres, O_{4C}^-, are detected as complexes $[O_{4C}^- \cdot O_2]$ (or $[Y \cdot O_2]$).

The results obtained allow us to propose the following formation mechanism for all the observable radical forms of oxygen photosorbed upon illumination of MgO and CaO in the presence of the acceptor molecules O_2 and N_2O:

$$(Me^{n+}-O^{2-}) \overset{h\nu}{\rightleftarrows} (Me^{(n-1)+}-O^-)^* \tag{1}$$

$$
\begin{array}{c}
\xrightarrow{+O_2} \quad O_2^- \;+\; [O_{st}^- \cdot O_2] \\
\uparrow{+O_2} \quad \updownarrow{+O_2 / -O_2} \\
(Me^{(n-1)+}-O^-)^* \longrightarrow [e]_{st} \;+\; O_{st}^- \\
\downarrow{+N_2O} \\
\xrightarrow{+N_2O} \quad O_{ads}^- \;+\; O_{st}^- \\
\updownarrow{+O_2/-O_2} \quad \updownarrow{+O_2/-O_2} \\
[O_{ads}^- \cdot O_2] \;+\; [O_{st}^- \cdot O_2]
\end{array}
\tag{13}$$

Depending on the experimental conditions (temperature, gas phase pressure) all the equilibria may be moved either towards primary electron and hole centers or towards their complexes with molecular oxygen.

It should be noted that the electron centres O_{ads}^- formed in the presence of N_2O on MgO and CaO have the same EPR parameters as the hole centres described above $(O_{st}^-)_{3C}$. This corresponds with the literature data for MgO.

At the same time photoadsorption of oxygen at low temperatures can result in the formation of a new type of hole centres stabilized on the MgO and CaO surface: $(O_{st}^-)_{4C}$, anion-radicals which are unobserved by EPR. Their emergence can be tested by the formation of the complexes with the molecular oxygen $[O_{4C}^- \cdot O_2]$ and by the formation of the anion-radicals CO_2^- by its interaction with CO molecules.

3. Formation of $C_6H_6^+$ Cation-Radicals on ZSM-5 Zeolites

Investigation of the diffuse reflection spectra of aromatic compounds gave the first evidence of their ionization upon adsorption over oxide catalysts in the 1950s [31]. Later, these processes were intensively studied by ESR. The processes resulting in cation-radical formation upon adsorption were shown to be widely spread [32–37].

The phenomenon of aromatic compound ionization under adsorption over wide-band dielectrics was quite unexpected, as the ionization potentials even of polyaromatic compounds are quite high: biphenyl 8.23 eV, naphthalene 8.12 eV, anthracene 7.38 eV, perylene 6.8 eV. This phenomenon could be explained only by the presence of very strong electron accepting sites on the surface.

It should be noted that a considerable amount of work on homogeneous systems has been performed in recent times. Many reactions of electrophilic substitution were shown to proceed through the formation of intermediate donor–acceptor complexes between aromatic molecules with electrophilic substituents [38–41]. In a number of cases one succeeds in obtaining direct proof of the formation of cation-radicals as intermediates in such reactions [41].

Therefore, the interest in those investigations which have been performed with variable intensity up till now is quite understandable. The high activity of alumosilicate systems in this process was noted, even in the first papers [31–37]. Zeolites appeared to be even move active [42, 43]. Thus, over high silica zeolites (mordenite, ZSM-5) even the appearance of benzene cation-radicals (ionization potential-9.24 eV) has been observed [44–46].

3.A. THERMALLY ACTIVATED PROCESSES ON ZSM-5 ZEOLITE WITH ADSORBED BENZENE

Benzene adsorption at 248–298 K on an initial sample is accompanied by the formation of two types of radicals with constants of hyperfine interactions: $a_1^H = 0.44$ mT and $a_2^H = 0.22$ mT (Figure 7, spectrum 1) which, in accordance with literature data [45–48], can be related to cation-radicals of benzene monomer and dimer, respectively.

Since the ionization potential of a benzene molecule is very high ($I = 9.24$ eV),

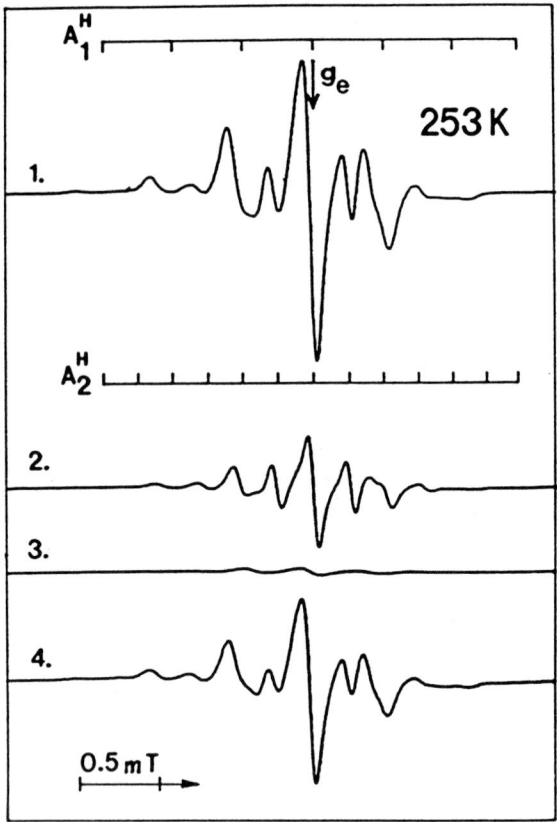

Fig. 7. EPR spectra occuring after benzene adsorption over the zeolite: (1) immediately after benzene adsorption ($1 \cdot 10^{21}$ mol/g); (2) after 4 h of evacuation at 253 K; (3) after 14 h of evacuation at 295 K; (4) after additional adsorption of benzene ($1 \cdot 10^{21}$ mole/g). The temperature of benzene adsorption and EPR spectrum recording was 253 K.

it would appear natural to assume that the formation of the cation-radicals results from an interaction with strong acceptor centres on the zeolite surface (A_s), according to the scheme:

$$C_6H_6 + A_s \rightarrow C_6H_6^+ \rightarrow \text{products} \tag{14}$$

the reaction being irreversible. However, the dependence of total concentration of cation-radicals on the quantity of adsorbed benzene has a form which is typical of an equilibrium process, i.e. presupposing the reversible character of the cation-radical formation:

$$C_6H_6 + A_s \rightleftarrows C_6H_6^+ + (A_s)^- ? \tag{15}$$
$$\tag{16}$$
$$C_6H_6^+ + C_6H_6 \rightleftarrows (C_6H_6)_2^+$$

Direct experiments with evacuation of adsorbed benzene confirm the presence of such reversibility. Thus, removal of benzene by evacuation at 295 K leads to practically complete disappearance of the EPR spectra of these cation-radicals (Figure 7, spectrum 3). The centres responsible for their formation and stabilization are retained, and subsequent benzene adsorption regenerates the initial spectrum (Figure 7, spectrum 4). A weak signal with $a^H = 0.33$ mT (Figure 7, spectrum 3), which remains after evacuation at 295 K, can be related to a biphenyl cation-radical. This signal disappears after evacuation at 395 K.

It should be noted that under evacuation the ratio between the quantity of cation-radicals of benzene monomer and dimer changes towards the latter (Figure 7, spectrum 2). However, with complete filling of zeolite channels the monomer form prevails.

3.B. PHOTOSTIMULATED FORMATION OF BENZENE CATION-RADICALS ON ZSM-5 ZEOLITES

It would appear reasonable that the process of charge transfer is preceded by the formation of some donor–acceptor complex of a benzene molecule with an active centre:

$$A_s + C_6H_6 \rightleftarrows [A_s \cdot C_6H_6] \tag{17}$$

followed by thermal ionization of this complex according to the scheme:

$$[A_s \cdot C_6H_6] \overset{kT}{\rightleftarrows} [A_s \cdot C_6H_6]^* \rightarrow A_s^- + C_6H_6^+ \tag{18}$$

The $[A_s \cdot C_6H_6]$ complex and the process of its ionization are difficult to study in the course of adsorption experiments described above, the latter requiring practically complete filling of zeolite channels with benzene. However, it appeared possible to use light in such systems to stimulate the charge transfer in the complex. This allows one to perform the experiments with a considerably smaller quantity of adsorbed benzene (less than 1 wt.-%) and opens a new avenue for obtaining information about the electronic structure of such a complex.

Figure 8 presents an EPR spectrum that occurs under irradiation of the sample containing 0.4 wt.-% of adsorbed benzene ($3 \cdot 10^{19}$ mol/g) with visible and near-UV light at 173 K.

If the photon energy is 2.8 eV and higher, the process of photostimulated formation of benzene cation-radicals occurs (Figure 9, curve 1). A study of dependence between this process and the light intensity has shown the former to proceed according to a one-quantum mechanism ($\lambda = 365$ nm). It should be noted that the photon energy is less than the energies of both singlet–singlet (4.84 eV) and singlet-triplet (4.7 eV) transfers in the benzene molecule [49]. It would be natural to relate this absorption to a charge transfer band in the initial adsorption complex:

$$[A_s \cdot C_6H_6] \overset{h\nu}{\rightleftarrows} [A_s \cdot C_6H_6]^* \rightarrow A_s^- + C_6H_6^+ \tag{19}$$

According to published data [50], such a considerable shift of the charge-transfer band to longer wavelengths was observed only for complexes of benzene with tetracyanoethylene in homogeneous systems. Thus, a correlation could be made between the strength of acceptor centres on ZSM-5 zeolite and the acceptor

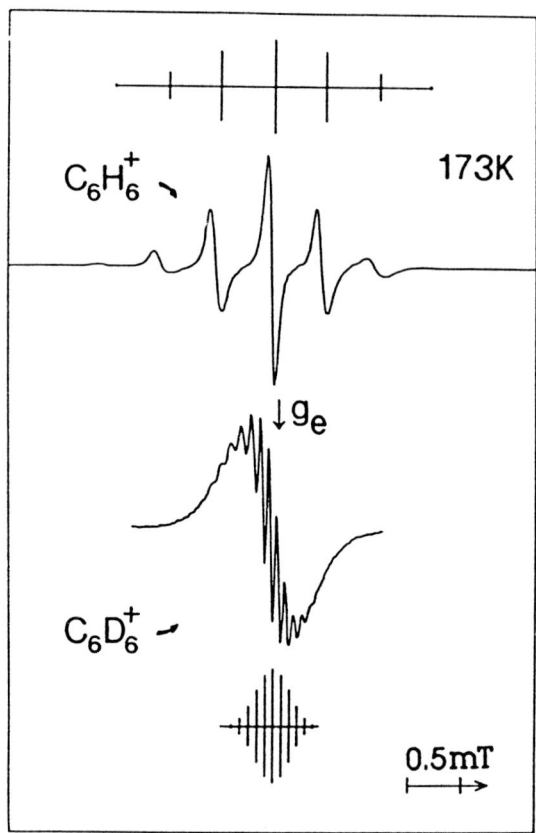

Fig. 8. EPR spectra appearing under illumination (λ = 365 nm) of the zeolite with adsorbed benzene and deuterobenzene. Temperature of illumination and spectrum recording was 173 K.

properties of the TCNE molecule. For comparison, the charge transfer band on chlorinated Al_2O_3, which possesses quite strong acceptor properties, lies at considerably shorter wavelengths (λ = 290 nm) [51].

The problem of observing the conjugate radical centre (counterion) on the formation of cation-radicals on oxide systems remains unsolved to date. Attempts to observe the conjugate anion-radical by spectral methods fail. Neither have we given the answer in the present paper to the question about the nature and subsequent transformations of the counterion. However, we succeeded in observing a stage preceding that of cation-radical formation. This stage is the formation of a new type of radical particles occurring immediately after the charge transfer but before the spatial separation of ion-radicals as a result of diffusion and chemical processes.

When illuminating the zeolite with benzene adsorbed at 93 K it has been possible to record a new paramagnetic particle R_x^*, the EPR spectrum of which is presented in Figure 10 (spectrum 1). A similar spectrum is obtained upon the adsorption of deuterobenzene (spectrum 2).

FORMATION OF RADICALS ON OXIDE SURFACES

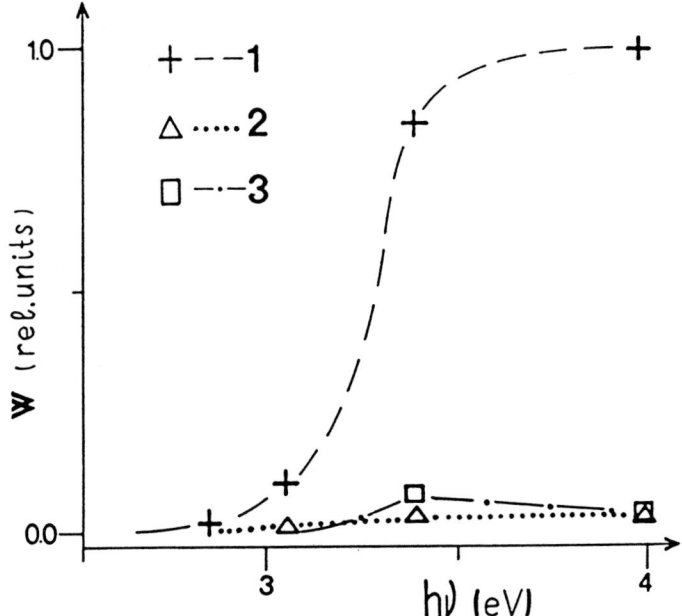

Fig. 9. Spectral dependencies of the initial rate of benzene cation-radical formation. (1) Initial sample; (2) the zeolite treated with hydrogen (823 K); (3) the sample after evacuation at 423 K. Temperature of illumination and spectrum registration was 173 K.

In the spectrum one can see two doublets symmetric in their position in relation to the g-factor of a free electron. Components of the external doublet (splitting 24 mT) have the shape of parallel components in anisotropic spectra, and those of the internal doublet (splitting 8 mT) have the shape of perpendicular components.

The ESR spectra of the R_x^- particles obtained by us resemble those of the radical pairs with the structure $[RO_2^- \cdots RO^{\cdot}]$ observed upon γ-irradiation of diphenylethane peroxide [52]. Radical pairs were also supposed, on indirect grounds, to be intermediates in ion-radical formation upon the interaction of aniline with zeolite surface active sites [53]. Thus, the paramagnetic particle R_x^- can also be supposed to be a radical pair. As a zeolite surface active centre can incorporate chemisorbed molecular oxygen [45, 54, 55], the radical pairs can be proposed to have the structure $[O_2^- \cdots C_6H_6^+]$.

As the temperature rises the spectral intensity of R_x^- radicals was observed to fall simultaneously with the benzene CR concentration growth. ESR spectra of $C_6H_6^+$ and $C_6D_6^+$ cation radicals are presented in Figure 8.

In order to discover a correlation between the disappearance of R_x^- radicals and the emergence of benzene cation-radicals, the following experiment was conducted. The zeolite with adsorbed benzene was irradiated at 93 K with subsequent momentary heating of the sample, the heating temperature varying from 123 K to 173 K. The sample was then cooled down to 93 K with subsequent recording of the ESR spectrum.

The results of the experiment are shown in Figure 11. One can see that the

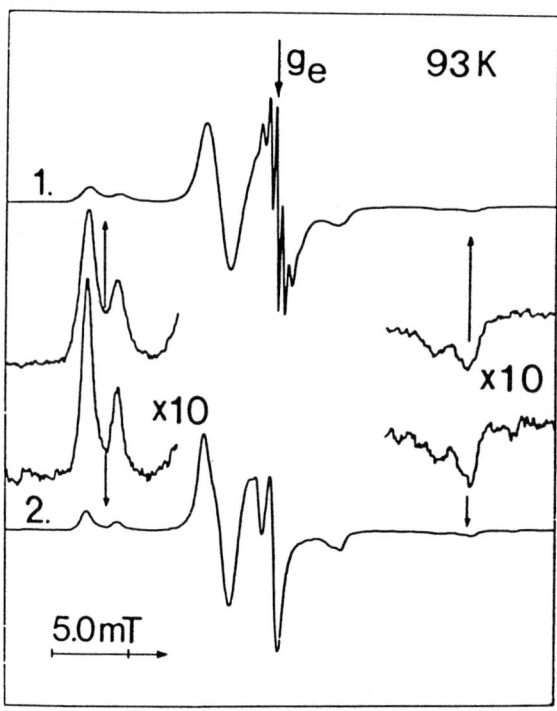

Fig. 10. EPR spectra appearing under illumination ($\lambda = 365$ nm) of zeolite with adsorbed benzene C_6H_6 (spectrum 1) and C_6D_6 (spectrum 2). Temperature of illumination and spectra registration is 93 K

relationship between the growth of benzene cation-radical concentration as the $R_x^{\cdot}$ spectral intensity falls is almost linear. That makes it possible to conclude that the radicals $R_x^{\cdot}$ discovered by us are precursors of the benzene CR in their photostimulated formation over H-ZSM-5 zeolites. Thus, appearance of the benzene CR in this process is caused by primary destruction of radical particles $R_x^{\cdot}$ and requires no irradiation.

$$R_x^{\cdot} \xrightarrow{T>120 \text{ K}} C_6H_6^+ \qquad (20)$$

If H-ZSM-5 zeolite with adsorbed benzene is irradiated at temperatures higher than 150 K, the $R_x^{\cdot}$ stationary concentration is low and only benzene cation-radicals are observed by ESR.

On the basis of the results obtained in the present paper and the analysis of known data we have suggested a formation mechanism of benzene cation-radicals and other compounds adsorbed over alumosilicates through the intermediate donor-acceptor complexes of the molecules with acceptor centres of the adsorbent (reactions 17, 18). It has been shown that photostimulated processes can proceed in such systems in the area of absorption band of the charge transfer of these complexes (reaction 19). The formation of radical pairs has been found to be precursors of cation radicals in the reaction studied.

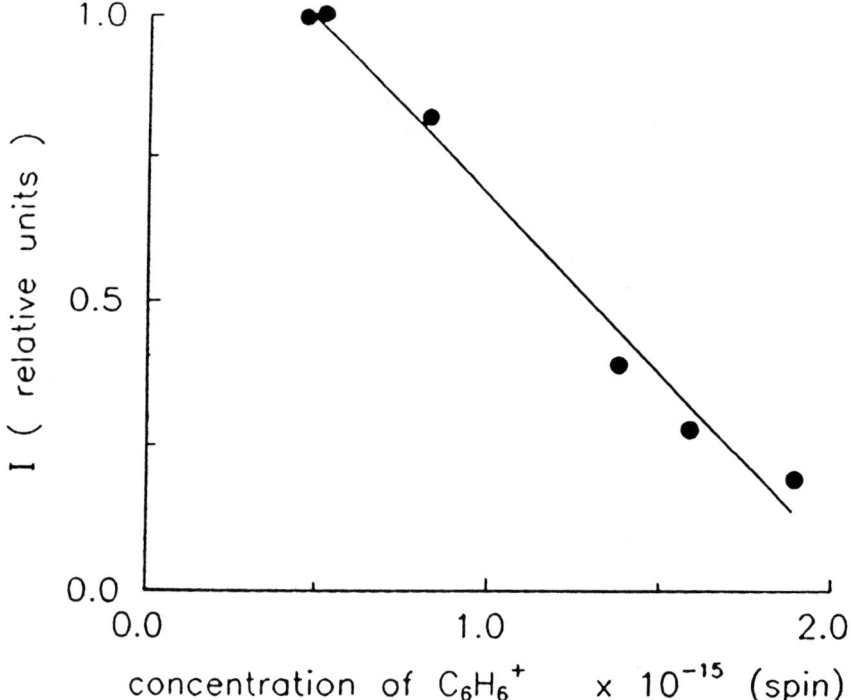

Fig. 11. Dependence of the concentration of benzene cation-radicals on the intensity of EPR spectra of radical pairs (I). Temperature of EPR spectrum recording 93 K.

4. Formation of Radical Particles from Aromatic Nitrocompounds Adsorbed on Alumina

The proposed scheme for the benzene cation-radical formation is quite universal and can be applied to the interaction between adsorbed acceptor molecules and surface donor centres proceeding through the formation of EDA complexes. The reaction between nitrobenzene and alumina surface active centres was chosen as an example of such a process.

The generation of paramagnetic particles under adsorption of nitrobenzenes (NB) on surface donor centers (D_s) of alumina and alumosilicates is well known from the literature [56–58], which also reports some examples of photoinduced processes in these systems [56, 57].

The adsorbed molecules of nitrocompounds do not possess any paramagnetism. Hence, the radicals appearing after such adsorption are usually assumed [56, 57] to be anion-radicals of the adsorbed compounds, resulting from their one-electron reduction on the surface donor centres D_s:

$$D_s + NB \rightarrow NB^- + D_s^+ \qquad (21)$$

The formation of counterions (D_s^+) in these reactions cannot be observed in the ESR spectra. In this case the concentration of donor centers D_s can be determined

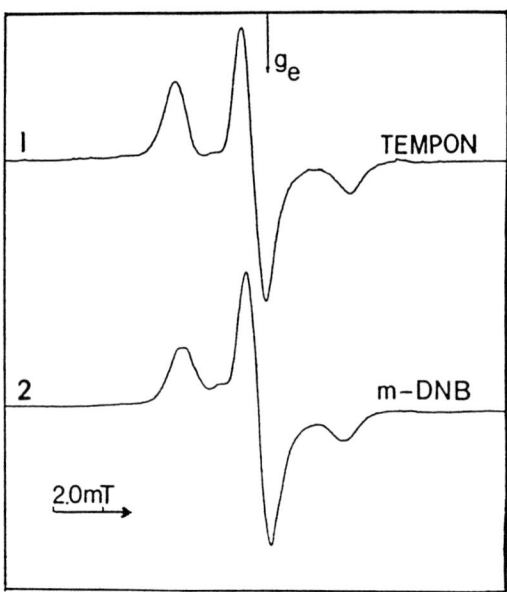

Fig. 12. EPR spectra for TEMPON radicals adsorbed on γ-Al_2O_3 (1) and radicals appearing after m-dinitrobenzene adsorption on the same catalyst (2). Recording temperatures were 193 K (1) and 295 K (2).

from the maximum concentration of radical particles (NB^-) produced upon adsorption. Note that the available experimental procedures [56, 57] imply the adsorption of nitrocompounds at concentrations much higher than the number of radical particles formed.

Figure 12 (spectrum 2) presents typical EPR spectra appearing after adsorption of nitrobenzenes on surface active sites of γ-Al_2O_3. Attention may be paid to the similarity of the EPR parameters of these particles with those of TEMPON (2,2,6,6-tetramethylpiperidine-4-oxo-N-oxyl) stable nitroxyl radical adsorbed on the same surface (spectrum 1). This makes it possible to suggest that the interaction of m-DNB with the donor centres of γ-Al_2O_3 can lead not only to electron transfer and the formation of ion-radicals (Equation 21) but also to chemical conversion to produce adsorbed nitroxyl radicals. Note that the reactions between donor (aromatic compound) and acceptor (nitrocompound) molecules to form nitroxyl radicals as intermediates are known for homogeneous systems [59, 60].

The available ESR parameters for the radicals appearing after the adsorption of nitrobenzenes on γ-Al_2O_3 (A_{iso} = 14.3 G [57]) are also typical of the isotropic h.f.s. constant of nitroxyl radicals (A_{iso} = 14 G [61]) rather than either for the anion-radicals of these nitrobenzenes (A_{iso} = 4.5 G for m-DNB and A_{iso} = 10.3 G for NB [62, 63]) or their ion pairs with cations of alkali metals (A_{iso} = 10.3, 9.0 and 7.0 G for mono-, di- and trinitrobenzenes, respectively [64]).

The results suggest that the appearance of the radical particles upon the adsorption of m-DNB on γ-Al_2O_3 is due to the formation of nitroxyl radicals stabilized on the surface after the interaction of m-DNB with the surface donor molecules

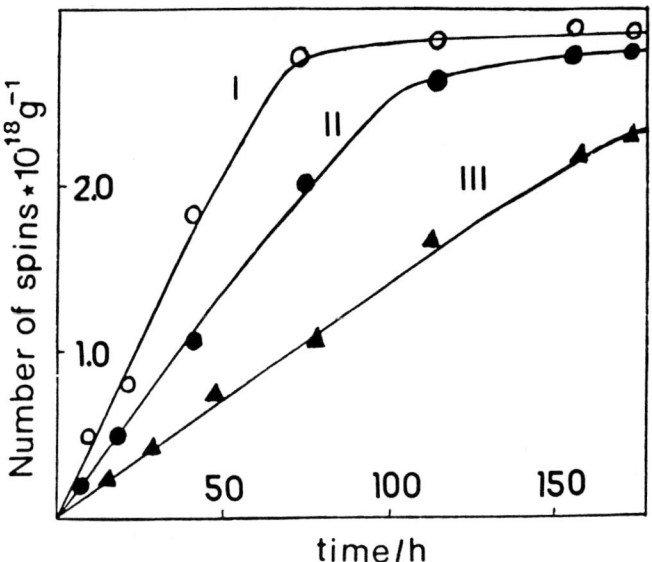

Fig. 13. Kinetic curves of radical accumulation under adsorption of m-DNB on γ-Al$_2$O$_3$ ($T = 295$ K) from o-xylene (I), benzene (II) and heptane (III).

[59, 60]. In our case the hydroxyl groups (H–O–⦃) of surface donor centres D_s could be the most probable donor of hydrogen [65]. Similar to homogeneous systems [59, 60], the following scheme for this reaction on γ-Al$_2$O$_3$ could be suggested

$$\text{Ar–NO}_2 + \text{H–O–⦃} \rightleftarrows [\text{Ar–}\overset{|}{\underset{}{\text{N}}}\text{–O} \cdots \text{H–O–⦃}] \rightarrow \text{Ar–}\overset{\uparrow}{\underset{}{\text{N}}}\text{–O–⦃} + \text{OH}^{\cdot}$$

In contrast to Equation (21), this scheme does not imply the formation of anion-radicals of the adsorbed nitrocompounds. Hydroxyl radicals OH˙ are highly reactive and can take part in further chemical conversions to produce nonparamagnetic products. Hence, under normal conditions they are not recorded in ESR spectra, whereas nitroxyl radicals formed on a surface are stable even at temperatures above 373 K.

Accumulation kinetic curves of the radicals (Fig. 13) indicate that the generation rate of radicals depends essentially on the solvent type, whereas their maximum concentration does not. The maximum concentration equals $(2-3) \cdot 10^{18}$ spin/g independently of the m-DNB concentration in the initial solution, and is determined by the concentration of the active donor centers (D_s) on γ-Al$_2$O$_3$.

One should note that kinetic curves of radical accumulation under m-DNB adsorption from solutions in heptane on the surface of certain alumosilicates have been studied previously [58]. It was found that the maximum concentration of radicals for these systems at 295° K is obtained after approximately 50 h and equals $2.9 \cdot 10^{18}$ spin/g.

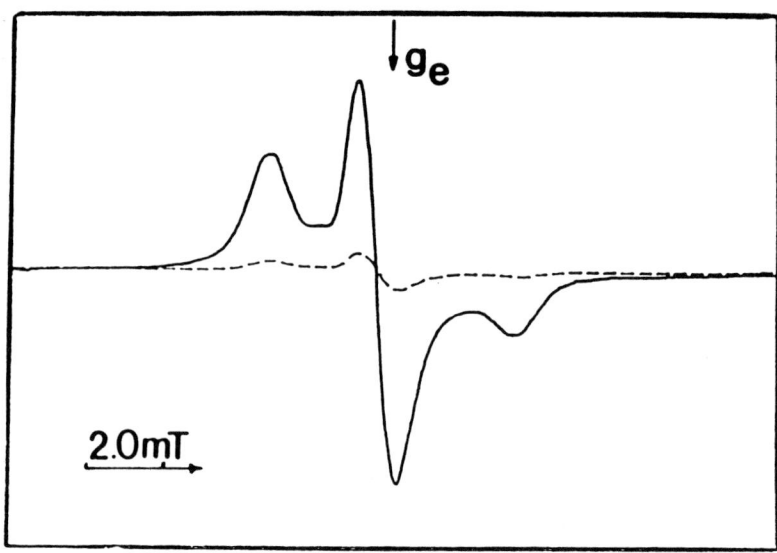

Fig. 14. EPR spectra 20 h after m-DNB adsorption on γ-Al₂O₃ (dashes) and after additional 30 min irradiation with λ = 365 nm at 295 K (solid line).

The effect of irradiation on the concentration of radicals produced is illustrated in Figure 14 (solid line). In this case their limiting concentration is determined by the amount of surface donor centers, whereas the energy characteristics of these processes essentially depend on the properties of solvent molecules. Since we used nonpolar solvents (heptane, benzene, o-xylene), it is natural to suggest that the factor accounting for the generation rate of radicals (and, hence, the activation energy of this process) is the donor properties of the solvent molecules. These properties depend on the ionization potential of molecules and differ essentially for heptane (I_p = 10.1 eV), benzene (I_p = 9.25 eV) and o-xylene (I_p = 8.56 eV). The band of charge transfer (the red edge for the occurrence of photoprocesses) essentially depends on the donor properties of the solvents and equals 2.3 eV (546 nm) for o-xylene, 2.8 eV (436 nm) for benzene and 3.4 eV (365 nm) for heptane. The spectral dependence of the initial generation rates of radicals measured in these solvents is presented in Figure 15.

The possibility to achieving photoinduced processes in the above systems, as in the case of the generation of cation-radicals, implies the occurrence of these reactions through the formation of EDA complexes.

Note that it was as early as 1962 that Lagercrantz [66] suggested that the radical particles produced under photolysis of trinitrobenzene in tetrahydrofuran are anion-radicals and are due to the formation of EDA complexes between the donor molecule of the solvent and the acceptor molecule of TNB. This concept, however, has not been further developed.

The observed shift of the absorption band for solvents with various donor properties is convincing evidence in favor of the fact that the processes under examination do occur through the formation of EDA complexes between the

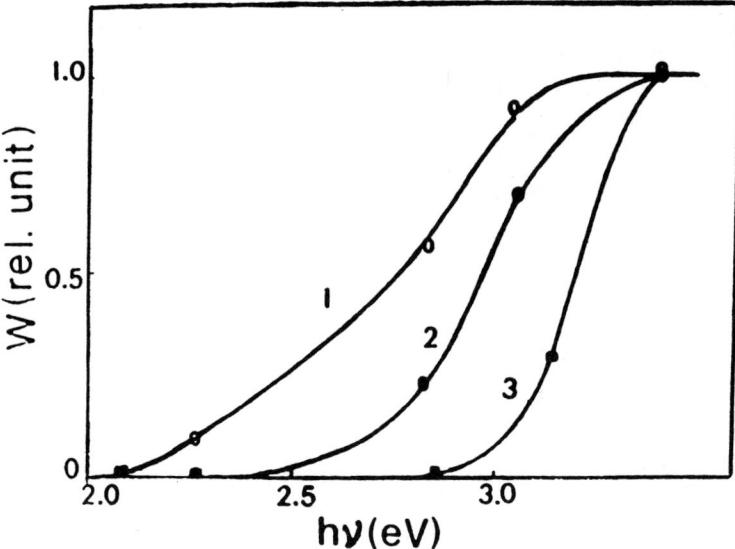

Fig. 15. Spectral dependence of the initial generation rate of radicals from m-DNB in o-xylene (1), benzene (2) and heptane (3).

molecules of m-DNB (NB) and the surface donor centers (D_s). It is essential that molecules of the solvent (D) enter into the composition of these complexes.

These results as well as investigation of the reaction by means of *in situ* EPR made it possible to isolate the following stages for the process of the formation of radical particles upon adsorption of nitrobenzene on alumina surface.

1. Nitrobenzene adsorption on the surface active sites:

$$\text{NB} + D_s \rightleftarrows [\text{NB} \cdot D_s] \tag{23}$$

This reaction is completely reversible and does not result in the formation of any paramagnetic particles neither upon illumination nor upon temperature rise. Nitrobenzene is desorbed at temperatures above 400 K.

2. Formation of EDA complexes with participation of solvent molecules:

$$[\text{NB} \cdot D_s] \underset{}{\overset{+D}{\rightleftarrows}} [D \cdot \text{NB} \cdot D_s] \tag{24}$$

The energetic characteristics of the complex depend substantially on the donor properties of the solvent.

3. Thermal or photoexcitation of the complex with participation of radical particles stabilized on the surface:

$$[D \cdot \text{NB} \cdot D_s] \underset{}{\overset{kT, h\nu}{\rightleftarrows}} [D \cdot \text{NB} \cdot D_s]^* \rightarrow \dot{R}_{st} \tag{25}$$

One may suppose the well known process of the formation of radicals upon nitrobenzene adsorption of MgO [65] to follow an analogous mechanism. Essen-

TABLE II

Photon energies (eV) required for the photostimulated formation of radicals from trinitrobenzene (TNB) in solution and from nitrobenzene adsorbed on alumina (NB/Al$_2$O$_3$) in different solvents.

	Toluene	o-xylene	Mesitylene
TNB in solution	3.4	3.04	2.85
NB$_{ads}$/Al$_2$O$_3$	2.85	2.3	–

tially, nitrobenzene itself can play the role of solvent if it is adsorbed in large quantities.

As stated above, photolysis of nitrobenzenes in homogeneous systems in donor solvents is accompanied by the formation of nitroxyl radicals as a result of recombination between a nitrobenzene and a solvent molecule [59, 60]. Earlier we have studied the photolysis of trinitrobenzene in donor solvents, resulting in the formation of nitroxyl radicals [67]. EDA complexes were shown to be key intermediates in the process, its energetic characteristics substantially depending on the solvent donor properties.

Comparing the results of trinitrobenzene photolysis in solution with the ones for adsorbed nitrocompounds, one can conclude that the photostimulated process on the surface requires much smaller photon energies.

Table II presents data for the red edge of the photoprocesses for trinitrobenzene in solution and nitrobenzene adsorbed over alumina. The data testify that nitrobenzene adsorbed on the surface active sites possesses far stronger acceptor properties than trinitrobenzene in solution.

5. Conclusion

Mechanistic investigations of the photostimulated processes over oxides presented in this paper reveal that these reactions are often accompanied by the formation of radical or ion-radical particles. Such a reaction can proceed both in the absorption band of intrinsic surface defects and in the band of EDA complexes formed upon adsorption of an adsorbate on surface active sites. Study of the spectral dependences of the photostimulated processes makes it possible to obtain information about the energetic characteristics of nonparamagnetic surface complexes participating in the absorption of photons.

Formation of EDA complexes as key intermediates in the reaction of adsorbed aromatic substances are also of great interest. The photon energy necessary for the photoinduced formation of radicals correlates well with the activation energy of the processes of thermal generation of the radicals. Energetic characteristics of photo- and thermally-stimulated reactions of adsorbed aromatic compounds are shifted towards lower energies than the same processes in homogeneous systems.

It should also be emphasized that in photoinduced processes on the surface, electron and hole centres are always generated together. But in the majority of cases only part of them can be observed by means of EPR. Special test reactions should be applied to discover and study other centres.

References

1. L. L. Basov, Yu. P. Solonitzin, and A. N. Terenin: *Dokl. Akad. Nauk SSSR* **164**, 122 (1965).
2. A. A. Lisachenko and F. I. Vilesov: *Uspekhi Fotoniki* **4**, 18 (1974).
3. L. L. Basov, G. N. Kuzmin, I. M. Prudnikov, and Yu. P. Solonitzin: *Uspekhi Fotoniki* **6**, 82 (1977).
4. E. Garrone, A. Zecchina, and F. S. Stone: *Phil. Mag. B* **42**, 683 (1980).
5. E. Garrone and F. S. Stone: *J. Chem. Soc. Farad. Trans. I* **83**, 1237 (1987).
6. M. Che and A. J. Tench: *Advances in Catal.* **31**, 77 (1982).
7. S. Coluccia, A. M. Deane, and A. J. Tench: *J. Chem. Soc., Faraday Trans. I* **74**, 2913 (1978).
8. M. Anpo, Y. Yamada, Y. Kubokawa, S. Coluccia, A. Zecchina, and M. Che: *J. Chem. Soc., Faraday Trans. I* **84**, 751 (1988).
9. V. N. Kuznetsov, A. O. Klimovsky, and A. A. Lisachenko: *Kinetika i Kataliz* **31**, 659 (1990).
10. V. A. Shvets, A. V. Kuznetsov, V. A. Fenin, and V. B. Kazansky: *J. Chem. Soc. Farad. Trans. I* **81**, 2913 (1985).
11. A. V. Kuznetsov and V. A. Shvets: *Zhurn. Fiz. Khimii* **64**, 717 (1990).
12. A. M. Volodin: *React. Kinet. Catal. Lett.* **28**, 189 (1985).
13. A. M. Volodin: *Khimicheskaya Fizika* **11**, 1054 (1992).
14. A. M. Volodin: *React. Kinet. Catal. Lett.* **44**, 171 (1991).
15. V. A. Bolshov and A. M. Volodin: *React. Kinet. Catal. Lett.* **46**, 337 (1992).
16. T. A. Konovalova and A. M. Volodin: *React. Kinet. Catal. Lett.* **51**, 227 (1993).
17. A. M. Volodin, A. E. Cherkashin, and V. S. Zakharenko: *React. Kinet. Catal. Lett.* **11**, 103 (1979).
18. A. M. Volodin and A. E. Cherkashin: *Kinetika i Kataliz* **22**, 598 (1981).
19. A. M. Volodin, V. S. Zakharenko, and A. E. Cherkashin: *React. Kinet. Catal. Lett.* **18**, 321 (1981).
20. I. M. Prudnikov and Yu. P. Solonitzin: *Kinetika i Kataliz* **13**, 426 (1972).
21. A. J. Tench and T. Lawson: *Chem. Phys. Lett.* **7**, 459 (1970).
22. N.-B. Wong and J. H. Lunsford: *J. Chem. Phys.* **55**, 3007 (1971).
23. A. J. Tench, T. Lawson, and J. F. Kibblewhite: *J. Chem. Soc. Farad. Trans. I* **68**, 1169 (1972).
24. M. Ivamoto and J. H. Lunsford: *Chem. Phys. Lett.* **66**, 48 (1979).
25. Yu. P. Solonitzin: *Kinetika i Kataliz* **7**, 43 (1977).
26. M. Anpo and Y. Yamada: *Material Chemistry and Physics* **18**, 465 (1988).
27. A. M. Volodin, A. E. Cherkashin, and K. N. Prokopiev: *React. Kinet. Catal. Lett.* **46**, 373 (1992).
28. A. M. Volodin, A. E. Cherkashin, and K. N. Prokopiev: *Kinetika i Kataliz* **33**, 1190 (1992).
29. M. Che, S. Coluccia, and A. Zecchina: *J. Chem. Soc. Farad. Trans. I* **74**, 1324 (1978).
30. D. Cordishi, V. Indovina, and M. Occhiuzzi: *J. Chem. Soc. Farad. Trans. I* **74**, 883 (1978).
31. A. Terenin and A. Sidorova: *Izv. Akad. Nauk. SSSR, Ser. Khim.* **2**, 152 (1950).
32. J. K. Fogo: *J. Phys. Chem.* **65**, 1919 (1961).
33. J. J. Rooney and R. C. Pink: *Trans. Faraday Soc.* **58**, 1632 (1962).
34. A. Terenin, V. Barachevsky, E. Kotov, and V. Kolmogorov: *Spectrochim. Acta.* **19**, 1797 (1963).
35. D. N. Stamires and J. Turkevich: *J. Am. Chem. Soc.* **86**, 749 (1964).
36. A. Terenin: *J. Chim. Phys.* 646 (1965).
37. B. D. Flockhart, J. A. N. Scott, and R. C. Pink: *Trans. Faraday Soc.* **62**, 730 (1966).
38. V. D. Pokhodenko, V. A. Khizhny, B. G. Koshetchko, and O. I. Shkreby: *Dokl. Akad. Nauk SSSR* **210**, 640 (1973).
39. A. S. Morkovnik, N. M. Dobaeva, O. Yu. Okhlobystin, and V. V. Bessonov: *Zhurn. Org. Khimii* **17**, 2618 (1981).
40. B. Reitstoen and V. D. Parker: *J. Am. Chem. Soc.* **113**, 6954 (1991).
41. J. K. Kochi: *Acta Chemica Scandinavica* **44**, 409 (1990).
42. A. J. Bard, A. Ledwith, and H. J. Shine: *Adv. Phys. Org. Chem.* **13**, 155 (1976).
43. M. B. Loktev and A. A. Slinkin: *Uspekhi Khimii* **45**, 1594 (1976).
44. Y. Kurita, T. Sonada, and M. Sato: *J. Catal.* **19**, 82 (1970).
45. J. C. Vedrine, A. Auroux, V. Bolis, P. Dejaifve, C. Naccache, P. Wierzchowski, E. G. Derouane, J. B. Nagy, J.-P. Gilson, J. H. C. van Hooff, J. P. van den Berg, and J. Wolthuizen: *J. Catal.* **59**, 248 (1979).

46. A. V. Kucherov, A. A. Slinkin, D. A. Kondratyev, T. N. Bondarenko, A. M. Rubinstein, and Kh. M. Minachev: *J. Mol. Catal.* **37**, 107 (1986).
47. M. K. Carter and G. Vincow: *J. Chem. Phys.* **47**, 292 (1967).
48. O. Edlung, P.-O. Kinel, A. Lund, and A. Shimizu: *J. Chem. Phys.* **46**, 3679 (1967).
49. T. Tanei: *Bull. Chem. Soc. Jap.* **41**, 833 (1968).
50. R. Foster: *Organic Charge-Transfer Complexes*, Academic Press, London (1969).
51. A. Ayame and T. Izumizava: *J. Chem. Soc., Chem. Commun.* 645 (1989).
52. A. V. Zubkov, A. T. Koritsky, and Ya. S. Lebedev: *Dokl. Akad. Nauk SSSR* **180**, 1150 (1968).
53. F. R. Chen and J. J. Fripiat: *J. Phys. Chem.* **96**, 819 (1992).
54. V. A. Bolshov, A. M. Volodin, A. G. Maryasov, G. M. Zhidomirov, and V. F. Yudanov: *Dokl. Akad. Nauk. SSSR* **316**, 141 (1991).
55. I. V. Yudanov, I. I. Zakharov, and G. M. Zhidomirov: *React. Kinet. Catal. Lett.* **48**, 411 (1992).
56. B. D. Flockhart, I. R. Leith, and R. C. Pink: *Trans. Faraday Soc.* **66**, 469 (1970).
57. B. D. Flockhart, M. C. Megarry, and C. Pink: *Advances in Chem.* **121**, 509 (1973).
58. H. Kashiwagi: *Bull. Chem. Soc. Jpn.* **64**, 423 (1991).
59. H. G. Aurich and W. Weiss: *Topics in Current Chem.* **59**, 65 (1975).
60. J. A. Menapace and J. E. Marlin: *J. Phys. Chem.* **94**, 1906 (1990).
61. A. L. Buchachenko and A. M. Vaserman: *Stable Radicals*, Khimiya, Moskva (1973).
62. D. H. Geske and A. H. Maki: *J. Am. Chem. Soc.* **82**, 2671 (1960).
63. A. H. Maki and D. H. Geske: *J. Chem. Phys.* **33**, 825 (1960).
64. R. L. Ward: *J. Am. Chem. Soc.* **83**, 1296 (1961).
65. R. M. Morris and K. J. Klabunde: *Inorg. Chem.* **22**, 682 (1983).
66. C. Lagercrantz and M. Yhland: *Acta Chem. Scand.* **16**, 1043 (1962).
67. A. M. Volodin, T. A. Konovalova and A. F. Bedilo: *XVIth International Conference on Photochemistry 'ICP'*, August 1–6, 1993, Vancouver, Canada, p. 321 (1993).

Part III: Trends in Modern Techniques

Fourier Transform Electron Paramagnetic Resonance Studies of Photochemical Reactions in Heterogeneous Media

HANS VAN WILLIGEN and PATRICIA R. LEVSTEIN[1]
Department of Chemistry, University of Massachusetts at Boston, Boston, MA 02125, U.S.A.

(Received: 15 December 1993; accepted: 5 July 1994)

Abstract. FT-EPR has proved to be an excellent spectroscopic technique for the study of free radicals generated in photochemical reactions. The high spectral and time resolution make it possible to identify transient free radicals. Analysis of the time dependence of the spectra gives data on the kinetics of radical formation and decay. Chemically induced dynamic electron polarization (CIDEP) effects give information on reaction mechanisms and radical pair characteristics. Relaxation and linewidths data provide an insight into molecular motion. This contribution reviews instrumental aspects of the technique and CIDEP mechanisms that can affect the spectra. The utility of the method is illustrated with a discussion of applications in the study of photoinduced electron transfer reactions in homogeneous and heterogeneous media.

Key words. EPR, photochemistry, radicals, triplets, CIDEP.

1. Introduction

This contribution is concerned with the application of Fourier Transform Electron Paramagnetic Resonance (FT-EPR) in studies of photochemical reactions. Of particular interest is the use of this new technique in investigations that deal with reactions in heterogeneous media. It should be noted at the outset that the number of applications dealing with this field of research so far is very limited. It is hoped that this review will stimulate further work.

To start it will be useful to consider FT-EPR in the context of alternative EPR measurement methodologies in so far as it concerns studies of photochemical reactions. It is of course well established that EPR is a powerful tool for the study of the mechanisms and kinetics of chemical reactions [1, 2]. In such applications the following features can be exploited.

1. Only paramagnetic species formed in a reaction give a signal contribution. As a consequence, spectra will be relatively simple so that it should be straightforward to use the time development of signal components to determine rate constants of radical formation and decay.
2. Information on electron spin – nuclear spin hyperfine interactions and g values can give unequivocal evidence for the identity of paramagnetic molecules.
3. The technique can be used to measure spin relaxation rates which can give an

---

[1] Present Address: FAMAF, Ciudad Universitaria, 5000 Cordoba, Argentina.

insight into molecular motion. (A feature that is of special interest in studies of heterogeneous systems.)
4. Populations of electron spin states of free radicals formed in chemical reactions initially will deviate from Boltzmann distribution. This can give rise to Chemically Induced Dynamic Electron Polarization (CIDEP) effects in the EPR spectra. Studies of CIDEP can provide unique insights into the mechanism of a chemical reaction [3-6].

Until about two decades ago studies of photochemical reactions were carried out almost exclusively with conventional cw EPR instruments. A continuous-wave (cw), fixed-frequency, microwave source is used in the measurements and signal detection involves application of field modulation combined with phase-sensitive detection. Spectra are recorded by sweeping the magnetic field. Kinetic measurements can be performed by monitoring growth and decay of a signal component at a fixed field setting [2]. In this mode of operation the optimum time resolution is determined by the field modulation frequency (typically 100 kHz) and is around 0.1 ms. It can be improved by an order of magnitude or so by increasing the modulation frequency [7]. Even then it may not be adequate for directly measuring reaction or spin-lattice relaxation rates.

Kinetic information on transient paramagnetic species formed in photochemical reactions can be obtained indirectly by applying sinusoidal modulation of light intensity combined with phase-sensitive detection of the EPR signal [8]. The dependence of amplitude and phase of the signal on light modulation frequency conveys information on the dynamics of the system. The drawback of this method is that whereas analysis of the data must be based on a detailed mechanistic model, the data provide little *a priori* guidance with regard to which model might be appropriate.

The limitations of the cw instrument can be removed to a large extent by a relatively simple modification that adapts it for use in a direct-detected, time-resolved (TREPR) mode [4,6,9,10]. In TREPR measurements, field modulation and phase sensitive detection are not used. Instead, following the initiation of a chemical reaction with a light pulse, the time dependence of the signal at a given magnetic field is measured directly using a fast data acquisition system. Alternatively, the spectrum of paramagnetic species present at a given time after the light pulse can be obtained by sweeping the magnetic field and sampling the signal with a boxcar integrator. The optimum time resolution of TREPR is about 50 ns [11]. Drawbacks associated with this method are the following. Firstly, the penalty for doing away with field modulation/phase-sensitive detection is a significant reduction in sensitivity. However, it has been shown that the TREPR spectrum from a $\sim 10^{-4}$ M stable nitroxide radical solution has a good signal-to-noise ratio [12]. Radical concentrations of this order of magnitude in many cases can be readily generated by pulsed-laser irradiation of a sample held in the cavity of a TREPR spectrometer. Secondly, the time development of the spectra is governed, in part, by the interaction between spin system and microwave field in the interval between radical formation and data acquisition. In the time domain where the inverse of the time delay between radical formation and signal detection (τ_d)

becomes comparable to or exceeds the intrinsic linewidth of the EPR signal, the perturbation gives rise to linebroadening. This can preclude signal detection or radical identification at early times. Apart from affecting the linewidth, the radiation field also influences the time evolution of the signal amplitude. For this reason, the determination of rate parameters must be based on a set of Bloch equations that account not only for chemical kinetics and CIDEP mechanisms, but also for the perturbation of the system by the microwave field [5]. It is clear, therefore, that it is no simple matter to interpret TREPR data quantitatively.

With pulsed-EPR techniques these difficulties can be avoided. The objective of time-resolved EPR measurements is to monitor the time dependence of magnetization due to the formation and decay of free radicals. This can be accomplished by turning the magnetization vector from the magnetic field axis (z) into the transverse (xy) plane with one or more short microwave pulses followed by measurement of the transverse magnetization. The spin system is not perturbed by the presence of a radiation field in the interval between radical formation and detection. Furthermore, the sensitivity of the technique can be much higher than that given by TREPR.

The first applications of pulsed EPR in the study of transient free radicals made use of electron spin echo (ESE) [4,13,14]. With this method, the magnetization is rotated in the xy plane with a $\pi/2$ microwave pulse. This is followed by a π pulse delivered after a short delay. The echo signal created by this pulse sequence is a measure of z magnetization existing at the time the first pulse is delivered. ESE can be applied in two modes. (1) The delay τ_d between laser pulse and $\pi/2$ pulse can be kept constant and the echo signal measured as function of magnetic field strength. This method gives the spectrum of the radicals present at time τ_d. It should be noted that, as in TREPR, off-resonance signal contributions can degrade spectral resolution. Line broadening becomes more pronounced with reduction in microwave pulse width and increase in microwave power (cf. Section 2). (2) The echo amplitude can be measured at a fixed field as function of τ_d to determine the time dependence of z magnetization.

Magnetization also can be measured via the transverse component created by a single $\pi/2$ pulse. This method is desirable because the ESE signal will not be as strong as the free induction decay (FID) signal given by a single pulse. Improvements in microwave and data acquisition hardware have removed the constraints that prevented application of this method a few decades ago. This first led to the application of the FID integration method by Trifunac *et al.* [15,16]. Here the integrated FID serves as a measure of magnetization present at the time of the microwave pulse. Measurement of this signal as function of magnetic field, at fixed τ_d, gives the EPR spectrum. Measurement at fixed field as function of τ_d gives information on chemical kinetics and spin dynamics.

Currently, instruments are available [17–20] with which it is possible to record spectra in a manner which is analogous to that used in FT-NMR. It involves measurement of an FID at fixed field and frequency followed by Fourier transformation of the time domain signal. FT-EPR has proven ideally suited for the study of photochemically generated transient paramagnetic species because it offers the ultimate in time and spectral resolution, and high sensitivity. The time evolution of spectra can be monitored over a time regime extending from nanoseconds to milliseconds. A complete analysis of the time dependence of the spectra gives [21]:

- the rate constants of formation and decay of free radicals,
- information on the characteristics of (paramagnetic) reaction precursors that cannot be observed directly,
- magnitudes of spin polarization generated by Chemically Induced Electron Polarization (CIDEP) mechanisms,
- data on spin relaxation times, and
- information on molecular motion.

As will be discussed in more detail in the next section, FT-EPR also has its limitations. For instance, the spectral width covered by the measurement is relatively small so that it may be impossible to get complete information with a measurement carried out with a single field/frequency setting. Furthermore, instrument deadtime sets a lower limit of about 0.5 μs on the decay of transverse magnetization. This precludes studies of systems giving broad EPR signals. Notwithstanding these restrictions, it can be expected that the number and diversity of FT-EPR applications will increase rapidly now that a commercial pulsed-EPR instrument has become available.

2. Instrumental Aspects

FT-EPR is the microwave analogue of FT-NMR so that discussions of general principles can be found in NMR textbooks. A good coverage of instrumental details and measurement methodology at the introductory level can be found in the book by Derome [22]. An excellent discussion of the specific instrumental requirements of FT-EPR is given in an article by Gorcester and Freed [19]. Schweiger gives a thorough presentation of pulsed-EPR principles and applications [23]. Instruments built by a number of research groups have been described in the literature [17–20].

As noted in the Introduction, in FT-EPR a short microwave pulse rotates the magnetization vector from the external magnetic field (z) axis into the transverse plane. Following the $\pi/2$ pulse, the time evolution of the magnetization in the xy plane (FID) is measured. Fourier transformation of the FID gives the frequency domain EPR spectrum. Two conditions must be fulfilled to capture all frequency components in a spectrum:

1. the inverse of the pulse width (τ_p) must exceed the spectral width, and
2. the data sampling rate must be at least twice the frequency of the highest frequency component in the FID.

With pulse width τ_p, the microwave field ($\mathbf{B}_1$) required for a $\pi/2$ pulse, assuming that the Larmor frequency of the spin corresponds to the microwave frequency, is given by

$$\mathbf{B}_1 = \pi/2\gamma_e\tau_p. \tag{1}$$

Here γ_e denotes the gyromagnetic ratio. The $\mathbf{B}_1$ field that can be generated is a function of transmitter power and the power-to-field conversion efficiency. With the low-Q resonators used in instruments described in the literature [17–20], 1 kW of transmitter power may give a $\mathbf{B}_1$ field of about 6 gauss (0.6 mT) only [19]. Then, according to Equation (1), τ_p must be at least 14 ns for a $\pi/2$ rotation. For minimal reduction in signal strength caused by an offset of the resonance frequency ($\omega_{res} = \gamma_e \mathbf{B}_0$, where $\mathbf{B}_0$ denotes the external magnetic field) from the applied microwave frequency (ω_{mw}), the microwave field strength $\mathbf{B}_1$ must meet the requirement

$$|\omega_{res} - \omega_{mw}| \ll \gamma_e \mathbf{B}_1. \tag{2}$$

It follows that, with $\mathbf{B}_1 = 6$ gauss, one can expect to measure an undistorted spectrum only if the spectral width does not exceed 12 gauss (34 MHz). This assumes that ω_{mw} corresponds to the center frequency of the spectrum. Actually, a considerably broader bandwidth is covered in the measurement. With $\mathbf{B}_1 = 6$ gauss, a resonance offset of about 16 gauss (45 MHz) gives a reduction in signal amplitude of about a factor of two [19]. One can easily make corrections for the effect the offset has on signal amplitude and phase, so that a broad range of free radicals can in fact be studied. In cases where the entire spectral range is not covered adequately, FIDs can be measured for a number of field settings. The Fourier transforms of the FIDs can then be assembled to give the complete EPR spectrum [24].

To realize the bandwidth coverage given by the microwave field, the resonator bandwidth must be at least as large. For this reason, a low-Q resonator is required. A bandwidth coverage of 100 MHz requires that $Q \sim 100$ [19]. Since the power-conversion efficiency is reduced by a reduction in Q, the requirement of a large $\mathbf{B}_1$ field on the one hand, and a wide resonator bandwidth on the other conflict with each other. In instruments that make use of cavities [17–20], the loss in power conversion efficiency is compensated for by the use of a high-power microwave amplifier (1 kW traveling wave tube amplifier, TWT). With the use of alternative devices, such as the loop-gap [25] and bridged-loop-gap [26,27] resonators, a combination of low Q and high microwave field can be attained at greatly reduced microwave power. Freed and coworkers have used the loop-gap resonator in studies of nitroxides. Their work demonstrates that a spectrum covering a frequency range of 90 MHz or more can be recorded without significant distortion [28]. In the applications discussed in this review, the use of standard cavities is preferred because measurements can be made with a broad range of sample conditions (dielectric constant and temperature) and there is ready access to the sample for light excitation. Other resonators may have to be tailor-made to fit the requirements of a particular measurement.

The FID is detected by amplification of the microwave signal generated by the sample followed by down conversion of the frequency by mixing with a reference signal from the microwave source. A quadrature IF mixer is used which gives two channel output [22]. The two signals represent the FID of the magnetization along two orthogonal axes in the xy plane. With this detection method ω_{mw} can be set to match the center of the bandwidth one wants to cover because positive frequency deviations can be distinguished from negative deviations. The result is that the bandwidth coverage of the measurement is optimized. A CYCLOPS phase-cycling routine is applied to correct for amplitude and phase errors introduced by the mixer [22]. The signals are sampled with a two-channel data acquisition system. A digital oscilloscope can be used for this purpose [19,20]. The frequency down conversion gives FIDs that can contain frequencies ranging from less than -50 MHz to more than $+50$ MHz. Hence, the data acquisition device should be capable of two-channel sampling at a rate of at least 100 Ms/s (10 ns per point). An FID signal may last as long as 10 or more microseconds so that sampling at 100 Ms/s can generate 2000 or more data points (1000 per detection channel). In general, signal averaging will be required to improve the signal-to-noise.

In FT-NMR, the spin-lattice relaxation rate generally dictates the rf pulse repetition rate. Since electron spins of free radicals in fluid solution relax to thermal equilibrium in microseconds, the maximum pulse repetition rate in an FT-EPR measurement typically is determined by capabilities of the data acquisition system. With two-channel measurement of 1000 data points at 10 ns/point the upper limit of the repetition rate may be $100\,\text{s}^{-1}$. In photochemical applications, constraints imposed by the sample or laser are likely to set an upper limit on the repetition rate.

The FID cannot be measured immediately after the $\pi/2$ pulse because of cavity ring down. For instruments using standard rectangular cavities [18–20], the deadtime is at least 100 ns. It can be reduced somewhat with the use of a bimodal cavity [17] or loop-gap resonators [28]. If the FID decays fast, the sensitivity of the measurement will be reduced by the deadtime. The rate of exponential damping of the FID (T_2^{-1}) is related to the full width at half height (ΔW, in Hz) of a Lorentzian line by $\Delta W = (\pi T_2)^{-1}$ [22]. Thus, the FID of a radical that gives a frequency domain spectrum with peak widths $\Delta W \approx 1$ gauss will decay to about $1/e$ of its initial amplitude during the spectrometer deadtime. This consideration reveals a serious limitation of FT-EPR. Without a significant reduction in deadtime, the technique cannot be used in studies of paramagnetic systems giving resonance peaks with widths much in excess of 1 gauss.

The method used in time-resolved studies of photo-generated paramagnetic molecules is presented in Figure 1 [18,29,30]. The chemical reaction is initiated by a laser pulse. After a delay τ_d, the $\pi/2$ microwave pulse is delivered and the FID measured. *The FID represents the time-domain spectrum of the paramagnetic species present at the time of the microwave pulse.* Free radicals formed during the time interval the FID is measured – which can cover 10 or more microseconds – only affect the signal indirectly through radical–radical interactions. Decay of free radicals during that time is reflected in the form of line broadening in the frequency-domain spectrum. By recording the FIDs for a range of τ_d settings – which can cover six orders of magnitude, from nanoseconds to milliseconds – the time evolution of the spectra from photo-generated free radicals can be monitored.

The high sensitivity and spectral resolution of the measurement is illustrated in Figure 2. It displays the spectrum given by the (perdeuterated) acetone ketyl radical, $(CD_3)_2\dot{C}OD$, formed in the reaction of photo-excited acetone($d6$) with 2-propanol($d8$)

$$(CD_3)_2CO \xrightarrow{h\nu} {}^1(CD_3)_2CO^* \xrightarrow{isc} {}^3(CD_3)_2CO^*$$

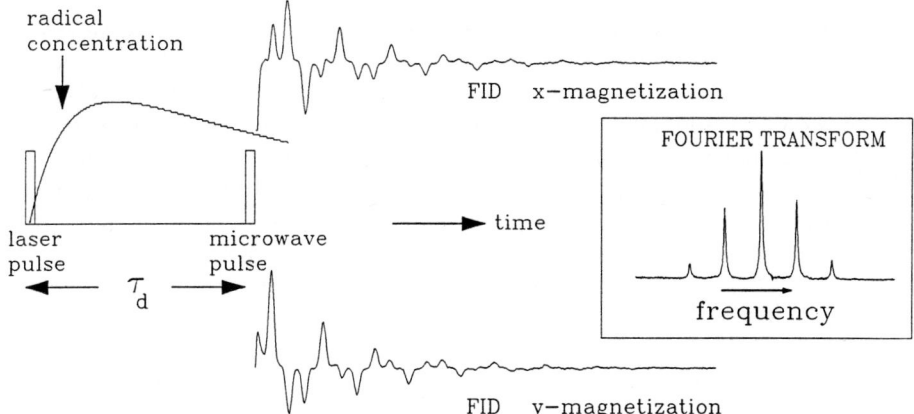

Fig. 1. Schematic diagram illustrating the sequence of events in measurements of photo-generated free radicals with FT-EPR.

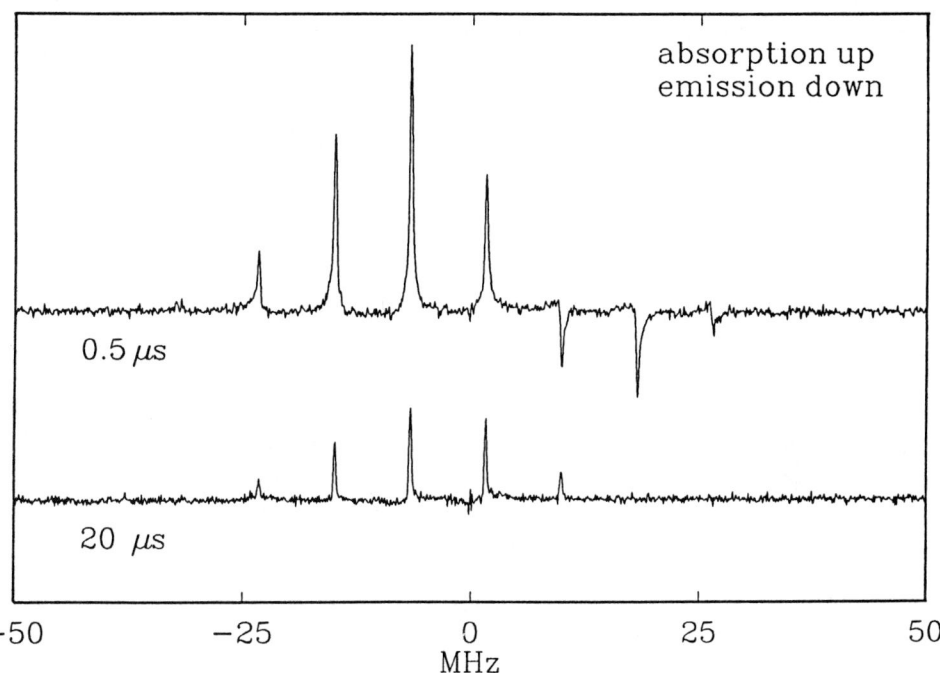

Fig. 2. FT-EPR spectrum of photogenerated $(CD_3)_2\dot{C}OD$ in acetone-d_6/2-propanol-d_8. Microwave pulse 0.5 µs and 20 µs after laser pulse (308 nm, ~50 mJ/pulse). Total number of laser shots 4 (one per phase in the CYCLOPS phase-cycling routine).

$$^3(CD_3)_2CO^* + (CD_3)_2CDOD \rightarrow [(CD_3)_2\dot{C}OD\text{---}(CD_3)_2\dot{C}OD]$$
$$\rightarrow 2(CD_3)_2\dot{C}OD$$

for delay times between laser pulse and microwave pulse of 0.5 and 20 µs. *The spectra were obtained by Fourier transformation of the* (average) *FID produced by four laser shots, total data acquisition time* 40 μs.

Of interest is the time dependence of all spectral parameters: frequencies, amplitudes, linewidths, and phases of all peaks. This information can be derived from the frequency-domain spectra obtained by Fourier transformation of the FIDs. However, it is more straightforward to extract the data directly from the FIDs with a linear prediction-singular value decomposition (LP-SVD) analysis routine [31,32]. The procedure is based on the assumption that, to a good approximation, the FID can be represented by the sum of a number of exponentially damped sinusoids. LP-SVD directly gives numerical data for the spectral parameters for each τ_d setting so that a computer analysis of the time dependence of the parameters is facilitated. If the data analysis method involves Fourier transformation of the FID as the initial step, the data points missed because of the instrument deadtime can cause artifacts in the frequency-domain spectrum which can make it difficult to extract reliable data [22]. Because LP-SVD can be applied on any set of data points in the FID, the deadtime does not pose a problem. The time-domain signal can be analyzed also with a non-linear least-squares routine that uses the Variable Projection (VARPRO) method. The VARPRO analysis can be applied not only on exponentially damped sinusoids, but on any function that describes the decay of the time-domain signal [32].

3. CIDEP Mechanisms

The spin system of photochemically generated radicals generally will not be at thermal equilibrium at the time of formation. If the EPR spectrum is recorded before Boltzmann equilibrium has been established, the time evolution of the spectrum will reflect chemical kinetics as well as spin dynamics. Chemical decay of free radicals also can give rise to deviations from thermal equilibrium. EPR spectra can show non-equilibrium spin polarization conditions for extended time periods.

The fact that signal amplitude is determined both by chemical processes and spin dynamics vastly complicates the analysis of time-resolved spectra. However, a complete analysis provides a great deal of information on the mechanisms of reactions. For the most part this information is not accessible via other spectroscopic techniques so that time-resolved EPR studies can give unique insights.

The processes that create non-Boltzmann electron spin polarization in paramagnetic molecules are collectively known as Chemically Induced Dynamic Electron Polarization (CIDEP) mechanisms. A number of reviews concerned with CIDEP effects have been published [3–6]). This section gives a brief overview of the various CIDEP mechanisms and their effects on EPR spectra.

3.1. TRIPLET MECHANISM

Triplet mechanism (TM) spin polarization can be generated if formation of radicals involves a precursor in a photoexcited triplet state [33,34]

$$^1M \xrightarrow{h\nu} {}^1M^* \xrightarrow{isc} {}^3M^*$$

$$^3M^* + Q \rightarrow {}^2R_1 + {}^2R_2.$$

Photoexcited triplets are formed by intersystem crossing (isc) from an excited singlet state. The isc process is spin selective and creates spin polarization in the triplet state molecules [35,36]. If the chemical reaction is fast enough to compete with triplet spin-lattice relaxation, triplet spin polarization will be carried over to the doublet radicals.

The requirement for the generation of TM CIDEP is that precursor triplets have populations in the T_{+1} and T_{-1} spin states that deviate from Boltzmann equilibrium at the time of the reaction. If the spin state remains conserved in the radical formation step, T_{+1} triplets will give α spin doublet radicals and T_{-1} triplets β spin radicals. Therefore, an excess population in the T_{+1} level will give excess population in the α level producing a stimulated-emission EPR signal. If the reverse is true, an enhanced absorption signal is observed. The mechanism is independent of nuclear spin state, so that relative intensities of peaks, given by hyperfine interactions between electron spin and nuclear spins, remain unaffected by TM CIDEP.

With a triplet spin-lattice relaxation time of T_1^T and a pseudo first-order radical formation rate constant k_f, the time development of the TM CIDEP signal contribution is given by:

$$S_{TM} = \frac{k_f P_T}{k_f + T_1^{T-1}}[1 - \exp\{-(k_f + T_1^{T-1})t\}], \qquad (3)$$

where P_T represents the initial difference in population of the T_{-1} and T_{+1} triplet states. The expression shows that transfer of triplet spin polarization requires that $k_f \geqslant 1/T_1^T$. T_1^T of triplet state molecules in fluid solution is determined by modulation of the dipole–dipole interaction between the unpaired electrons due to molecular motion. Values are expected to be in the nanosecond range [37, 38]. Hence, reaction rates of the order of 10^8 to $10^9\,s^{-1}$ are required for creation of TM CIDEP. This is a stringent requirement that may not be fulfilled [39]. Theory predicts a strong dependence of P_T on the rotational correlation time of the triplet state molecules [37]. Moreover, if the environment imposes some orientational order (i.e. triplets adsorbed on films or dissolved in liquid crystals), this will also have a strong effect on the magnitude of the effect.

Given the short spin-lattice relaxation times of triplet state molecules, the rise time of the TM signal contribution will be fast. This, together with the fact that it is independent of nuclear spin state, can be used to identify TM CIDEP. With FT-EPR, the time evolution of spectra during the first few hundred nanoseconds following a laser pulse can be monitored readily [29,30,40,41] so that it can provide unambiguous evidence of TM contributions. Quantitative analysis gives the values of T_1^T and P_T. Both parameters depend on the rotational correlation time of the triplets and, therefore, can be used to study media effects on molecular motion. This can, for instance, be useful in studies of photochemical reactions in microheterogeneous environments [42].

It should be noted that reactions involving triplet precursors may give rise to hyperfine-independent, non-Boltzmann, spin polarization in doublet radicals for other reasons as well. Firstly, the population difference, at thermal equilibrium, between T_{-1} and T_{+1} spin states is approximately twice as large as that between the β and α spin states in a doublet radical. Hence, a reaction involving triplets at thermal equilibrium will produce free radicals that initially should give enhanced-absorption EPR signals. Secondly, spin polarization can be produced by a triplet-sublevel-dependent reaction rate [43]. In either case, the polarization can be distinguished from that produced by spin-selective isc on the basis of the difference in kinetics. TM CIDEP develops with a rate determined by chemical kinetics and T_1^T. Contributions from the other two mechanisms reflect chemical kinetics only.

3.2. RADICAL PAIR MECHANISM

The formation of free radicals in a photochemical reaction involves a transient radical pair. For instance, in an excited-state electron transfer reaction:

$$D^* + A \rightarrow {}^{1,3}[D^+ \cdots A^-] \rightarrow D^+ + A^-.$$

The radical pair, $[D^+ \cdots A^-]$, can be formed in a singlet or triplet state depending on the spin state of the excited-state precursor D^*. The spin state of the pair evolves in time because of the difference in precession frequencies of the two unpaired electrons [44]. This gives rise to what is known as Radical Pair Mechanism (RPM) CIDEP. The difference in precession frequencies can be caused by a difference in g value and/or the effect of hyperfine interactions. Since the mixing terms are small compared to the Zeeman interaction, appreciable singlet-triplet mixing during the time the two radicals stay correlated, generally, can occur only between the singlet (S) and triplet T_0 states of the pair. Even for these two states the mixing terms are small compared with the $S-T_0$ splitting caused by the exchange interaction (J) between the unpaired electrons so that little polarization can develop during the brief time the radicals are in close contact. Instead, the generation of RPM spin polarization in low-viscosity solvents is accounted for in terms of a three-step process [45,46]. In the first step the reaction produces a contact radical pair. In the second, the radicals diffuse apart so that $J \sim 0$ and ST_0 mixing is effective. In the third, the radicals reencounter and the effect of the mixing is expressed in the form of excess α spin character for one radical and a concomitant excess in β spin character for the other. The originally proposed [44] one-step (encounter followed by cage escape) mechanism of ST_0 CIDEP can play a role in systems where radical pair lifetime is long, for instance in high viscosity solvents or micelles [6].

The theoretical expression for ST_0 RPM spin polarization in radical 1 in nuclear spin state a because of interaction with radical 2 in nuclear spin state b is given by [46]:

$$P_1^{ab} = C[Q_{ab}^{1/2} - \gamma Q_{ab}]. \tag{4}$$

Here, Q_{ab} represents half the difference in resonance frequencies of radicals 1 and 2, and is given by:

$$Q_{ab} = \frac{1}{2}\left\{\mu_B B_0 \Delta g + \sum_m a_{1m} m_{1m}^a - \sum_n a_{2n} m_{2n}^b\right\}. \tag{5}$$

The g value difference is given by Δg, hyperfine interactions in radicals 1 and 2 by a_{1m} and a_{2n}, respectively. The spin quantum number of the mth (nth) nucleus of radical 1 (2) in overall nuclear spin state a (b) is given by m_{1m}^a (m_{2n}^b). In a calculation of the *RPM CIDEP* spectrum of radical 1 one sums over contributions from all possible nuclear spin states of radical 2 taking nuclear spin-state degeneracy (d_b) into account:

$$P_1^a = \sum_b d_b P_1^{ab}.$$

The first term in Equation (4) represents the contribution from the re-encounter process, the second the encounter-dissociation sequence. Under most conditions $\gamma \sim 0$. The value of the proportionality constant C is a function of the mean time between relative diffusive displacements of the radicals. The sign of C depends on the sign of J and the spin state of the excited-state precursor molecule. Generally, $J < 0$ in which case, for triplet-state precursors, low-field hyperfine peaks will be in emission and high-field peaks in absorption. For reactions involving singlet excited-state precursors the pattern is reversed.

Overall, the ST_0 RPM does not create net spin polarization. Under special conditions, ST_{-1} mixing can make a contribution and in that case RPM CIDEP will have a net polarization component. As pointed out earlier, this will generally not play a role because the energy gap between S and T_{-1} states, induced by the external magnetic field, is large compared to the mixing terms. However, if hyperfine coupling is strong or the radical pair is captured in a configuration so that there is near degeneracy of the S and T_{-1} states, ST_{-1} CIDEP contributions can be observed [47–49]. Theory predicts that this mechanism gives an emission spectrum if the excited-state precursor is a triplet, an absorption spectrum is given by a singlet excited-state precursor [4–6]. The method of calculating the polarization given by this mechanism is described by McLauchlan [6].

Since the perturbation terms producing singlet-triplet mixing depend on the nuclear spin state, relative intensities of hyperfine components differ from those given by nuclear spin state degeneracy. In this respect, RPM CIDEP differs from TM CIDEP. TM CIDEP also can be distinguished from RPM CIDEP on the basis of the time evolution of the two contributions. According to Equation (3), the development of the TM signal component depends, in part, on the value of T_1^T. The rate of RPM spin polarization generation equals the rate of free radical formation.

So far the focus has been on RPM spin polarization involving geminate radical pairs. However, chemical decay caused by random encounters of free radicals also generates RPM spin polarization [4–6]. Encounters giving singlet-state pairs are reactive leaving triplet radical pairs which will produce the same spin polarization as triplet geminate pairs. The contribution to RPM CIDEP produced by random encounters of radicals is labeled F-pair spin polarization. An FT-EPR study of acetone ketyl radical formation in the reaction of 2-propanol with *tert*-butoxy radicals formed by photocleavage of *tert*-butylperoxide provides a good illustration of F-pair CIDEP [24]. RPM and TM polarization cannot be observed during the time domain that the radicals are formed. On the other hand, spectra clearly show the effect of RPM CIDEP at longer delay times.

In conclusion, it is clear that RPM CIDEP can be a source of information on the mechanism of radical formation. It can be used to determine the spin state of precursor molecules and the characteristics of radical pairs.

3.3. SPIN-CORRELATED RADICAL PAIRS

The preceding discussion has been concerned with the effects of spin dynamics in paramagnetic precursors on the spectrum of free radicals. In special cases radical pairs can live long enough so that they can be observed directly with time-resolved EPR [42,51–55]. TREPR [51–53] and FT-EPR [42,55] spectra of these spin-correlated radical pairs (SCRP) are characterized by a derivative-like lineshape. This unusual spectral feature is accounted for as follows [51, 52]. It is assumed that the radicals in the pair are weakly coupled so that the exchange interaction (J) between the unpaired electrons is small compared to the difference in resonance frequency ($\Delta\omega = \omega_1 - \omega_2$) of the two radicals. In that case the spectrum of the pair – ignoring hyperfine splittings for the moment – will consist of two pairs of lines centered at ω_1 and ω_2 with a doublet splitting of $2J$. One line of each doublet will be in absorption whereas the other will be in emission. Whether the pattern is low field absorption/high field emission or vice versa is determined by the spin state of the precursor. A triplet will give an *e/a* pattern, a singlet an *a/e* pattern. With the value of J of the order of the linewidth – and also taking into account that there may be a distribution of J values [52] – the doublet splitting will not be resolved. Instead, the partial merging of absorption and emission peaks will produce a derivative-like signal for each hyperfine component in the spectrum.

A special case arises in pulsed-EPR measurements if SCRPs dissociate into free radicals in the interval between delivery of the microwave pulse and recording of the ensuing signal [20,56–58]. In this case the exchange interaction between the unpaired electrons produces an out-of-phase (or dispersion) contribution to the FID of the cage escape products. The growth and decay of these species can be determined with a measurement of the time dependence of the amplitude of the dispersion signal [20,57,58].

The derivative-like signals produced by direct observation of SCRPs and the dispersion component carried over to the signals from *free* radicals are similar in appearance. In principle, the two can be distinguished on the basis of linewidths since peaks in spectra from SCRPs contain the unresolved exchange interaction splitting. With TREPR, which carries no phase information, only direct observation of SCRPs is possible. Hence, a comparison of results obtained with cw and pulsed EPR techniques can be useful for interpretation of derivative-like signal contributions. It must be stressed that both kinds of radical pair signals can only be observed with FT-EPR if the turning angle of the magnetization vector is $<\pi/2$ [55].

3.4. POLARIZATION FROM DOUBLET–QUARTET MIXING

The interaction between excited-state molecules and free radicals can produce spin polarization as well [59–65]. The first reports were concerned with the effect of photo-excited triplets on the EPR spectrum of doublet radicals [59,61,62]. The interaction between triplet- and doublet-state molecules in fluid solution gives an emissive signal contribution, which in some cases can be strong enough to produce a net emission signal [12,65]. It is important to note that TREPR measurements generally give a light-on minus light-off signal so that only the light-induced (emissive) polarization is observed [12,65]. FT-EPR measurements [12] demonstrate that the polarization builds up with a rate matching the rate of diffusion-controlled encounters between triplet and doublet radicals. This finding is in accord with the proposal that polarization is produced by spin state evolution of the doublet-triplet encounter complex [65,66].

Encounters between molecules in the singlet excited state and free radicals in a similar way can give rise to enhanced-absorption signals [63]. Since the relative magnitudes depend critically on singlet excited-state lifetime, both effects can be observed in the spectra of suitably chosen systems. Enhanced absorption is found shortly after laser excitation because of the interaction of the radicals with singlet excited-state molecules. An emission signal may be observed for longer delays, at which time the contribution from the longer-lived triplets is at a maximum [63].

While the polarization is not produced in a chemical reaction, it is clear that it can play a role in studies of photochemistry with time-resolved EPR. Efforts to enhance the signal from photo-generated free radicals (high solute concentration, high laser power) may inadvertently lead to conditions where the mechanism gives a significant signal contribution.

4. Applications

Most FT-EPR studies of photochemical reactions have been concerned with photo-induced electron transfer [29,30,40,41,55,57,58,67–75]. A few investigations focused on photoinduced hydrogen abstraction reactions [20,24]. As noted in the Introduction, so far only a few applications have dealt with photochemistry in heterogeneous media. Studies dealing with excited state electron transfer from porphyrins to quinones (see structural formulas) in micellar solutions have been reported [42,76–78]. The electron transfer reaction involving porphyrins and quinones adsorbed in the pores of silica gel has been studied as well [78].

In the following the information derived from spectra given by porphyrin/quinone in homogeneous solution will be considered first. This provides the necessary background for the subsequent discussion of the investigations dealing with heterogeneous media.

4.1. PORPHYRIN/QUINONE IN ETHANOL

Figure 3 shows a series of spectra obtained in a study of ZnTPP (**1**, 5×10^{-4} M)/DQ (**2**, 5×10^{-3} M) in ethanol at $-28\,°C$ [29]. The spectra are from the duroquinone anion radical (DQ^-) formed by reversible electron transfer from ZnTPP triplets produced by laser excitation (600 nm, 2 mJ/pulse). The cation radical is not observed because of the fact that its signal contribution decays within the deadtime of the spectrometer due to a short T_2 (cf. Section 2). The normal EPR spectrum of DQ^- consists of 13 lines with binomial intensity distribution because of hyperfine coupling (1.9 gauss [79]) with 12 equivalent protons. Figure

ZnTPP X = H ; ZnTPPS X = SO$_3^-$

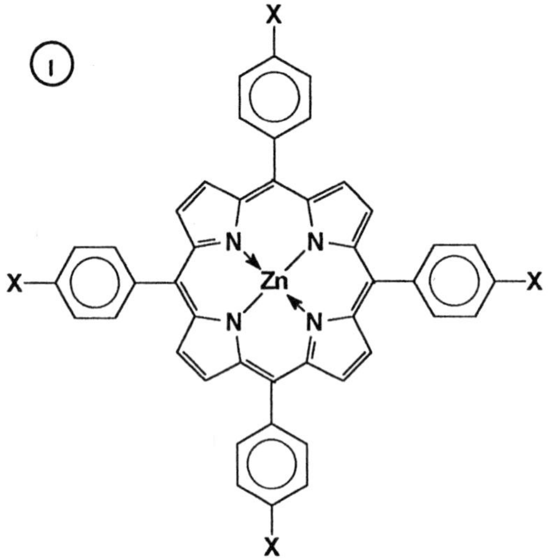

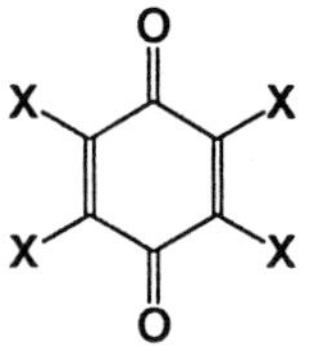

BQ X = H ; DQ X = CH$_3$

3 shows that the spectrum from the transient free radical depends strongly on delay time between microwave pulse and laser pulse both in terms of the overall signal amplitude as well as the relative amplitudes of hyperfine components. For delay times up to about 1 μs there is rapid overall signal growth. In this time period, the spectrum, which initially is in absorption, gradually turns into a spectrum in which the high-frequency (low-field) peaks are in emission. With τ_d increasing from 1 μs to 80 μs the signal intensity diminishes and emission peaks turn back into absorption, eventually giving the normal binomial intensity distribution. A number of papers have addressed the question of the interpretation of the spectra [29,30,40,41,50,67] and it has been shown that the results can be accounted for within the framework of the following reaction mechanism:

$$\text{ZnTPP} \xrightarrow{h\nu} {}^1\text{ZnTPP}^* \xrightarrow{\text{isc}} {}^3\text{ZnTPP}^*$$

$$^3\text{ZnTPP}^* + \text{DQ} \xrightarrow{k_{et}} [\text{ZnTPP}^+ \cdots \text{DQ}^-] \xrightarrow{k_{ce}} \text{ZnTPP}^+ + \text{DQ}^-$$

$$\text{ZnTPP}^+ + \text{DQ}^- \xrightarrow{k_d} \text{ZnTPP} + \text{DQ}.$$

Signal growth depends, in part, on the kinetics of electron transfer from ^{3}ZnTPP* to DQ. Since ZnTPP triplets are formed in a spin-selective isc process that produces excess population in the T_{-1} spin state [80], DQ$^-$ formed before the triplet spin system is at thermal equilibrium will show an enhanced absorption (TM CIDEP) component. Electron transfer involving the T_0 triplet state will generate ST_0 RPM polarization which is associated with a low-field-emission/high-field-absorption pattern. From the data presented in Figure 3 it can be concluded that TM CIDEP dominates at early times ($\tau_d < 200$ ns). As the triplet spin system

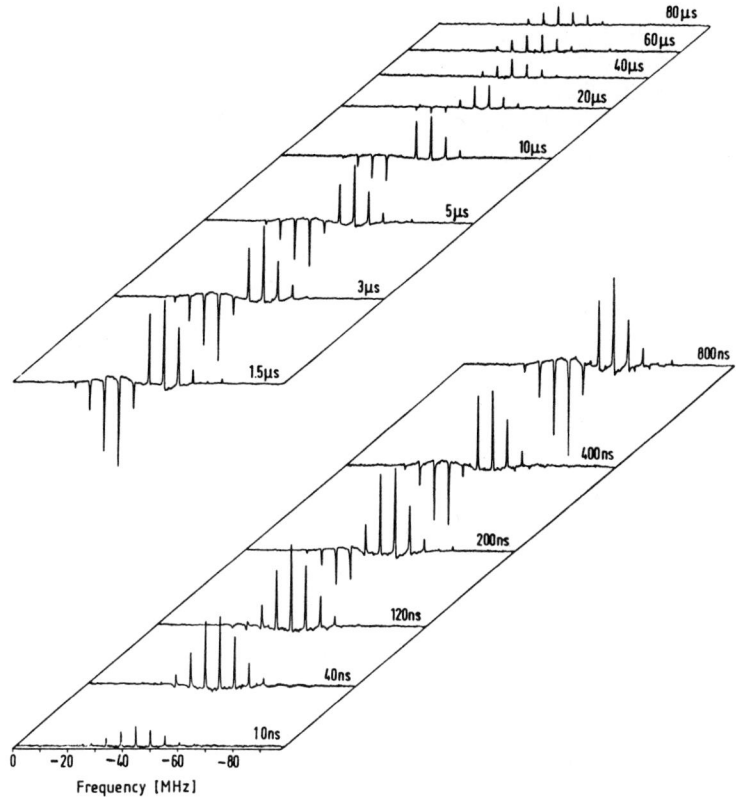

Fig. 3. FT-EPR spectrum of DQ$^-$ produced by photo-induced electron transfer from ZnTPP to DQ in ethanol at -28 °C. Delay time between laser pulse and microwave pulse ranges from 10 ns to 80 μs. The spectra represent Fourier transforms of the sum of 10000 (10 μs long) FIDs acquired with 40 Hz repetition rate. Excitation wavelength 600 nm, laser power 2 mJ, laser pulse width 15 ns. From Reference [29].

relaxes, the RPM signal contribution gains in relative importance. Consequently, low-field hyperfine lines which are initially in absorption turn into emission. The conclusion that the TM gives a significant contribution at early times is illustrated convincingly by the finding that the polarization pattern changes when the donor molecule is MgTPP [40]. In this case the isc process gives a triplet system with excess population in the T_{+1} level so that the TM gives rise to an emission signal.

A complete analysis of the time dependence of signal intensities in terms of chemical and spin dynamics can be performed in a step-wise fashion by separating TM signal contributions from RPM contributions. TM CIDEP contributes equally to pairs of hyperfine lines with overall nuclear spin states M and $-M$. Therefore, the time evolution of the RPM contribution is given by the difference in signal amplitudes (ΔS_M) of M and $-M$ hyperfine lines. In the time regime from 0 to 0.5 μs the RPM signal, to a good approximation, reflects the formation of DQ$^-$ so that the time dependence of ΔS_M can be used to determine its rate. This method of analysis gave pseudo first-order rate constants of 7.7×10^6, 1.5×10^7, and $2.2 \times 10^7 \, \text{s}^{-1}$ at -23, 0, and 27 °C, respectively [67]. Within experimental error the rate is proportional to temperature/viscosity establishing that the reaction rate is diffusion controlled.

The rate of ΔS_M decay is determined by spin-lattice relaxation and back-electron transfer. For $\tau_d < 20 \, \mu\text{s}$ the first process is the dominant decay channel and it was determined that the spin-lattice relaxation time (T_1^R) of DQ^- ranges from 4.5 μs at 27 °C to 10.4 μs at -23 °C [67].

The value of the spin-lattice relaxation time T_1^T of the triplet and the magnitude of the TM polarization were determined from the time dependence of the $M = 1$ hyperfine peak for $\tau_d < 150 \, \text{ns}$. The analysis is based on the conclusion that the RPM contribution to this peak must be small, so that signal growth is represented by Equation (3) (Section 3.1). A least-squares fit of the experimental data to Equation (3) gave 50, 40, and 20 ns for T_1^T at -23, 0, and 27 °C, respectively [67]. The theoretical expression for TM polarization relative to the Boltzmann value, P_{eq}, is [34]:

$$\frac{P_T}{P_{\text{eq}}} = \frac{4kT\omega_{\text{zfs}}}{30g\beta B_0 \omega_{\text{mw}}} \frac{T_1^T}{(T_1^T + k_{\text{et}}^{-1})} [2P_z - P_x - P_y]. \tag{6}$$

In this equation B_0 denotes the field strength, ω_{mw} and ω_{zfs} the microwave frequency (9.5 GHz) and zero-field-splitting (1 GHz), respectively. P_x, P_y, and P_z give the relative rates of population of the zero-field triplet spin states. For ZnTPP isc almost exclusively populates the $|z\rangle$ level so that $P_z \approx 1$, $P_x = P_y \approx 0$ [36,81]. Equation (6) gives $P_T/P_{\text{eq}} = 5$, a value that is in good agreement with the value of 6 derived from the experimental data [67]. Under conditions that the polarization is not attenuated by relaxation, i.e. $T_1^T \gg k_{\text{et}}^{-1}$, TM spin polarization would amount to $22 \times P_{\text{eq}}$.

An LP-SVD analysis analysis (cf. Section 2) of the FIDs given by ZnTPP/DQ in ethanol shows that there is a pronounced out-of-phase signal component for $\tau_d < 300 \, \text{ns}$ [57, 58]. The effect was attributed to the presence of spin correlated radical pairs (SCRP) at the time the microwave pulse is delivered. These radical pairs have vanished for the most part by the time the FID is measured. However, as pointed out in Section 3.3, the exchange interaction between the unpaired electrons of the pair in the interval between microwave pulse and signal detection generates a characteristic phase shift in the FID of the free radicals [56–58]. A study of the time evolution of the out-of-phase signal [58] gave the rate constants of formation and decay of the SCRPs. The rate of radical-pair generation was found to match the rate of free radical formation. The rate of dissociation k_{ce} ranged from $3.2 \times 10^6 \, \text{s}^{-1}$ at -48 °C to $1.0 \times 10^7 \, \text{s}^{-1}$ at 25 °C. In a low viscosity solvent like ethanol, free translational diffusion would be expected to give a much shorter SCRP lifetime (4–40 ns) than measured experimentally. The unusually long lifetime has been interpreted in terms of a model that assumes that an attractive potential of the order of a few kT restricts relative motion of the radicals [58,82]. In [ZnTPP$^+\cdots$DQ$^-$], the electrostatic interaction between the redox ions obviously can account for restricted diffusive separation.

A pronounced solvent effect on SCRP characteristics was found in an FT-EPR study of the anthraquinone/triethylamine photoredox system [71]. In this case electron transfer kinetics was monitored via the spectrum of the antraquinone anion radical (AQ^-). An order of magnitude reduction in the rate constant of AQ^- formation was found upon going from methanol to 2-propanol as solvent. The increase in solvent viscosity cannot be solely responsible for this effect. The strong solvent dependence was attributed to the change in cation radical-anion radical binding potential induced by a change in solvent dielectric constant. As will be shown further on, the fact that EPR spectra can show the effect of radical-radical interactions can be used to get information on the spatial distribution of transient paramagnetic molecules in heterogeneous media.

Instead of the step-wise data analysis method described above, the evolution of the intensities of hyperfine peaks, over a time span from nanoseconds to tens of microseconds, also can be analyzed with comprehensive kinetic models that account for radical formation and decay as well as spin dynamics (i.e. TM CIDEP, RPM CIDEP, and spin-lattice relaxation) [20,68]. Apart from giving numerical data on chemical rates and spin-lattice relaxation rates, such an analysis also gives the magnitudes of spin polarization generated by RPM and TM CIDEP.

An analysis [68,69] of the time dependence of the intensities of three hyperfine peaks in spectra given by ZnTPP/benzoquinone (BQ, **2**) in ethanol gave $T_1^T = 28 \pm 10$ ns. The initial spin polarization of the ^{3}ZnTPP* triplets relative to thermal equilibrium was found to be 22 ± 2. The result is in excellent agreement with the value predicted by theory and the result derived from the study of ZnTPP/DQ [67]. The relative amplitudes of hyperfine components due to RPM CIDEP derived from the spectra are $-2.1 : -6.9 : -6.0 : -0.88 : +0.39$ (the results are given in order of increasing field, negative values denote emission signals). Relative intensities given by ST_0 spin polarization can be calculated with Equations (4) and (5) (Section 3.2). For $\gamma = 0$, the calculated values are $-1.8 : -5.9 : -6.0 : -1.1 : +0.5$. The systematic deviation between experimental and theoretical values suggests that the second term in Equation (5) cannot be neglected. This conclusion is supported by the finding that a reduction in temperature enhances the deviations. A TREPR study of ZnTPP/BQ in ethanol by Schlüpmann et al. finds evidence of an ST_{-1} signal contribution [83]. The values of the RPM spin polarization relative to Boltzmann polarization are $-42, -35, -20, -4.1$, and $+8.1$ [69].

4.2. PORPHYRIN/QUINONE IN MICELLAR SOLUTIONS

A study has been made of ZnTPP/BQ in a micellar solution of cetyl trimethyl ammonium chloride (CTAC) [76] and of ZnTPPS (**1**)/DQ in micellar solutions made up of anionic, cationic, or neutral surfactant molecules [42,77,78]. Interest in these systems stems in part from the potential relevance to applications in conversion and storage of solar energy [84,85]. Also, studies of these systems can contribute to the understanding of the factors that play a role in electron transfer across boundaries between hydrophobic and hydrophilic domains. FT-EPR measurements can give a unique insight into structure and dynamics of these systems because of the fact that CIDEP contributions, lineshapes, linewidths, and relaxation times are sensitive to the spatial distribution of reactants and products, and motional correlation times.

In the study of ZnTPP/BQ in CTAC, a comprehensive analysis of the time evolution of the FT-EPR spectrum of the BQ anion radical yielded data on the rate constants of radical formation and decay as well as the magnitudes of spin polarization. Additional information was provided by linewidths data. It was found that the rate of BQ$^-$ formation is small compared to that found in homogeneous solution at comparable concentrations of donor and acceptor. By contrast, the rate of the back reaction is similar to that found in ethanol. Furthermore, the linewidths of the hyperfine components were similar to those found for BQ$^-$ in aqueous solution. The slow forward rate indicates that electron transfer mainly takes place from ZnTPP solubilized in the micelles to BQ in the aqueous phase [76]. Linewidth data show that the anion radicals formed in the electron transfer process remain in the aqueous phase. However, to account for the relatively fast back electron transfer, notwithstanding the compartmentalization of the redox products, one has to assume that BQ$^-$ must remain trapped at the micelle-water interface, i.e. close to its redox partner ZnTPP$^+$ in the micelle.

In the case of ZnTPPS/DQ, the acceptor molecules are expected to be, almost exclusively, inside the micelles because of the low water solubility of the quinone. The location of the anionic porphyrin depends on the charge of the polar headgroup of the surfactant molecules [86]. As a result, changes in surfactant molecules are found to have a profound effect on the FT-EPR spectra [42,77,78].

Representative FT-EPR spectra from of DQ$^-$ in a Triton X-*100* (TX*100*) solution generated by photo-induced electron transfer from ZnTPPS are shown in Figure 4. The spectra are qualitatively similar to those found for ZnTPP/DQ in ethanol (cf. Figure 3) [67]. At early times ($\tau_d < 100$ ns) a TM CIDEP contribution dominates which establishes that the reaction involves ZnTPPS triplets. For $\tau_d > 100$ ns the RPM signal dominates. The low-frequency absorption/high-frequency emission pattern observed in this time domain also is consistent with triplet excited state electron transfer. However, contrary to what is found in homogeneous solution, the growth of the RPM signal cannot be represented by a single exponential. Instead, it appears to involve two consecutive steps with first-order rate constants of 1.7×10^7 and $1.4 \times 10^6 \, \text{s}^{-1}$. This suggests that the kinetics of *free* DQ$^-$ formation is determined by the radical pair formation step as well as the cage-escape step. Even the fastest of the two steps has a rate constant that is smaller than the value of $2.2 \times 10^7 \, \text{s}^{-1}$ found for ZnTPP/DQ ([DQ] = 5×10^{-3} M) in ethanol. The result is surprising because acceptor as well as donor molecules are associated with the micelles [86]. It is estimated that the average distance between reactants is less than 20 Å compared to ~38 Å for the ZnTPP/DQ in ethanol system. With such a short distance between reactants one would expect fast electron transfer, with a significant singlet excited state contribution. Instead, the spectral data show a relatively slow reaction with no evidence of DQ$^-$ formation via the singlet excited state. However, the fluorescence of ZnTPPS in TX100 is quenched appreciably by the addition of DQ. It is concluded, therefore, that electron-transfer quenching of the singlet excited state does occur, but does not yield free DQ$^-$ because of fast back electron transfer. If singlet state quenching inactivates closely spaced reaction partners, the average spacing between ^{3}ZnTPPS* and DQ can be substantially larger than 20 Å. This would account for the relatively slow rate of triplet excited state electron transfer.

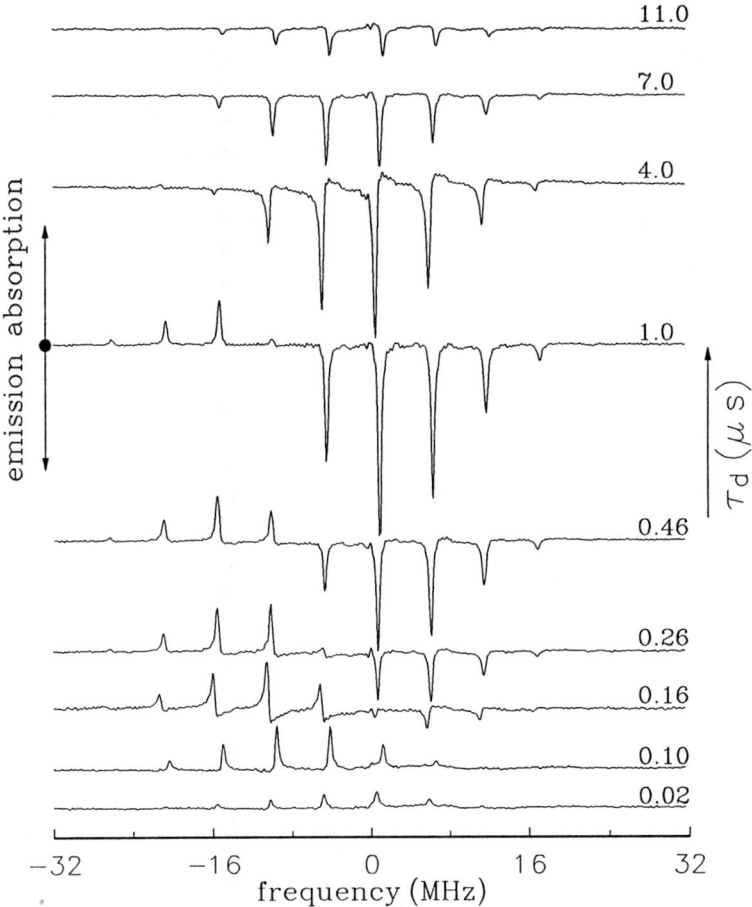

Fig. 4. FT-EPR spectra of photo-generated DQ$^-$ in TX100 solution for delay times between laser excitation of ZnTPPS and microwave pulse ranging from 20 ns to 11 μs. The central hyperfine line ($M = 0$) is at ≈ -4.5 MHz. From Reference [42].

An analysis of the time evolution of the TM CIDEP contribution shows that T_1^T is 100–150 ns, a factor of 4 longer than in homogeneous solution [67]. The result is consistent with the conclusion that ZnTPPS is associated with the micelles. Molecular motion in the micelles will be inhibited which is expected to increase the spin-lattice relaxation time. By contrast, T_1^R of DQ$^-$ is similar to the value found in ethanol. In addition, the linewidths of the hyperfine lines match those found in homogeneous solution (cf. Figure 5). Restricted molecular motion inside the micelle and interaction with the porphyrin radical would cause severe line-broadening. Hence, the findings indicate that the anion radicals move into the aqueous phase. A slight reduction in linewidth with increase in delay between laser pulse and microwave pulse, apparent for $\tau_d < 100$ ns may reflect the movement of DQ$^-$ across the micelle/water interface.

The spectra given by ZnTPPS/DQ in sodium dodecylsulfate (SDS) micellar

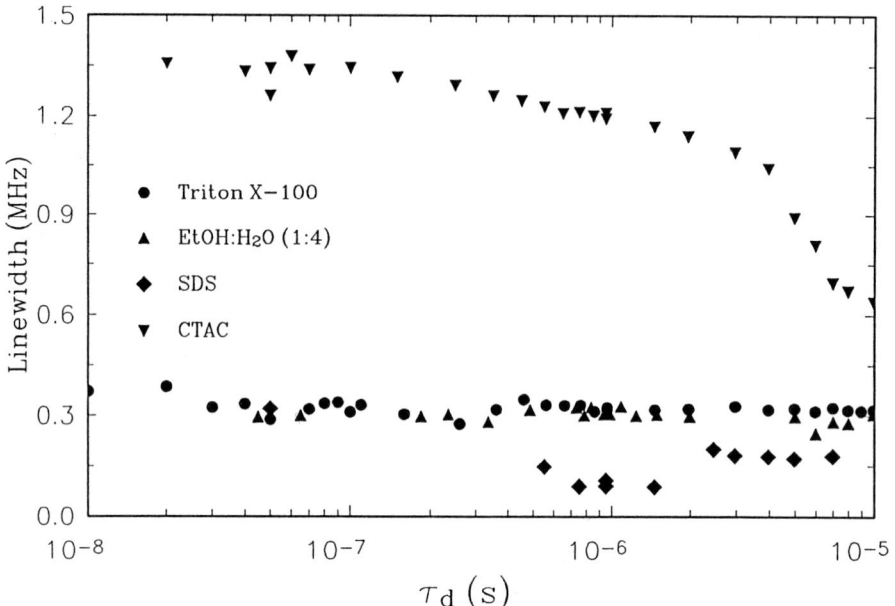

Fig. 5. Time evolution of the linewidth ($\Delta\nu = 1/\pi T_2$) of the central line in the spectra from DQ^- in TX100 (●), CTAC (▼), and SDS (◆). The linewidth of the $M = 0$ line in the spectrum from BQ^- in H_2O/ethanol (4:1) (▲) is given for comparison. From Reference [42].

solution displayed in Figure 6 are very similar to those found for ZnTPPS/DQ in homogeneous solution as well. The main differences with the results obtained with ZnTPPS/DQ in TX100 are (1) a significant reduction in signal and (2) a reduced rate of DQ^- formation. The findings reflect the difference in reactant distribution in the two media. In SDS the ZnTPPS molecules are found in the aqueous phase rather than inside the micelles [86]. Hence, electron transfer must involve encounters of ^{3}ZnTPPS* molecules with the few DQ molecules that are in the aqueous phase as well. The compartmentalization of reactants leads to a reduction in the pseudo first-order reaction rate. Fluorescence measurements show that in SDS solution ^{1}ZnTPPS* quenching by DQ is significant. Optical absorption spectra indicate that this is due to ground-state complexation. Apparently, singlet excited state quenching does not generate free DQ^- and, by diminishing the ^{3}ZnTPPS* concentration, reduces the EPR signal intensity.

Figure 7 depicts spectra from ZnTPPS/DQ in CTAC solution. It is evident that the spectra recorded for $\tau_d < 5\ \mu s$ differ fundamentally from those given by the other systems. All peaks have a derivative-like lineshape in this time regime. The lineshape bears a striking resemblance to that found in the FT-EPR spectrum from the photoreduced quinone moiety in a covalently-linked donor-acceptor tetrad [55]. A similar lineshape has been found in TREPR spectra of photo-generated radicals in viscous and micellar solutions [51–53]. The unique lineshape is a manifestation of the presence of SCRPs ([ZnTPPS^{3-}···DQ$^-$]) present at the time the FID is recorded. As explained in Section 3.3, a weak exchange interaction – in this case of the order of 1 MHz or less – between the unpaired electrons splits

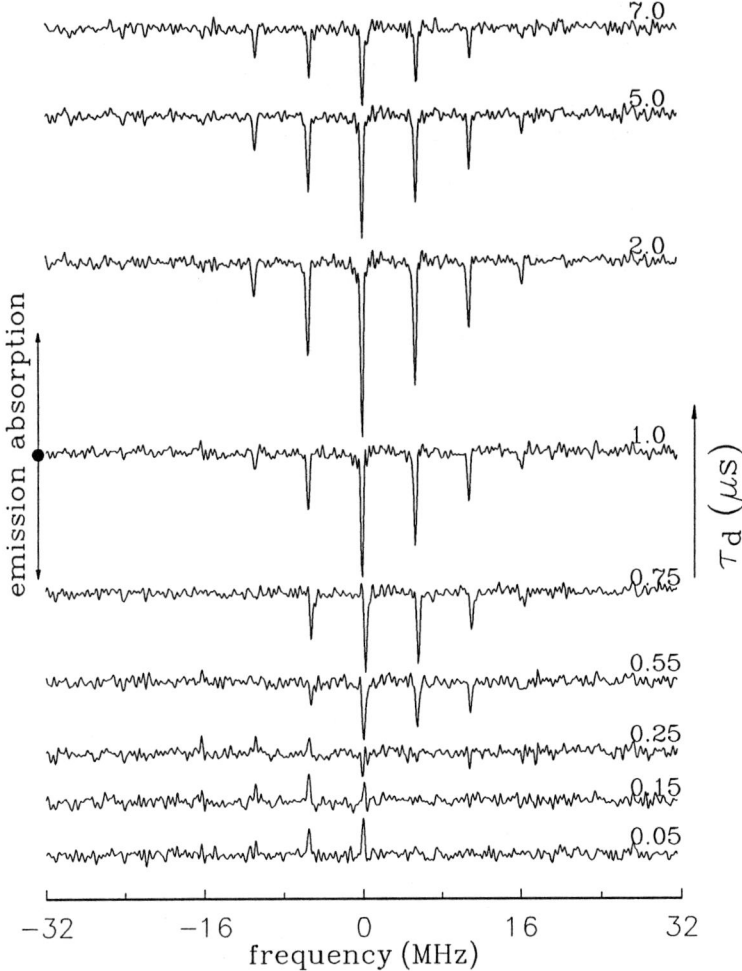

Fig. 6. FT-EPR spectra of photogenerated DQ^- in SDS solution for delay times between laser excitation of ZnTPPS and microwave pulse ranging from 50 ns to 7 μs. The central hyperfine line ($M = 0$) is at ≈0 MHz. From Reference [42].

every hyperfine line into an unresolved doublet. The doublets exhibit a low-frequency absorption/high-frequency emission pattern, consistent with formation via a triplet-state precursor. By contrast, in the covalently-linked donor-acceptor system electron transfer involves the singlet excited state and the radical pair spectrum displays an emission/absorption pattern [55]. The formation and decay of the radical pair spectrum can be described reasonably well with single exponentials ($k_1 = 3.6 \times 10^6 \, s^{-1}$, $k_2 = 5.0 \times 10^5 \, s^{-1}$). In CTAC, as in TX100, electron transfer is between reactants associated with the micelles [86]. The long lifetime of SCRPs – the maximum signal amplitude is given for $\tau_d \approx 0.8 \, \mu s$ – is attributed to interaction between the anionic reaction products and the positive electrostatic

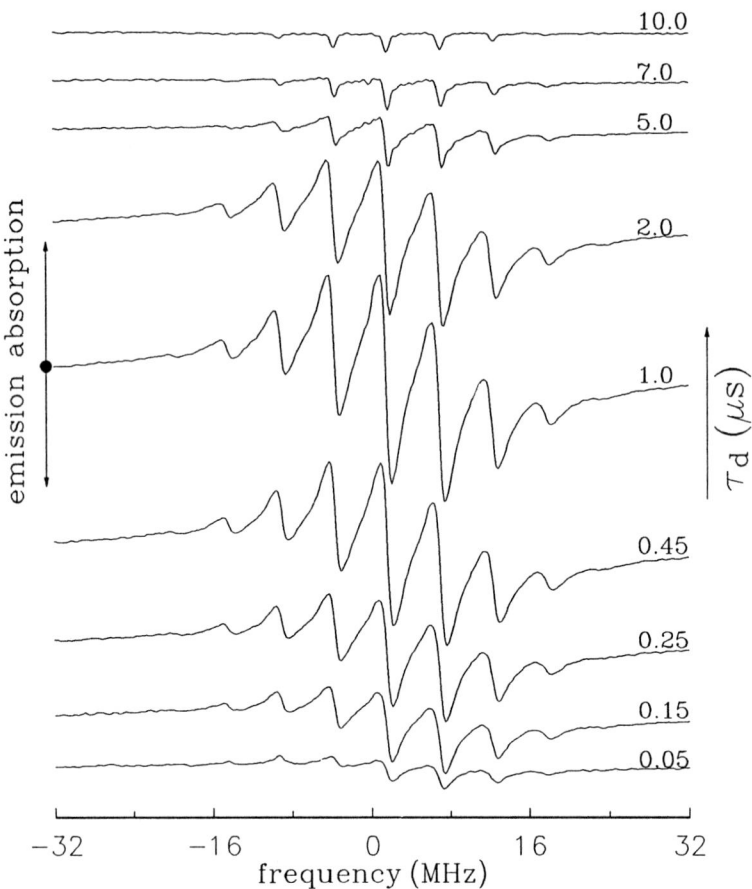

Fig. 7. FT-EPR spectra of photogenerated DQ$^-$ in CTAC solution for delay times between laser excitation of ZnTPPS and microwave pulse ranging from 50 ns to 10 µs. The central hyperfine line ($M = 0$) is at ≈7 MHz. From Reference [42].

field generated by the cationic headgroups of surfactant molecules. The interaction inhibits translational diffusion of the redox products.

The spectrum of free DQ$^-$ is observed when $\tau_d > 5$ µs. At that time the spectrum still reflects spin polarization (most hyperfine components are in emission, cf. Figure 7). From the decay of the polarization it is deduced that T_1^R of DQ$^-$ in CTAC is ~8 µs as compared to 4.5 µs in TX100 solution. Figure 5 shows that the linewidth in the spectrum of 'free' DQ$^-$ in CTAC solution also is significantly larger than in TX100 and SDS. These results indicate that the anion radicals remain trapped in the cationic micelles.

Figure 8 shows that the linewidths in the spectrum of the SCRP are a function of nuclear spin state (M) of DQ$^-$. The M dependence can be described by the polynomial:

$$\Delta \nu_M = a + bM + cM^2. \tag{7}$$

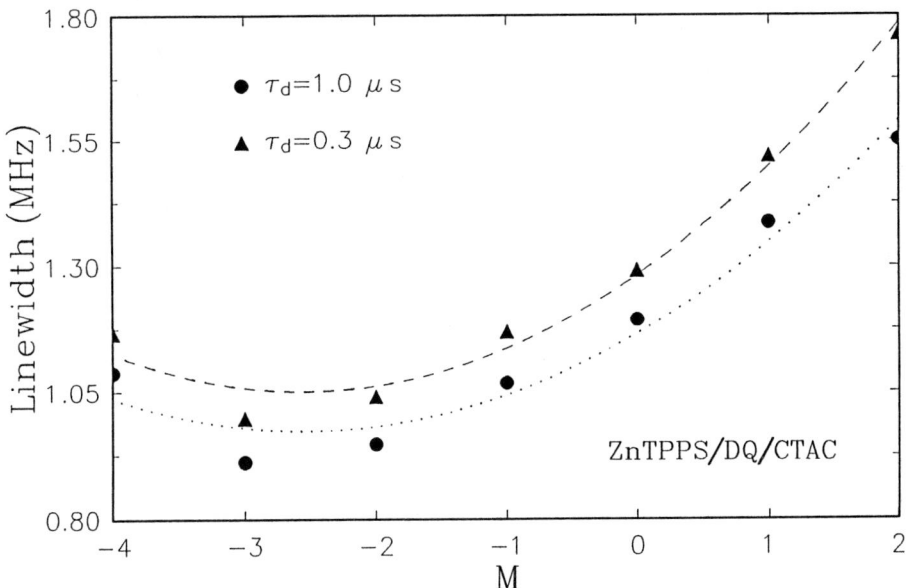

Fig. 8. Dependence of the linewidth ($\Delta \nu_M = 1/\pi T_2(M)$) on the total nuclear spin quantum number for DQ^- in the ZnTPPS/DQ/CTAC system for τ_d settings of 0.3 μs (▲) and 1.0 μs (●). The dashed and dotted lines represent the least-squares fittings of the data points to Equation (7). The coefficients (in MHz) derived from the fits are: $a = 1.28$ (1.16), $b = -0.18$ (-0.15), and $c = 0.035$ (0.03) for $\tau_d = 0.3$ (1.0) μs. From Reference [42].

Values of the coefficients a, b, and c derived from a least-squares analysis of the linewidths data are given in the caption of Figure 8. The M dependence is attributed to incomplete averaging of g and hyperfine anisotropies by rotational motion [87,88]. The rotational correlation time, τ_R, can be derived from the coefficients with the aid of the theoretical expressions given in the paper by Suga et al. [88]. The τ_R values derived from b and c are 10.5 and 3.3 ns, respectively. The paper by Leniart et al. gives the experimentally determined values of b and c as function of τ_R for DQ^- in ethanol [87]. An extrapolation of their results gives values of 24 and 10 ns. It is concluded that the linewidth data point to strongly inhibited rotational motion of the radicals. By comparison, $\tau_R = 2.3$ ns for DQ^- in ethanol at $-70\,°C$ [87]. The systematic deviation between the rotational correlation times derived from the two coefficients can be due to non-random rotational motion of DQ^- as a result of the interaction with the electrostatic field inside the micelles.

4.3. PORPHYRIN/QUINONE ADSORBED ON SILICA GEL

Photochemistry involving reactants adsorbed on solid supports – such as silica gels and zeolites – is of considerable current interest [89]. This is because the constraints imposed on spatial organization and motion of reactants can enhance product yield and suppress undesired side reactions. Time-resolved EPR is an ideal spectroscopic technique for this field of research particularly because it can provide information on molecular motion of transient paramagnetic species and on the characteristics of radical pairs. The value of the technique in this field of research has been illustrated in recent publications by Forbes et al. [90,91].

FT-EPR work has been concerned with photo-induced electron transfer from ZnTPP to BQ adsorbed in the pores of silica gel (Davisil 634, 60 Å pore) [77, 78]. Because the electron transfer process is reversible, this is a convenient system to investigate as it does not require sample flow through the cavity. Sample preparation followed the procedure given by Johnston et al. [92]. If it assumed that all the donor and acceptor molecules end up in the solvent captured in the silica gel pores, concentrations are estimated to be 6 and 60 mM, respectively.

In the presence of oxygen only a weak spectrum of ^{3}ZnTPP* can be obtained with TREPR and this required cooling of the sample to $-100\,°C$ or below. After purging of ZnTPP/silica samples with argon, TREPR spectra of triplet-state ZnTPP could be observed even at room temperature. Modulation of the electron spin–spin dipolar interaction by molecular motion would lead to line broadening which would prevent the detection of the triplet EPR spectrum. It is concluded, therefore, that rotational motion of the pophyrin is essentially stopped. TREPR measurements of ZnTPP/silica samples purged with solvent-saturated argon reveal a triplet EPR spectrum only at reduced temperature. Apparently, adsorption of solvent molecules restores the motion of the porphyrin. This finding is of importance, since it implies that at room temperature donor and acceptor molecules have some motional freedom in silica gel pores as long as solvent molecules are present.

'Dry' ZnTPP/BQ/silica samples and ZnTPP/BQ/silica samples saturated with hexane did not give photoinduced FT-EPR spectra. If excited-state electron transfer does occur, the back reaction must be so fast that redox ion products cannot be detected. By contrast, samples exposed to methanol or ethanol gave well-resolved FT-EPR spectra from photogenerated BQ^-. Spectra from BQ^- obtained from a ZnTPP/BQ/silica sample saturated with methanol are given in Figure 9. The time dependence of peak intensities is displayed in Figure 10. It has the following characteristics: (1) the rise time of the signal is close to instrument limited (50 ns); (2) the spectrum at early times is in enhanced absorption; (3) signals persist into the millisecond time domain; and (4) for τ_d values up to tens of microseconds, the FIDs display unusual phase effects and deviations from the binomial intensity pattern are observed even in the millisecond time domain (cf. insets of Figures 9 and 10).

From the fact that the signal initially is in enhanced absorption it is deduced that redox products are formed by triplet excited state electron transfer. The rise time of the TM signal component (~ 50 ns) shows that the spin-lattice relaxation time of ^{3}ZnTPP* is similar to that found in homogeneous solution [67,68] which is indicative of free rotational motion of the donor. The relatively small RPM contribution grows in with a rate of about $1.5 \times 10^7\,s^{-1}$ when the solvent is methanol and $0.8 \times 10^7\,s^{-1}$ with ethanol as solvent. These values are approximate because the time evolution of the signals is not purely exponential. The unusually fast signal rise time shows that BQ^- formation must involve reaction partners that are closely spaced at the time of laser excitation of ZnTPP.

The change from methanol to ethanol as solvent also is accompanied by a

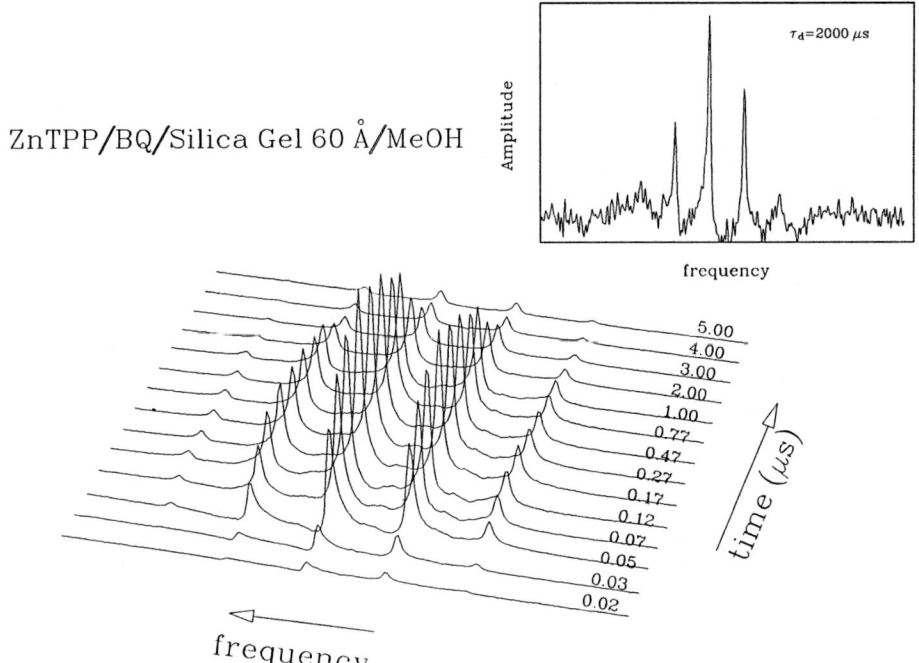

Fig. 9. FT-EPR spectra from BQ$^-$ formed by electron transfer from ZnTPP to BQ adsorbed in silica gel (pore size 60 Å, pores filled with methanol). The inset depicts the spectrum obtained with τ_d = 2 ms. From Reference [78].

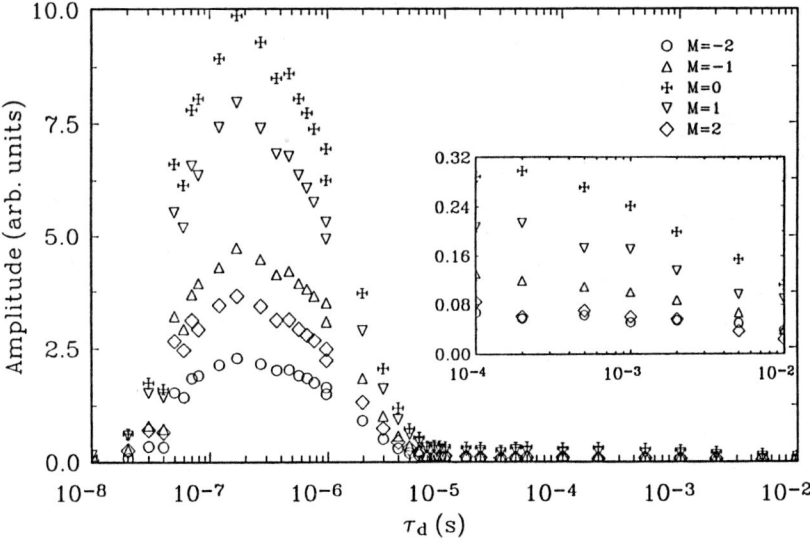

Fig. 10. Time dependence of the intensities of the five hyperfine components in the spectra from BQ$^-$ given in Figure 9. The inset highlights the dependence for long delay times. From Reference [78].

substantial increase in linewidth (from ~1 to 3 MHz). By comparison, the linewidth found for ZnTPP/BQ in ethanol is 0.26 MHz [68]. The large linewidth in the silica/alcohol environment can be due to a number of factors: there can be a linewidth contribution from spin–spin interaction between redox partners, incomplete averaging of hyperfine and g value anisotropies may play a role, and homogeneous electron transfer

$$BQ^- + BQ \xrightarrow{k_{et}} BQ + BQ^-$$

is expected to affect the linewidth.

A striking result is that the signal from BQ^- is generated with a rate which is appreciably higher than that found for ZnTPP/BQ in ethanol [68]. On the other hand, a fraction of the free radicals formed persists for a very long time. These apparently contradictary results can be explained by considering sample topology. First, the acceptor concentration is high so that homogeneous electron transfer can play an important role. Electrostatic interaction between redox partners and restrictions imposed by the heterogeneous medium are expected to inhibit charge separation, but electron hopping from BQ to BQ may counteract these effects. Second, ZnTPP molecules near the surface of the silica gel particles will be preferentially excited so that most electron transfer events will occur near the surface as well. Following electron transfer, translational diffusion and homogeneous electron transfer can move a fraction of the anion radicals to the interior of the particles. Back electron transfer involving these anions is expected to be a slow process.

Signal decay in the time interval from 200 ns to 10 μs is reasonably well described by a single exponential with a time constant of ≈ 1 μs. The decay is attributed to establishment of thermal equilibrium by spin-lattice relaxation. The fact that T_1^R is more than a factor of two shorter than that found in homogeneous solution [68] probably is due to spin-spin interaction with the cation radical.

It must be stressed that in the application of FT-EPR in the study of photochemistry in heterogeneous media the standard methods of analysis of FIDs may not reveal all the information contained in the time-domain signal. For instance, during the recording of the FID from DQ^- generated by electron transfer from ZnTPPS to DQ in micellar solution, radical pairs may dissociate into free radicals or DQ^- may move from the lipophilic to the aqueous phase. Evidently, the FIDs contain a complete record of these changes in characteristics of the paramagnetic species. The extraction of the information from the time-domain signal will require development of additional methods of data analysis.

Acknowledgement

Financial support for this work was provided by the Division of Chemical Sciences, Office of Basic Energy Sciences of the U.S. Department of Energy (DE-FG02-84ER-13242).

References

1. J. E. Wertz: *Chem. Rev.* **55**, 829 (1955).
2. J. A. Weil, J. R. Bolton, and J. A. Wertz: *Electron Paramagnetic Resonance*, Wiley, New York (1993).
3. K. M. Salikhov, Y. N. Molin, R. Z. Sagdeev, and A. L. Buchachenko: in Y. N. Molin (ed.), *Spin Polarization and Magnetic Effects in Radical Reactions*, Elsevier, Amsterdam (1984).
4. A. D. Trifunac, R. G. Lawler, D. M. Bartels, and M. C. Thurnauer: *Progress in Reaction Kinetics* **14**, 43 (1986).
5. P. J. Hore: in A. J. Hoff (ed.), *Advanced EPR: Applications in Biology and Biochemistry*, Elsevier, Amsterdam, pp. 405–440 (1989).
6. K. A. McLaughlan: in L. Kevan and M. K. Bowman (eds.), *Modern Pulsed and Continuous-Wave Electron Spin Resonance*, Wiley, New York, pp. 285–364 (1990).
7. B. Smaller, J. R. Remko, and E. C. Avery: *J. Chem. Phys.* **48**, 5174 (1968).
8. H. Paul: *Chem. Phys.* **40**, 265 (1979); **43**, 294 (1979).
9. N. C. Verma and R. W. Fessenden: *J. Chem. Phys.* **58**, 2501 (1973).
10. S. S. Kim and S. I. Weissman: *J. Magn. Res.* **24**, 167 (1976).
11. M. D. E. Forbes, J. Peterson, and C. S. Breivogel: *Rev. Sci. Instr.* **62**, 2662 (1991).
12. J. N. Turro, I. V. Koptyung, H. van Willigen, and K. A. McLauchlan: *J. Magn. Res.* **A109**, 121 (1994).
13. A. D. Milov, M. D. Schirov, V. E. Khmelinskii, and Y. D. Tsvetkov: *Dokl. Acad. Nauk SSR* **218**, 878 (1974).
14. A. D. Trifunac and J. R. Norris: *Chem. Phys. Lett.* **59**, 140 (1978).
15. A. D. Trifunac and R. G. Lawler: *Chem. Phys. Lett.* **84**, 515 (1981).
16. D. M. Bartels, R. G. Lawler, and A. D. Trifunac: *J. Chem. Phys.* **83**, 2686 (1985).
17. O. Dobbert, T. Prisner, and K. P. Dinse: *J. Magn. Res.* **70**, 173 (1986).
18. R. J. Massoth: Thesis, University of Kansas (1987).
19. J. Gorcester and J. H. Freed: *J. Chem. Phys.* **88**, 4678 (1988).
20. P. R. Levstein and H. van Willigen: *J. Chem. Phys.* **95**, 900 (1991).
21. For a recent review see H. van Willigen, P. R. Levstein, and M. H. Ebersole: *Chem. Rev.* **93**, 173 (1993).
22. A. Derome: *Modern NMR Techniques for Chemistry Research*, Pergamon Press, Oxford (1987).
23. A. Schweiger: *Angew. Chem., Int. Ed.* **30**, 265 (1991).
24. P. R. Levstein, P. Doering, and H. van Willigen: *Chem. Phys. Lett.* **197**, 265 (1992).
25. W. Froncisz and J. S. Hyde: *J. Magn. Res.* **47**, 515 (1982).
26. S. Pfenninger, J. Forrer, A. Schweiger, and Th. Weiland: *Rev. Sci. Instrum.* **59**, 752 (1988).
27. A. Schweiger: in L. Kevan and M. K. Bowman (eds.), *Modern Pulsed and Continuous-Wave Electron Spin Resonance*, Wiley, New York, pp. 43–118 (1990).
28. J. H. Freed: in L. Kevan and M. K. Bowman (eds.), *Modern Pulsed and Continuous-Wave Electron Spin Resonance*, Wiley, New York, pp. 119–194 (1990).
29. T. Prisner, O. Dobbert, K. P. Dinse, and H. van Willigen: *J. Am. Chem. Soc.* **110**, 1622 (1988).
30. K. P. Dinse, H. van Willigen, O. Dobbert, and T. Prisner: in C. P. Keijzers, E. J. Reijerse, and J. Schmidt (eds.), *Pulsed EPR: A New Field of Applications*, North Holland, Amsterdam, pp. 89–95 (1989).
31. H. Barkhuijsen, R. de Beer, W. M. M. J. Bouvée, and D. van Ormondt: *J. Magn. Res.* **61**, 465 (1985).
32. R. de Beer and D. van Ormondt: in A. Hoff (ed.), *Advanced EPR: Applications in Biology and Biochemistry*, Elsevier, Amsterdam, pp. 135–173 (1989).
33. J. K. S. Wan, S. K. Wong, and D. A. Hutchinson: *Acc. Chem. Res.* **7**, 58 (1974).
34. F. J. Adrian: *J. Chem. Phys.* **61**, 4875 (1974).
35. M. A. El-Sayed: *Ann. Rev. Phys. Chem.* **26**, 235 (1975).
36. J. H. van der Waals, W. G. van Dorp, and T. J. Schaafsma: in D. Dolphin (ed.), *The Porphyrins*, Vol. 4, Academic Press, New York, pp. 257–312 (1979).
37. P. W. Atkins and G. T. Evans: *Mol. Phys.* **27**, 1633 (1974).
38. P. W. Atkins, A. J. Dobbs, and K. A. McLauchlan: *Chem. Phys. Lett.* **29**, 616 (1974).
39. S. Basu, A. I. Grant, and K. A. McLauchlan: *Chem. Phys. Lett.* **94**, 517 (1983).

40. M. K. Bowman, M. Toporowicz, J. R. Norris, T. J. Michalski, A. Angerhofer and H. Levanon: *Isr. J. Chem.* **28**, 215 (1988).
41. A. Angerhofer, M. Toporowicz, M. K. Bowman, J. R. Norris, and H. Levanon: *J. Phys. Chem.* **92**, 7164 (1988).
42. P. R. Levstein and H. van Willigen: *Chem. Phys. Lett.* **187**, 415 (1991).
43. D. A. Leinwand, S. M. Lefkowitz, and H. C. Brenner: *J. Am. Chem. Soc.* **107**, 6179 (1985).
44. R. Kaptein and L. J. Oosterhoff: *Chem. Phys. Lett* **4**, 214 (1969).
45. F. J. Adrian: *J. Chem. Phys.* **54**, 3918 (1971).
46. J. H. Freed and J. B. Pedersen: *Adv. Magn. Res.* **8**, 1 (1976).
47. A. D. Trifunac: *Chem. Phys. Lett.* **49**, 457 (1977).
48. A. D. Trifunac and D. J. Nelson: *Chem. Phys. Lett.* **46**, 346 (1977).
49. T. J. Burkey, J. Lusztyk, K. U. Ingold, J. K. S. Wan, and F. J. Adrian: *J. Phys. Chem.* **89**, 4286 (1985).
50. H. van Willigen, M. Vuolle, and K. P. Dinse: *J. Phys. Chem.* **93**, 2441 (1989).
51. C. D. Buckley, D. A. Hunter, P. J. Hore, and K. A. McLauchlan: *Chem. Phys. Lett.* **135**, 307 (1987).
52. G. L. Closs, M.D. E. Forbes, and J. R. Norris: *J. Phys. Chem.* **91**, 3592 (1987).
53. K. Tominaga, S. Yamauchi, and N. Hirota: *Chem. Phys. Lett.* **149**, 32 (1988).
54. D. Stehlik, C. H. Bock, and J. Petersen: *J. Phys. Chem.* **93**, 1612 (1989).
55. K. Hasharoni, H. Levanon, J. Tang, M. K. Bowman, J. R. Norris, D. Gust, T. A. Moore, and A. L. Moore: *J. Am. Chem. Soc.* **112**, 6477 (1990).
56. M. C. Thurnauer and J. R. Norris: *Chem. Phys. Lett.* **76**, 557 (1980).
57. K. P. Dinse, M. Plüschau, G. Kroll, T. Prisner, and H. van Willigen: *Bull. Magn. Res.* **11**, 174 (1989).
58. G. Kroll, M. Plüschau, K. P. Dinse, and H. van Willigen: *J. Chem. Phys.* **93**, 8709 (1990).
59. T. Imamura, O. Onitsuka, and K. Obi: *J. Phys. Chem.* **90**, 6741 (1986).
60. W. S. Jenks and N. J. Turro: *Tetrahedron Lett.* **30**, 4469 (1989).
61. C. Blättler, F. Jent, and H. Paul: *Chem. Phys. Lett.* **166**, 375 (1990).
62. A. Kawai, T. Okutsu, and K. Obi: *J. Phys. Chem.* **95**, 9130 (1991).
63. A. Kawai and K. Obi: *J. Phys. Chem.* **96**, 52 (1992).
64. C. Blättler and H. Paul: *Res. Chem. Intermed.* **16**, 201 (1991).
65. G. H. Goudsmit, H. Paul, and A. Shushin: *J. Phys. Chem.* **97**, 13243 (1993).
66. A. Shushin: *Chem. Phys. Lett.* **208**, 173 (1993).
67. M. Plüschau, A. Zahl, K. P. Dinse, and H. van Willigen: *J. Chem. Phys.* **90**, 3153 (1989).
68. M. Ebersole, P. R. Levstein, and H. van Willigen: *J. Phys. Chem.* **96**, 9310 (1992).
69. M. Ebersole: Thesis, University of Massachusetts (1992).
70. A. Berman, A. Michaeli, J. Feitelson, M. K. Bowman, J. R. Norris, H. Levanon, E. Vogel, and P. Koch: *J. Phys. Chem.* **96**, 3041 (1992).
71. D. Beckert, M. Plüschau, and K. P. Dinse: *J. Phys. Chem.* **96**, 3193 (1992).
72. M. Plüschau, G. Kroll, K.-P. Dinse, and D. Beckert: *J. Phys. Chem.* **96**, 8820 (1992).
73. G. Zilber, V. Rozenshtein, M. Rabinovitz, and H. Levanon: *Chem. Phys. Lett.* **196**, 255 (1992).
74. V. Rozenshtein, G. Zilber, M. Rabinovitz, and H. Levanon: *J. Am. Chem. Soc.* **115**, 5193 (1993).
75. A. Michaeli, A. Regev, Y. Mazur, J. Feitelson, and H. Levanon: *J. Phys. Chem.* **97**, 9154 (1993).
76. P. R. Levstein, H. van Willigen, M. H. Ebersole, and F. W. Pijpers: *Mol. Cryst. Liquid Cryst.* **194**, 123 (1991).
77. P. R. Levstein, M. H. Ebersole, and H. van Willigen: *Proc. Indian Acad. Sci. (Chem. Sci.)* **104**, 681 (1992).
78. P. R. Levstein and H. van Willigen: *Colloids and Surfaces A* **72**, 43 (1993).
79. B. Venkataraman, B. G. Segal, and G. K. Fraenkel: *J. Chem. Phys.* **30**, 1006 (1959).
80. B. Hoffman: *J. Am. Chem. Soc.* **97**, 1688 (1975).
81. I. Y. Chan, W. G. van Dorp, T. J. Schaafsma, and J. H. van der Waals: *Mol. Phys.* **22**, 741, 753 (1971).
82. A. I. Shushin: *Chem. Phys. Lett.* **162**, 409 (1989).
83. J. Schlüpmann, K. M. Salikhov, M. Plato, P. Jaegerman, F. Lendzian, and K. Möbius: *Applied Magn. Res.* **2**, 117 (1991).
84. K. Kalyanasundaram: *Photochemistry in Microheterogeneous Systems*, Academic Press, New York (1987).

85. M. Grätzel: in V. Balzani (ed.), *Supramolecular Photochemistry*, NATO ASI Series, Reidel, Dordrecht, p. 435 (1987).
86. K. M. Kadish, G. B. Maiya, C. Araullo, and R. Guillard: *Inorg. Chem.* **28**, 2725 (1989).
87. D. S. Leniart, H. D. Connor, and J. Freed: *J. Chem. Phys.* **63**, 165 (1975).
88. K. Suga, K. Maemura, M. Fujihira, and S. Aoyagui: *Bull. Chem. Soc. Jpn.* **60**, 2221 (1987).
89. L. J. Johnston: in V. Ramamurthy (ed.), *Photochemistry in Organized and Constrained Media*, VCH Publishers, New York, pp. 359–386 (1991).
90. M. D. E. Forbes, K. E. Dukes, T. L. Myers, H. D. Maynard, C. S. Breivogel, and H. B. Jaspan: *J. Phys. Chem.* **95**, 10547 (1991).
91. M. D. E. Forbes, T. L. Myers, K. E. Dukes, and H. D. Maynard: *J. Am. Chem. Soc.* **114**, 353 (1992).
92. S. Kazanis, A. Azarani, and L. J. Johnston: *J. Phys. Chem.* **95**, 4430 (1991).

Muon Spin Resonance of Radicals on Surfaces

EMIL RODUNER and MARTINA SCHWAGER
Physikalisch-Chemisches Institut der Universität Zürich, Winterthurerstrasse 190, CH-8057 Zürich, Switzerland

MEE SHELLEY
TRIUMF, 4004 Wesbrook Mall, Vancouver, B.C. V6T 2A3, Canada

(Received: 15 December 1993; accepted: 5 July 1994)

Abstract. Polarized positive muons are incorporated as spin labels in organic free radicals adsorbed on large-area surfaces. Two muon spin resonance techniques are introduced which allow the detection of the muonated species, either in transverse or near avoided crossings of energy levels in longitudinal magnetic fields. The radicals are characterized by their hyperfine interactions, and dynamic information is obtained from the extent of averaging of the hyperfine anisotropy. Because of the high spin polarization the method is extremely sensitive and allows the study of radicals at concentrations down to a single radical in the sample at a given time, and therefore under conditions of high mobility where conventional techniques often fail due to radical termination reactions.

Key words. Radicals, surface diffusion, silica, zeolite ZSM-5, muon spin resonance technique.

1. Introduction

The diffusion and desorption dynamics of adsorbed reactants, intermediates and products is of key importance for the function of processes in heterogeneous catalysis since it determines not only the turnover frequency but often also the balance between unimolecular and bimolecular transformations. Several excellent techniques are available for investigations on clean and well-defined single crystal surfaces. While such studies are of great importance for a detailed understanding of *fundamental* questions it is at the same time true that catalysts of *practical* importance are never clean and well-defined in the same sense, and that catalytic activity is even often related in an essential way to impurities and defects in the active material. It is therefore necessary also to have techniques available which allow studies of adsorbed species in large-surface area microcrystalline or amorphous materials which are not transparent or for other reasons just not accessible and tractable for high vacuum techniques. Nuclear magnetic resonance and infrared spectroscopy may be a good choice for the study of reactants and products, but the transient intermediates which are the key species of catalytic processes are much more elusive. Electron paramagnetic resonance can be used for paramagnetic transients at lower temperatures, but under catalytic conditions they are mobile and disappear by radical termination reactions so that their concentrations are often below the detection limit. It is for large-surface area materials and in particular also at ambient or elevated temperatures where the muon spin resonance (μSR) method can be an extremely sensitive alternative for the study of structure and dynamics of adsorbed organic radical intermediates.

The muon is a short-lived elementary particle which occurs in the positive as well as in the negative charge state. It constitutes the main component of cosmic rays on the surface of our planet. For experimental use it is available at the ports of certain accelerators in the form of energetic beams with a spin polarization of close to 100%. Of interest here is the positive muon (μ^+) which in a chemical sense is a lightweight proton. Implanted in matter it thermalizes at the end of its ionization track. During the slowing down process it can capture an electron from the environment to form a bound state which has been dubbed muonium (Mu $\equiv \mu^+ e^-$). This one-electron atom has within 0.5% the same ionization potential and Bohr radius and therefore the same chemical properties as the hydrogen atom; consequently it is regarded as a light hydrogen isotope with a mass of one-ninth the mass of H [1]. This reacts with unsaturated compounds by addition as in

Mu + C₆H₆ ⟶ C₆H₆(H)(Mu) [cyclohexadienyl radical]

to form a free radical, thereby leaving the muon as a polarized spin label in the β-position to the unpaired electron or delocalised system. Since the muon is a spin-$\frac{1}{2}$ particle like the proton it acts as a microscopic magnetic probe, and the information of interest is obtained in magnetic resonance type experiments [2] with signal detection based on a single particle counting technique. The method of μSR has found a wide variety of applications in solid state physics and in chemistry. For details, the reader is referred to corresponding reviews [3–8].

It is the scope of this article to introduce the μSR technique in view of its applications to the study of radicals on surfaces and to explore its potential and its limitations. The systems reported are organic radicals adsorbed on fully hydroxylated amorphous silica, both porous and non-porous, and on silica supported metal catalysts and on zeolites. It is of interest to see whether the binding of radicals to the surface is primarily by van der Waals interaction as for comparable diamagnetic molecules, or whether the open shell nature provides an additional chemical interaction. The structure of the adsorbed species compared with those which are free, and the mobility both on and off the surface depend critically on this interaction.

2. The Basics of Experiments Using Positive Muons

2.1. MUON PRODUCTION AND DECAY

Particles colliding with the nuclei of a suitable target with an energy above the threshold of *ca.* 140 MeV induce the production of π-mesons which then decay according to the sequence:

$$\pi^+ \to \mu^+ + \nu_\mu \qquad (\tau_\pi = 26 \text{ ns}) \tag{1}$$

$$\mu^+ \to e^+ + \bar{\nu}_e + \nu_\mu \qquad (\tau_\mu = 2.2 \text{ μs}) \tag{2}$$

The pion is a particle without a spin. Given that the neutrino has negative helicity, conservation of angular momentum ensures that the muon is polarized with its spin antiparallel to its momentum in the pion rest frame. Pions decaying at rest at the surface of the target are thus the source of a monoenergetic muon beam with 100% spin polarization.

Conservation of angular momentum again ensures that in the three-body decay of the muon the positron is emitted preferentially along the instantaneous muon spin direction. The orientation of the muon spins can thus be monitored via the observation of their decay positrons in a given direction, and if desirable as a function of time.

2.2. IMPLANTATION IN MATTER

The muons have an energy of 4 MeV which corresponds to a stopping range in matter of $0.15 \, \text{g cm}^{-2}$ and to a penetration depth in water of $ca.$ 1.5 mm. This limits the thickness of any material such as counters and windows in the muon path. The energy is dissipated in an ionization track, and the muon comes to rest at the end of this track in an environment which depends on the chemical composition of the material. A significant fraction is normally found in diamagnetic environments, i.e. as free muons or chemically bound in diamagnetic molecules in place of a proton. A further fraction forms Mu which in an inert environment is long-lived so that it can be observed directly. However, if double bonds are present it adds readily, following the same reactivity pattern as hydrogen atoms [8], to place the muon in a free radical. A third fraction of polarization is often lost in depolarizing encounters of Mu with other paramagnetic species present at the end of the track. The distribution process between the different species follows the rules of hot atom and radiation chemistry, and it can be influenced by addition of electron scavengers in particular [3,9].

For the experiments, the silica powders are dried in vacuum at a temperature of 385 K in stainless steel cells equipped with 50 μm windows, and a controlled amount of the organic material is adsorbed before the cells are sealed.

2.3. PERFORMING EXPERIMENTS WITH MUONS

2.3.1. *Time Resolved Experiments*

A generalised experimental set-up is shown as a schematic drawing in Figure 1. Individual muons in the polarized beam with their spins pointing in the backward direction enter from the left and stop in the experimental sample S. A thin scintillation counter M in front of the sample detects the passage of the particle and triggers the *START* signal for a clock which measures the muon life-time. The decay positron, detected either in the counter F in the forward or B in the backward direction, provides the *STOP* for the clock. This is a signature which serves to increment the proper channel of a histogramming memory where good events are accumulated as a function of the life-time of the corresponding muons.

In the absence of a time-evolution of muon polarization the histograms, one for each counter, are exponentials characterized by the muon life-time. The presence of a perpendicular magnetic field ($B_0 \| x$) leads to muon Larmor precession which

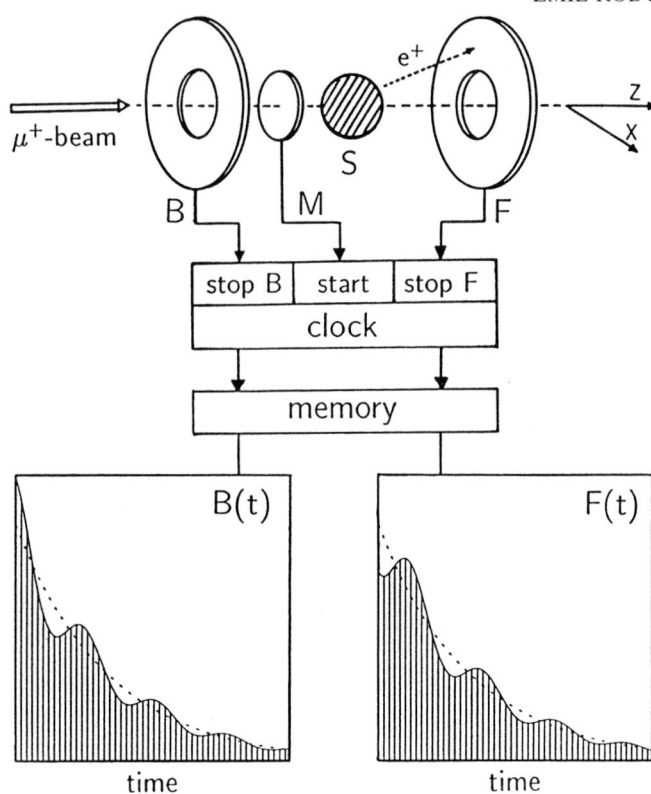

Fig. 1. Schematic drawing of the experimental set-up for muon spin resonance experiments. Spin polarized muons stop in the sample S which is surrounded by scintillation counters M triggered by the incoming muons, and F and B to detect the decay positrons. Experiments are done either in a longitudinal ($B_0 \| z$) or in a transverse ($B_0 \| x$) external magnetic field.

appears as superimposed oscillations in the histograms. This time-dependent muon polarization $P_\perp(t)$, akin to a free induction decay following a $\pi/2$ pulse in a magnetic resonance type experiment, represents the information about the magnetic environment of our probe. For experiments in longitudinal fields ($B_0 \| z$) the corresponding signal $P_\|(t)$ is often simply an exponential defined by the longitudinal relaxation time of the muon, but in special cases such as near avoided crossings of magnetic energy levels it may also reflect oscillation or precession of spin polarization.

The slightly idealised general form of the histograms may be written:

$$B(t) = N_0 \exp(-t/\tau_\mu)[1 + aP_{\perp,\|}(t)] \tag{3}$$

and

$$F(t) = N_0 \exp(-t/\tau_\mu)[1 - aP_{\perp,\|}(t)]. \tag{4}$$

N_0 is a normalization, a an asymmetry constant related to muon beam parameters and to the geometry of the experimental set-up, and the signal of interest is given by a Fourier series:

$$P_{\perp,\parallel}(t) = \sum_i p_i \exp(-\lambda_i t) \cos(\omega_i t + \phi_i), \tag{5}$$

where ω_i represents the frequency of the component i which has an initial phase ϕ_i, an amplitude p_i as permitted by the Hamiltonian of the species and the selection rules for the magnetic transitions [8], and a relaxation rate λ_i. For a radical in a high transverse field there are two allowed muon nuclear transitions which obey the ENDOR condition [8],

$$\omega_i/2\pi = |\nu_\mu \pm \tfrac{1}{2} A_\mu|, \tag{6}$$

where $\nu_\mu = B_0 \times 13.55$ kHz/G, is the muon Zeeman frequency and A_μ the muon–electron hyperfine interaction. From the experiment the frequencies are obtained by Fourier analysis of $P_\perp(t)$, and the parameters by fitting either the histogram directly in time space or individual frequencies in Fourier space [10].

2.3.2. Time Integral Experiments

In order to have well-defined events in time resolved experiments no more than one muon at a time can be present in the sample. This limits the number of acceptable muons to $<10^5$ per second. Modern accelerators provide beams with a flux which can be two orders of magnitude higher. In order to benefit from this situation one may sacrifice the direct information of the time dependence and simply count the number of positrons in the detectors F and B. This is often done in longitudinal magnetic fields and recorded as the experimental decay asymmetry $\mathcal{A}$ as a function of field:

$$\mathcal{A} = \frac{N_B - N_F}{N_B + N_F}, \tag{7}$$

where N_B (and similarly N_F) is the integral number of counts

$$N_B = \frac{\int_0^\infty B(t)\,dt}{\int_0^\infty \exp(-t/\tau_\mu)\,dt} = \frac{1}{\tau_\mu} \int_0^\infty B(t)\,dt \tag{8}$$

represented by the hatched area in the histograms in Figure 1. Given that by definition $\phi_i = 0$ for longitudinal fields we obtain for the normalised time integrated polarization

$$\bar{P}_\parallel(B_0) = \frac{\mathcal{A}}{a} = \sum_i p_i \frac{\lambda_i \tau_\mu + 1}{(\lambda_i \tau_\mu + 1)^2 + (\omega_i \tau_\mu)^2}$$

$$= \sum_i \frac{p_i}{\lambda_i \tau_\mu + 1} \left\{ 1 - \frac{(\omega_i \tau_\mu)^2}{(\lambda_i \tau_\mu + 1)^2 + (\omega_i \tau_\mu)^2} \right\}. \tag{9}$$

In high fields outside avoided-level-crossings the eigenstates of a system with isotropic hyperfine interactions are pure Zeeman states so that all transition frequencies with $\omega_i \neq 0$ have zero amplitude. For $\omega_i = 0$, p_i equals unity, so that the maximum value for the polarization is obtained in the absence of relaxation as $\bar{P}_\parallel = 1$.

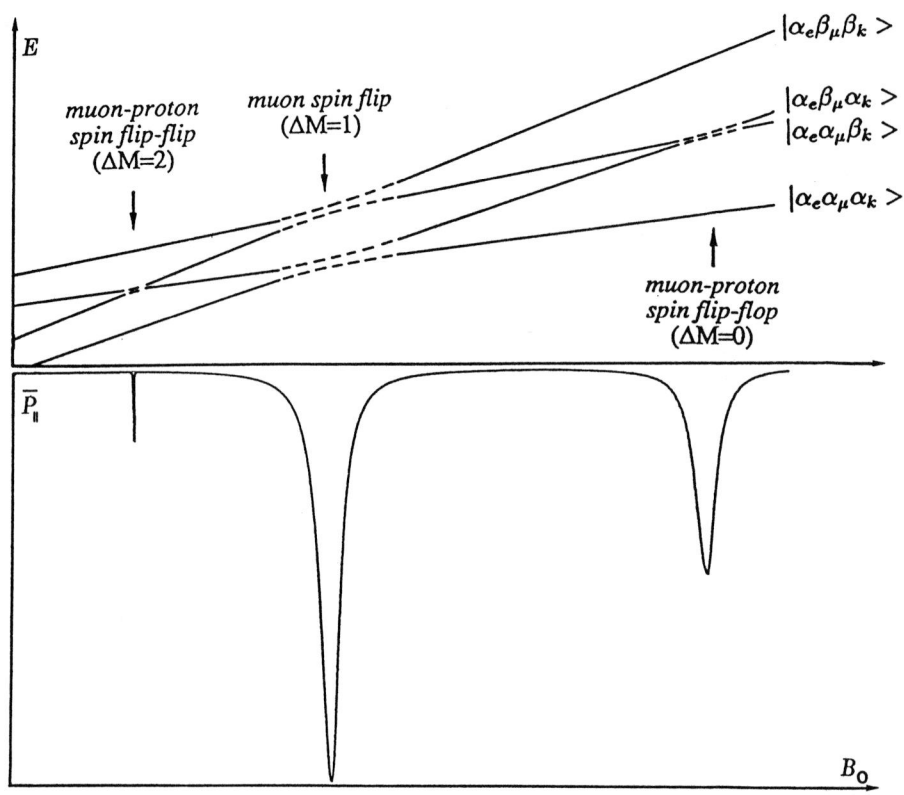

Fig. 2. Principle of avoided-level-crossing (ALC) muon spin resonance, where characteristic absorption lines are observed in the time integrated polarization at fields where there is an avoided crossing of energy levels. The simulation of the ALC spectrum is based on realistic parameters for the muonated cyclohexadienyl radical in a given orientation with respect to the external field.

2.4. AVOIDED-LEVEL-CROSSING (ALC) EFFECTS

In reality the time integrated polarization $\bar{P}_\parallel(B_0)$ is not constant in high fields but shows distinct resonances at fields B_r where there is an avoided crossing of magnetic energy levels. The situation is displayed in Figure 2. Near these avoided crossings, the eigenstates are mixtures of two Zeeman states, which causes the system to oscillate. Furthermore, if one of them belongs to muon spin α and the other one to β this results in oscillatory behaviour of the muon polarization on an angular frequency ω_i given by the energy gap, and with a fractional polarization p_i determined by the mixing coefficients [2,8,11]. According to the selection rules which are given by the change of the total magnetic quantum number M the resonances are denoted $\Delta M = 0$, ± 1, and ± 2. The $\Delta M = 0$ line in which the spin polarization oscillates between the muon and another magnetic nucleus is governed mainly by the *isotropic* hyperfine interactions. It is therefore present also in isotropic environments, whereas the other two are driven by the hyperfine *anisotropy* and thus not observed in liquid-like situations. In fact, the $|\Delta M| = 1$ resonance is the strongest in solids but absent in liquids and therefore an extremely sensitive probe for the presence of small anisotropies. It has been used as an indicator for the onset of reorientational dynamics for radicals in solid norbornene [12] and in fullerene C_{70} [13] in their plastic phases and is also of key importance to radicals on surfaces. The $|\Delta M| = 2$ line is always weak and has not been of great practical use so far. The resonance positions B_r are given to first order by

$$B_r = \left| \frac{A_\mu + (|\Delta M| - 1)A_k}{2[\gamma_\mu + (|\Delta M| - 1)\gamma_k]} \right|. \tag{10}$$

Here, A_k are the isotropic nuclear coupling constants and $\gamma_{\mu,k}$ the muon and nuclear gyromagnetic ratios. The formula makes clear that muon as well as nuclear coupling constants are obtained from ALC spectra. This contrasts with experiments in high transverse fields which reveal A_μ alone (see Equation (6)).

3. Spectroscopy of Surface Adsorbed Radicals

3.1. THE TRIMETHYLALLYL RADICAL ON SILICA

The first muonated radical on a surface was detected upon muon irradiation of spherical silica grains (Cab-O-Sil EH-5) with two different coverages of 2,3-dimethylbutadiene (DMBD) [14]. The transverse field spectra of the 1,1,2-trimethylallyl radical as a function of temperature are displayed in Figure 3. The two lines are centered at $\frac{1}{2}A_\mu$ as given by Equation (6) for $A_\mu > 2\nu_\mu$. A_μ obviously increases with decreasing temperature. This is a consequence of the preferred conformation of the muonated methyl group with the C—Mu bond eclipsing the p_z orbital containing the unpaired electron [16]. For the surface adsorbed species the lines broaden continuously as temperature decreases, but they remain isotropic down to 150 K. For the bulk, in contrast, one observes narrow lines in the liquid range, but below the freezing point there is a clear powder spectrum (Figure 3b). The near-axial anisotropy with $D_\| \simeq 14$ MHz is typical for the muon substituted in the rotating methyl group of the radical which is frozen in its orientation. Translational diffusion on the surface of the spherical SiO_2 particle leads to a dynamic averaging of the hyperfine anisotropy. The line width parameter which is plotted in Figure 4 for both coverages therefore contains the dynamic information. A theoretical treatment which relates to the reorientational correlation time τ_R has been worked out for ENDOR spectroscopy [15]. It applies here as well, as has been discussed by Heming et al. [14]. An Arrhenius fit for τ_R gives 13.6(8) kJ mol^{-1} for the activation energy in the monolayer and 15.9(1.4) kJ mol^{-1} for the 10% layer. Obviously, the radicals are less mobile under conditions of low surface coverage. This reflects the fact that there are sites with different energies on an amorphous surface, and that low energy sites are populated preferentially.

3.2. THE CYCLOHEXADIENYL RADICAL ON SILICA

The system which has been most extensively studied by μSR among the surface adsorbed species so far is the cyclohexadienyl radical on silica. Figure 5 gives the transverse field spectrum obtained with a monolayer of benzene on Cab-O-Sil EH-5. It demonstrates a principal limitation of this variant of μSR: the higher fre-

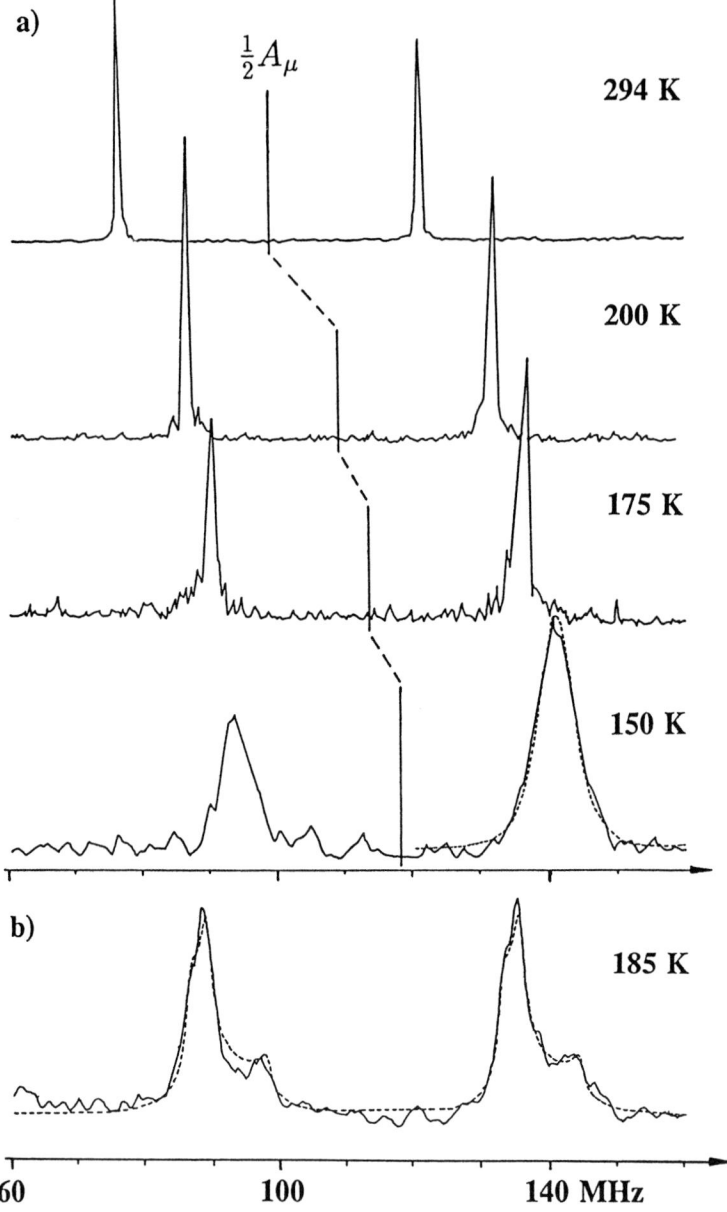

Fig. 3. Fourier power spectra obtained by μSR in a transverse field of 2 kGauss with DMBD. (a) In a monolayer adsorbed on spherical silica grains of 7 nm diameter; (b) with the bulk frozen polycrystalline compound (experiment and simulation).

quency, around 300 MHz, carries much less polarization than its low frequency partner line. With decreasing temperature, the latter disappears as well. This is a consequence of loss of coherence in the muon precession and due to the finite lifetime of the Mu precursor which has a hyperfine interaction of *ca.* 4.5 GHz and

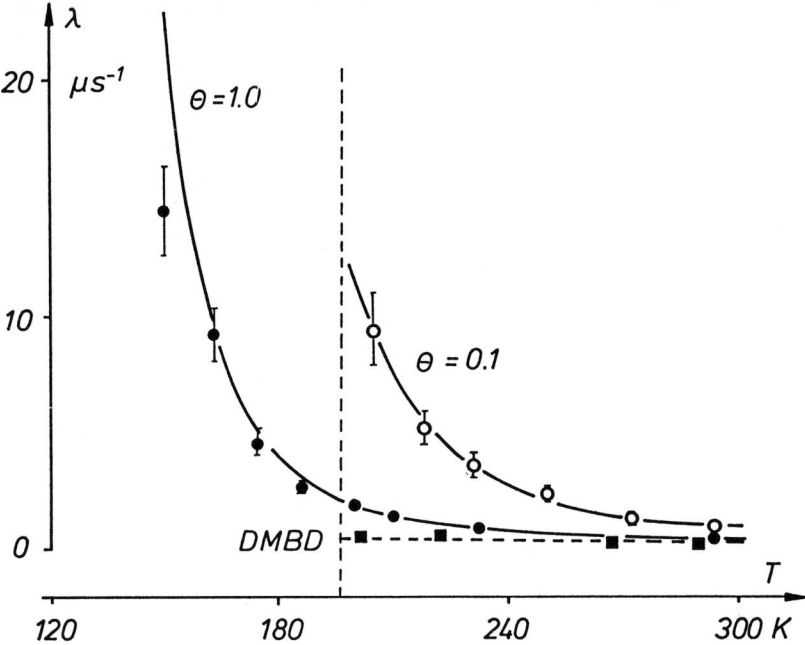

Fig. 4. Lorentzian line width parameter λ as a function of temperature for two coverages θ and for neat DMBD. The solid lines are theoretical curves assuming that isotropic modulation of the muon hyperfine anisotropy is the dominating origin of relaxation [14]. The vertical broken line gives the freezing point of bulk liquid DMBD.

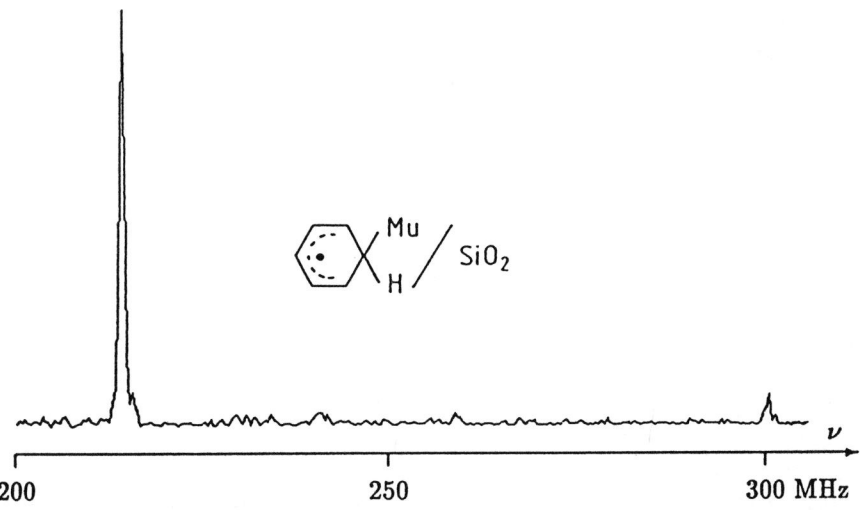

Fig. 5. Transverse field μSR spectrum of the cyclohexadienyl radical adsorbed on silica.

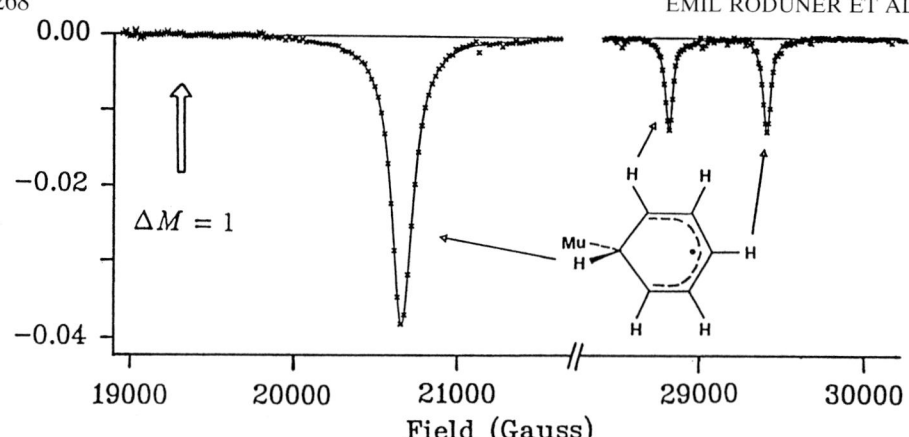

Fig. 6. ALC-μSR spectrum of the cyclohexadienyl radical observed at 334 K with a monolayer of benzene adsorbed on silica (Cab-O-Sil EH-5).

thus a rapid time evolution of spin polarization [8]. Mu has to add to benzene in a time ≪ 1 ns in order to ensure full polarization transfer, but addition is activated in the case of benzene much more than for DMBD.

The same system served for the first observation of a surface-adsorbed radical by means of the avoided-level-crossing resonance technique [17]. A $\Delta M = 0$ resonance was observed down to a temperature of 139 K, demonstrating that the above limitation due to slow radical formation is much less stringent in longitudinal compared with transverse fields. Instead of within one nanosecond the radical has to be formed to a significant amount *within the muon lifetime* as long as the energy gap at the avoided crossing is large enough to allow oscillation of the muon spin within τ_μ as well. The time window for radical formation is thus extended by more than two orders of magnitude, but formation is usually still the limiting factor for observation. This is shown by the fact that with this technique, the Mu adduct radical to DMBD on the same type of silica is observed down to far lower temperatures.

An example for the quality of an ALC spectrum that can be obtained is given in Figure 6. In this experiment, the counting time per point amounts to *ca.* 30 seconds. The three resonances at *ca.* 20 700 G, 28 850 G and 29 450 G are all of the type $\Delta M = 0$, whereas the $|\Delta M| = 1$ line, which is expected to occur approximately at the position indicated by the arrow, is absent. The resonance positions B_r are slightly shifted compared with those observed with bulk benzene at the same temperature, which confirms that the radicals are in an adsorbed state [18,19]. Based on an isotropic muon hfc of *ca.* 513 MHz we evaluate B_r and obtain the isotropic proton hfcs 126 MHz, −25 MHz, and −36 MHz, which are readily assigned to the methylene, the *ortho* and the *para* protons, respectively [20]. The ALC signal amplitude depends strongly on the magnitude of the hyperfine couplings [8], and since the *meta* proton coupling amounts to only 7.5 MHz the corresponding signal was not detected. Both the absence of the $|\Delta M| = 1$ line and the Lorentzian shape of the observed resonances indicate isotropic conditions. The

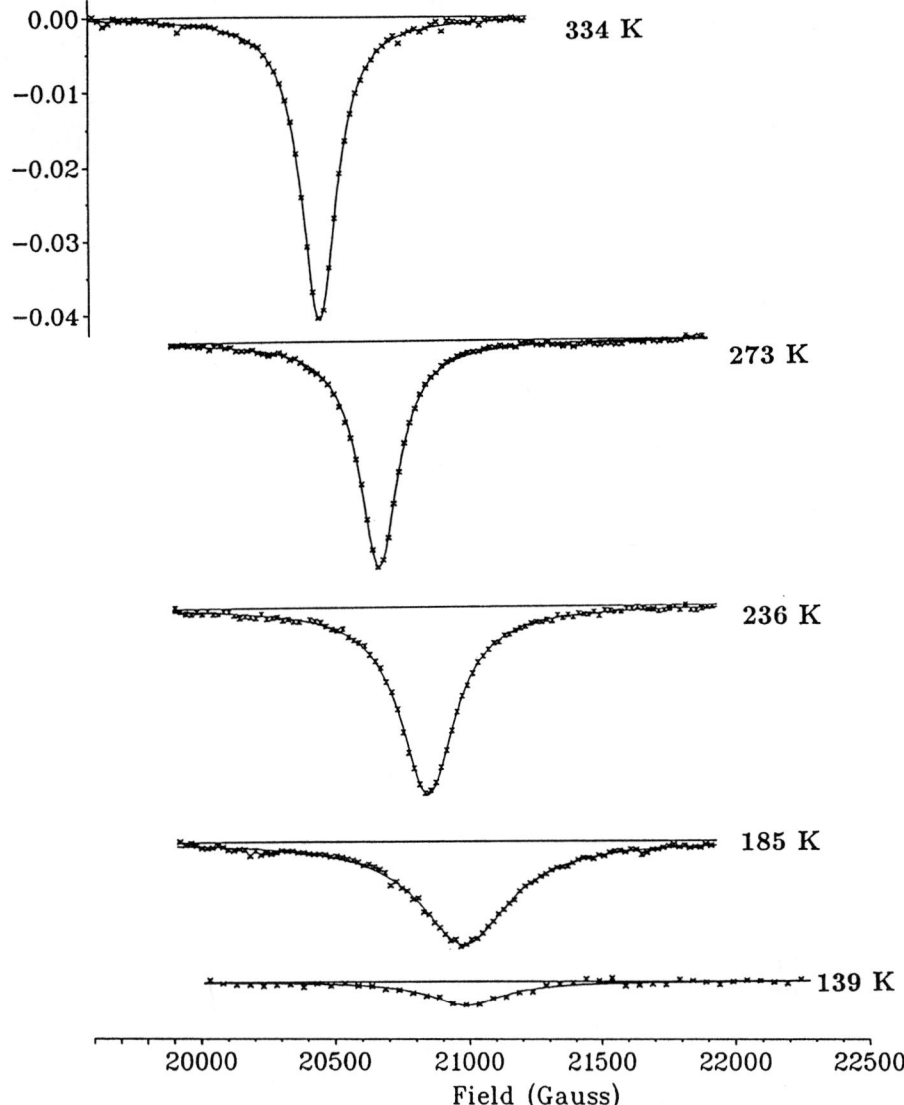

Fig. 7. Methylene proton $\Delta M = 0$ ALC-μSR resonance of the cyclohexadienyl radical observed as a function of temperature with a monolayer of benzene adsorbed on silica (Cab-O-Sil EH-5).

lines are almost as narrow as in bulk liquid benzene, which shows that the reorientational mobility is quite comparable with that in the liquid state.

The most prominent resonance is shown in Figure 7 as a function of temperature down to 139 K. The line width changes considerably but its shape remains Lorentzian to a good approximation over the entire range. This demonstrates that the system remains in the motional narrowing regime although mobility is reduced at the lower temperatures. The shift of the resonance position reflects the tempera-

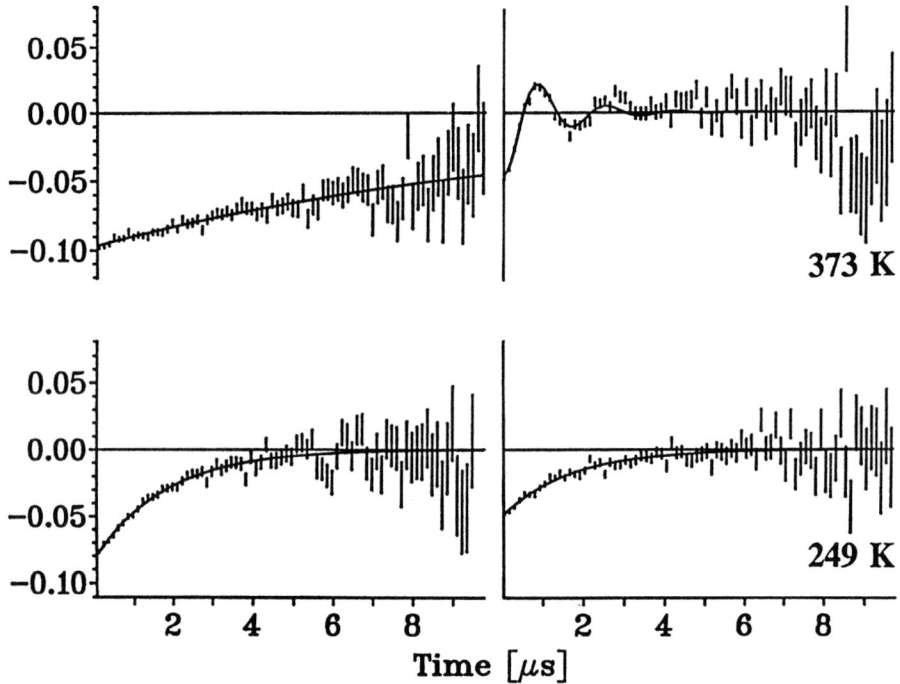

Fig. 8. Time differential measurements of the muon polarization function on resonance of the $\Delta M = 0$ muon–methylene proton spin flip-flop line (right) and of the $|\Delta M| = 1$ muon spin flip line (left) for a nominal 5% coverage of benzene adsorbed on silica (Cab-O-Sil EH-5).

ture dependence of A_μ and of A_p. The $|\Delta M| = 1$ resonance was not observed in these early experiments even at the lowest temperature, although in bulk frozen benzene this transition has double the intensity of the strongest $\Delta M = 0$ line [21] (compare Figure 2). The absence of the resonance is striking and has important implications regarding the dynamics of the systems. It is believed that frequent flipping over of the radical on the surface is mainly responsible for this [18].

Similar experimental series were obtained with coverages corresponding to 30% and to 5% of a monolayer [18]. The $|\Delta M| = 1$ resonance grew in at the lower coverages where the higher adsorption energy works against the flipping over of the adsorbed species. In contrast, a nominal fivefold layer gave spectra which had to be interpreted as superpositions of radicals in bulk polycrystalline benzene and of a surface adsorbed state [19].

Time integrated ALC-μSR has the advantage of allowing to accept as many muons as there are available in the beam. This is, however, at cost of the explicit time dependence of the muon relaxation function. In certain cases it may therefore be advisable to resort to time-dependent measurements in longitudinal fields on and off resonance. Figure 8 demonstrates the information obtained directly in experiments on-resonance for the $\Delta M = 0$ and the $|\Delta M| = 1$ lines for a 5% coverage of silica with benzene. At 373 K, on resonance of the $\Delta M = 0$ line, one finds a characteristic oscillation as drawn schematically for $F(t)$ in Figure 1. The frequency corresponds to the energy gap at the avoided crossing, which is 0.56 MHz for the cyclohexadienyl radical in the fully isotropic case. At 249 K this oscillation is overcritically damped. Here, and for the $|\Delta M|=1$ line at both temperatures, the signal simply decays exponentially. At 373 K, on resonance of the $|\Delta M|=1$ transition, the relaxation is very small so that only a weak ALC signal can build up during the muon life-time in a time-integrated experiment. At 249 K, both resonances were clearly observed [18].

The hyperfine coupling constants can change significantly when the radical is brought from the bulk solution onto the surface. For the cyclohexadienyl radical one observes at 298 K for the muon in the methylene position 514.5 MHz in benzene solution [20]; on the fully hydroxylated surface of silica this increases to 516.7 MHz in a monolayer of benzene (i.e. $\theta = 1$), to 516.7 MHz ($\theta = 0.3$), and to 517.5 MHz ($\theta = 0.05$) [18]. The methylene proton shows a comparable relative increase. The coverage dependence reveals that the shift grows when the adsorption energy increases, which is compatible with small geometric distortions. Comparatively larger effects were found for the cyclohexadienyl type radicals formed by Mu addition to solid naphthalene and ascribed to distortions due to the different crystal fields [22]. A similar effect was also reported for the ethyl radical adsorbed on silica, where the trend of the α-proton coupling to more positive values was interpreted in terms of a somewhat increased $2s$ character of the atomic orbital containing the unpaired electron, and thus to a slightly bent radical centre of the adsorbed species [23].

Shape, width, and intensity of the ALC resonances depend on the hyperfine anisotropies. Reorientation dynamics partly averages this anisotropy, and isotropic reorientation leads to $\Delta M = 0$ lines of Lorentzian shape whereas the $|\Delta M| = 1, 2$ resonances disappear. Line shape analysis thus obviously reveals dynamic information. A theory has been developed which treats isotropic rotational diffusion within the framework of a stochastic Liouville formalism [24]. Translational Brownian diffusion over the surface of spherical silica particles corresponds to small angle rotational diffusion. The theory has therefore been applied to study surface diffusion of adsorbed radicals.

Figure 9 displays the rotational diffusion coefficients extracted by fitting the above theory to the $\Delta M = 0$ transition of the cyclohexadienyl radical at three different coverages of benzene on spherical silica grains [18]. They show Arrhenius behaviour, except at the lowest temperatures where the theoretical line shape is no longer quite appropriate. The activation energies are around 9 kJ mol^{-1} and thus lower by a factor of 4–5 than the heat of adsorption of benzene [25]. This demonstrates that diffusion occurs as an adsorbed species and not via jumps off the surface. The fact that the activation energies are also close to the same for the cyclohexadienyl radical and for the benzene molecule shows that the nature of the interaction is the same in both cases. Obviously, the radical has a similar van der Waals interaction with the hydroxylated surface, and the unpaired electron provides no extra chemical bonding.

3.3. THE CYCLOHEXADIENYL RADICAL ON A SILICA SUPPORTED METAL CATALYST

Silica is often used as a support for metal catalysts. It is therefore of particular interest to also observe adsorbed radicals as potential transient intermediates in

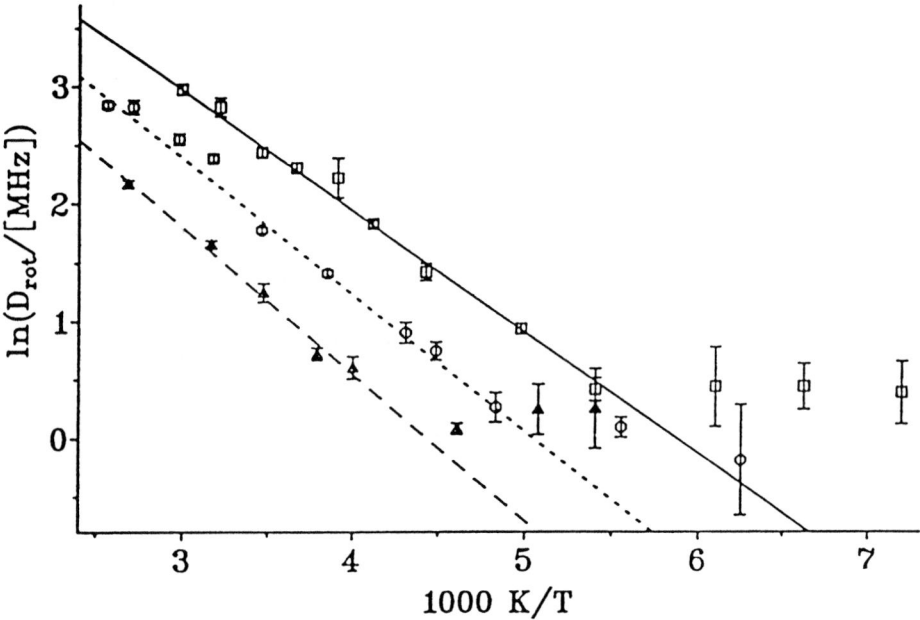

Fig. 9. Arrhenius plot for the rotational diffusion coefficient D_{rot} for the cyclohexadienyl radical adsorbed on the surface of silica (Cab-O-Sil EH-5) with benzene coverages of $\theta = 0.05$ (triangles), 0.3 (circles), and 1.0 (squares).

heterogeneous catalysis in real catalytic systems. Observation by ALC-μSR was indeed successful, in both Pt and Pd loaded silica catalysts [18] with a metal particle size of the order of 20 nm. A sample spectrum is displayed in Figure 10. It was obtained with a 5% benzene coverage on a Pt (2.5% wt) catalyst at 303 K. Because of the low coverage, the spectrum is weak, and therefore the derivative is displayed. It is striking that the resonances are the same as those in Figure 6, but in addition a weak $|\Delta M| = 1$ feature is observed. Also the temperature dependence of the line width is similar to that observed with the 5% coverage of benzene on plain silica. We conclude that the observed species obviously sits on the surface of silica, and that it does not take much notice of the metal particles. An estimate of the radicals' root-mean-square displacement shows that the metal is indeed out of range under the conditions of the above measurements [18].

The affinity of Pt and Pd metals to benzene is high [26], so that the metal particles are expected to be fully covered with the organic phase. This should lead to the formation of metal adsorbed radicals with significantly different hyperfine interaction. Indeed the signal relating to the silica adsorbed cyclohexadienyl radical was found to be diminished to ca. 50% in the Pd catalyst relative to a silica sample with comparable coverage. Despite this, no additional resonance which could have been attributed to the metal adsorbed radical was observed. This may have to be attributed to scattering of metal conduction electrons which gives rise to radical electron spin flips and provides an efficient mechanism for muon depolarization.

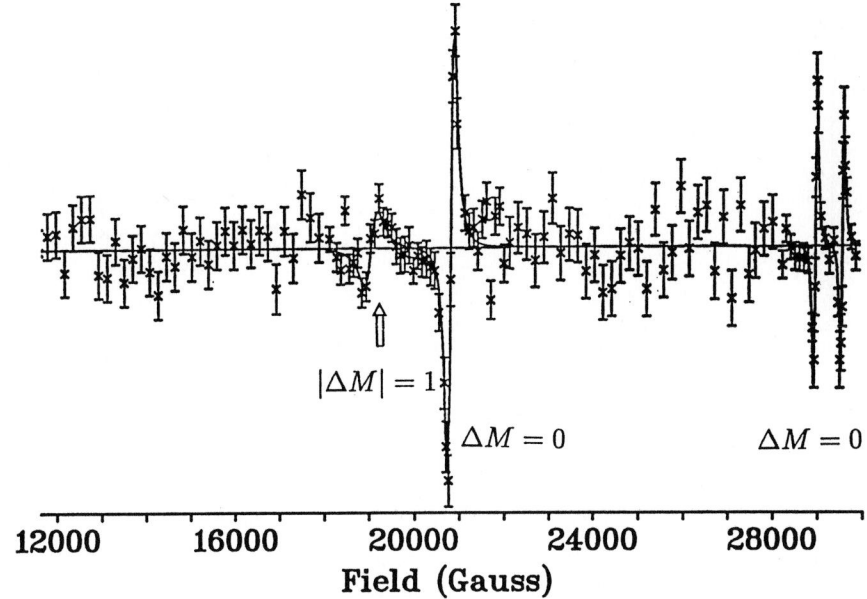

Fig. 10. Differential ALC-μSR spectrum obtained at 303 K with a 5% benzene coverage on Pt/silica (2.5 wt.-% on Cab-O-Sil EH-5).

3.4. DYNAMICS IN A ZEOLITE

Zeolites are high surface area catalysts of very high practical interest. While organic radical cations in these materials are routinely observed by electron spin resonance there are only few claims of magnetic resonance detections of *neutral* radicals at catalytically relevant temperatures despite significant efforts in this direction. In contrast, there has been a clear observation of the cyclohexadienyl radical by transverse field μSR in NaX at ambient temperature [27]. The greatly increased line width for the zeolitic environment compared with the bulk liquid indicated the presence of considerable anisotropy.

ALC-μSR has also been successful in zeolites. Figure 11 displays a spectrum obtained with benzene in Na-ZSM-5. While the high-field part of the spectrum is quite comparable with the one obtained with silica (Figure 6), the low-field side contrasts with the clear presence of the $|\Delta M| = 1$ resonance in the zeolite. The prominent line at 20 600 G has a Lorentzian shape and a width of 216 G FWHM. This is less than half the maximum width observed for the same radical on silica, and in terms of spherical rotational diffusion [24] it would translate into a reorientational correlation time of 30 ns. The $|\Delta M| = 1$ resonance, with a width of 425 G, is also narrower than in the static limit. The value is roughly compatible with uniaxial rotation of the radical about an axis perpendicular to the molecular plane, but the symmetric line shape and the amplitude which is reduced to *ca.* 50% of the maximum value disagree with this type of motion which is commonly found for the benzene molecule in zeolites [28]. Isotropic rotational diffusion is also excluded since, based on a correlation time of 30 ns as inferred from the

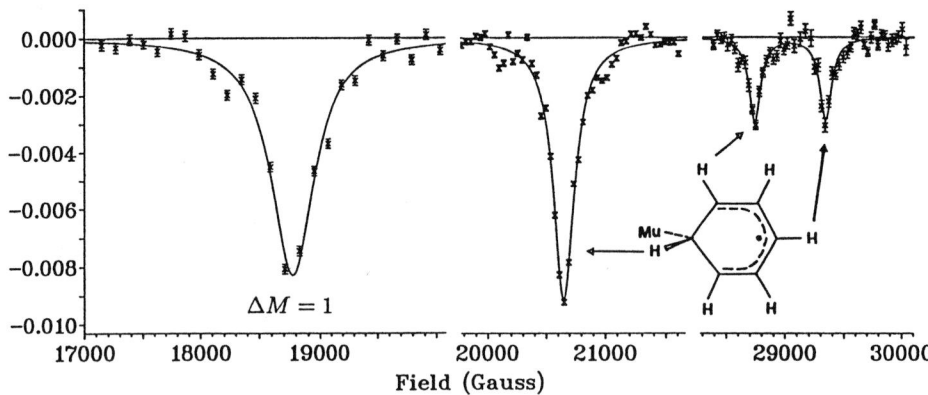

Fig. 11. ALC-μSR spectrum obtained with benzene in Na-ZSM-5 at a temperature of 344 K and a loading of 13 wt.-%.

$\Delta M = 0$ resonance, this would lead to a width of *ca.* 2300 G, and to a higher amplitude. By comparison with the case of the same radical on the surface of silica it is anticipated that flip motions of *ca.* 180° play a role in diminishing the amplitude, and that in addition a wobbling type motion exists which partly averages the hyperfine anisotropy.

Preliminary experiments with benzene in several different faujasites reveal a behaviour which is distinctly different from that observed with ZSM-5 [29]. In NaX at room temperature, one can clearly distinguish two types of cyclohexadienyl radicals with slightly different hyperfine couplings. Also, the absence of the $|\Delta M| = 1$ transition demonstrates isotropic conditions. On cooling, this resonance appears at once, indicating a phase transition to non-isotropic but not completely frozen conditions. Further cooling leads to a broadening of the line and then to a splitting which is, especially in NaUSY, too large to be explained by different orientations of the same radical, hence it must be a consequence of inequivalent sites.

4. Concluding Remarks

The μSR techniques have been shown to be suitable for studying neutral organic radicals derived by Mu addition to unsaturated molecules adsorbed on surfaces. For experiments in transverse fields it is required that the radicals are formed within <1 ns while for longitudinal field ALC experiments formation within τ_μ is sufficient. The radicals are characterized by their hyperfine coupling constants. Dynamic processes which occur on a timescale of the inverse muon hyperfine anisotropy, usually *ca.* 30 ns, are revealed from the narrowing of the resonances of selection rule $\Delta M = 0$. Particularly sensitive is the $|\Delta M| = 1$ muon spin flip transition which is the strongest under static conditions but absent when the radical orientations are averaged isotropically. Depending on the type of motion and on the corresponding correlation times the resonance can disappear via broadening (for spherical rotational diffusion [24]) or via narrowing (in the 'pseudo-static' case [30]) or it can simply lose amplitude [18]. It has been demonstrated elsewhere that the temperature behaviour of width and shape of this resonance can be drastically different even for molecules as similar as the two fullerenes C_{60} and C_{70} when they reorient in their plastic crystalline phases [30]. The main advantage of the μSR techniques is their sensitivity which allows detection of radicals under extremely low concentrations and in non-transparent and polycrystalline or amorphous environments of large surface area materials which are often of particular importance to heterogeneous catalysis.

Acknowledgements

We are indebted to I. D. Reid, H. Dilger, S. R. Kreitzman, P. W. Percival and A. Baiker for their help with the measurements, and for contributing in discussions to the understanding of the work presented here. Financial support by the Swiss National Science Foundation and through an NSERC (Canada) International Fellowship are gratefully acknowledged. We thank CU Chemie Uetikon AG for providing us with the zeolites.

References

1. D. C. Walker: *J. Phys. Chem.* **85**, 37 (1978).
2. E. Roduner: *Chem. Soc. Rev.* **22**, 337 (1993).
3. D. C. Walker: *Muon and Muonium Chemistry*, Cambridge University Press, Cambridge (1983).
4. J. Chappert and R. I. Grynszpan (eds.): *Muons and Pions in Materials Research*, North-Holland, Amsterdam (1984).
5. B. C. Webster: *Annual Reports C*, Royal Society of Chemistry (1984).
6. A. Schenck: *Muon Spin Rotation Spectroscopy*, Adam Hilger, Bristol (1985).
7. S. F. J. Cox: *J. Phys. C: Solid State Phys.* **20**, 3187 (1987).
8. E. Roduner: *The Positive Muon as a Probe in Free Radical Chemistry. Potential and Limitations of the μSR Techniques*, Lecture Notes in Chemistry, Vol. 49, Springer, Heidelberg (1988).
9. P. W. Percival, J.-C. Brodovitch, and K. E. Newman: *Chem. Phys. Letters* **91**, 1 (1982).
10. P. Burkhard, E. Roduner, J. Hochmann, and H. Fischer: *J. Phys. Chem.* **88**, 773 (1984).
11. M. Heming, E. Roduner, B. D. Patterson, W. Odermatt, J. Schneider, H. Baumeler, and H. Keller: *Chem. Phys. Letters* **128**, 100 (1986).
12. E. Roduner, I. D. Reid, M. Riccó, and R. De Renzi: *Ber. Bunsenges. Phys. Chem.* **93**, 1194 (1989).
13. T. J. S. Dennis, K. Prassides, E. Roduner, L. Cristofolini, and R. De Renzi: *J Phys. Chem.* **97**, 8553 (1993).
14. M. Heming and E. Roduner: *Surface Science* **189/190**, 535 (1987).
15. J. R. Wilson and D. Kivelson: *J. Chem. Phys.* **44**, 154 (1966): S. A. Goldman, G. V. Bruno, C. F. Polnaszek, and J. H. Freed: *J. Chem. Phys.* **56**, 716 (1971).
16. E. Roduner, W. Strub, P. Burkhard, J. Hochmann, P. W. Percival, H. Fischer, M. J. Ramos, and B. C. Webster: *Chem. Phys.* **67**, 275 (1982).
17. I. D. Reid, T. Azuma, and E. Roduner: *Nature* **345**, 328 (1990).
18. M. Schwager, H. Dilger, E. Roduner, I. D. Reid, P. W. Percival, and A. Baiker: *Chem. Phys.* in press.
19. I. D. Reid, T. Azuma, and E. Roduner: *Hyperfine Interactions* **65**, 879 (1990).
20. P. W. Percival, R. F. Kiefl, S. R. Kreitzman, D. M. Garner, S. F. J. Cox, G. M. Luke, J. H. Brewer, K. Nishiyama, and K. Venkateswaran: *Chem. Phys. Letters* **133**, 465 (1987).
21. E. Roduner: *Hyperfine Interactions* **65**, 857 (1990).
22. I. D. Reid and E. Roduner: *Struct. Chem.* **2**, 419 (1991).
23. M. Schwager, E. Roduner, I. D. Reid, P. W. Percival, J.-C. Brodovitch, S. Wlodek, and R. F. Marzke: *Hyperfine Interactions* **87**, 859 (1994).
24. S. R. Kreitzman and E. Roduner: *Chem. Phys.* in press.

25. A. V. Kiselev and V. I. Lygin: *Infrared Spectra of Surface Compounds*, Wiley, New York (1975).
26. G. A. Somorjai: *J. Phys. Chem.* **94**, 1013 (1990).
27. C. J. Rhodes, I. D. Reid, and E. Roduner: *J. Chem. Soc., Chem. Commun.* 512 (1993).
28. B. Boddenberg and R. Burmeister: *Zeolites* **8**, 488 (1988).
29. M. Shelley, D. J. Arseneau, J. J. Pan, R. Snooks, S. R. Kreitzman, D. G. Fleming, and E. Roduner: 'Muon Spin Relaxation Studies of Cyclohexadienyl Radicals in NaUSY', in J. Weitkamp, H. G. Karge, H. Pfeifer, and W. Hölderich (Eds.), *Zeolites and Related Microporous Materials: State of the Art 1994*, Studies in Surface Science and Catalysis, Vol. 84, Elsevier Science, Amsterdam, 469 (1994).
30. E. Roduner, K. Prassides, R. Macrae, I. M. Thomas, C. Niedermayer, U. Binninger, C. Bernhard, A. Hofer, and I. D. Reid: *Chem. Phys.*, in press.

Investigation of Radical Ions with Time-Resolved Surface Enhanced Raman Spectroscopy

RONALD L. BIRKE and JOHN R. LOMBARDI
Department of Chemistry, The City College of The City University of New York, Convent Avenue at 138th Street, New York, NY 10031, U.S.A.

(Received: 15 December 1993; accepted: 5 July 1994)

Abstract. Time-resolved surface enhanced Raman scattering (TRSERS) spectroscopic methods are discussed for the study of radical ions produced photochemically and electrochemically at silver or gold metal surfaces. Both single shot and pump-probe TRSERS experimental methods are illustrated which use an optical multichannel analyzer, OMA, for ms (single shot) to ns (pump-probe) time resolution. Fundamental chemical and physical processes for photochemically and electrochemically induced radical ion formation are described for adsorbed molecules at the metal-solution interface. Emphasis is given to the possibility of laser photoinduced radical ion formation by a direct molecule-to-metal charge transfer process. Applications of TRSERS techniques are discussed for the study of radical ions formed by various photochemical and electrochemical reactions at the surface of SERS active metals. These adsorbed reaction systems encompass electroreduction processes of adsorbed alkylviologens, p-nitrobenzoate, 4-cyanopyridine, 4-pyridine carboxaldehyde, 4-hydroxymethylpyridine, and direct photoinduced radical cation formation from flavin mononucleotide, FMN.

Key words. Surface enhanced Raman scattering, SERS, time-resolved surface enhanced Raman scattering, TRSERS, pump-probe TRSERS, photoinduced, photoelectrochemical, metal-solution interface, alkylviologens, p-nitrobenzoate, 4-cyanopyridine, 4-pyridine carboxaldehyde, 4-hydroxymethylpyridine, and flavin mononucleotide, FMN.

1. Introduction

1.A. RADICAL IONS FORMED AT ELECTRODES AND SERS

Free radical ions can be produced electrochemically at electrode surfaces by one-electron transfer either from or to the neutral species, producing the cation radical (oxidation) in the former case and the anion radical (reduction) in the latter case. The driving force for these electrochemical processes is a potential perturbation causing a potential drop between the metal phase and the solution phase at the site of the reacting molecule. On exhaustive electrolysis a sufficient concentration of these radical ions can be produced to use optical (UV-vis) absorption or ESR (electron spin resonance) absorption techniques for their investigation. On the other hand, surface enhanced Raman scattering (SERS), resonance Raman scattering (RRS), and surface enhanced resonance Raman scattering (SERRS) can be used to investigate a monolayer or less of radical ions produced at the electrode surface. This situation is somewhat similar to electrochemical perturbation methods such as pulse or linear sweep voltammetry where the current–potential curve senses the concentration perturbation and related dynamics close to the electrode surface. Raman spectroscopy provides a molecular structure method capable of investigating events which occur directly at the metal/solution interface after an electrode potential induced charge transfer process takes place. Since the Raman process occurs on the time scale of femtoseconds, the SERS method could, in principle, be used to investigate extremely rapid molecular transitions.

As with all resonance Raman methods for investigating time-resolved spectroscopy, the limiting factor in pushing the method to shorter times is the detection sensitivity. Resonance Raman scattering can be used at smooth electrodes of any metal but it is not as sensitive or surface specific as is SERS. On the other hand, SERS and SERRS require roughened surfaces of the coinage metals. In fact, we have recently developed methods for time-resolved SERS in the ns and μs time domains [1], and it has routinely been used in the ms and s time domains.

All electrochemical perturbation methods for studying charge transfer (CT) are limited in their time response by the requirement that the interfacial (double-layer) capacity be charged in response to any change of the potential drop across the interface. This leads to an RC_{dl} time constant which at its smallest value is several hundred ns. On the other hand, electromagnetic radiation can induce femtosecond CT, as well as chemical bond breaking reactions, and this EM perturbation can be directed at the interfacial region of the metal/solution interface. Flash photolysis [2] and pulse radiolysis [3] have been used to generate radicals and radical ions in the vicinity of electrode surfaces. The electrochemical dynamics for these radicals can be detected by either current-time measurements or UV-vis optical absorption-time measurements [3]. In such a case, the current-time response involves double-layer charging and thus the kinetic response for interfacial CT is, in part, controlled by this charging process. On the other hand, the optical absorption vs. time response will not be limited by the RC_{dl} time constant.

In the same way, SERS (probe) measurements following either pulsed laser excitation or pulsed electrochemical perturbation (pump) of adsorbed molecules on the surface will respond within femtoseconds of the pulse. The beauty of the SERS measurement is that using optical multichannel analyzer detection, the vibrational spectrum can be obtained which gives molecular details of the radical ions produced. As mentioned, the method depends on generating enough sensitivity to get a reasonable signal-to-noise ratio for the pump-probe SERS spectrum. One question is whether it is possible to generate radical ions with low energy (UV-vis) optical excitation at metal surfaces where excited state damping is very efficient. Because the photoinduced excitation of molecules at surfaces can involve electronic transitions to or from metal states, the photoionization energy required can be red shifted from the values necessary for photoionization of molecules in homogeneous solution. Recently, we have shown that UV-laser pulse excitation can indeed produce radical cations at surfaces under certain circumstances [1]. Thus, two incisive means for investigating radical ions at metal surfaces are possible with Raman scattering detection: (i) electrochemical generation of radical ions followed by SERS detection and (ii) UV-vis optical laser pulse generation of radical ions followed by SERS detection.

The study of radical ion transients with time-resolved SERS (TRSERS) could provide fundamental knowledge of primary photochemical and electrochemical processes which can occur at the surface of SERS active metals like Ag and Au. Shifts of vibrational bands and formation of new bands in the TRSERS spectrum following primary photochemical or electrochemical events have proven to be a powerful high resolution tool to elucidate the mechanism of fundamental kinetics. Such steps might include photoionization, photoisomerization, photodecomposition, surface and solvent interactions of photoproducts, and chemical kinetics following electron transfer for molecules either adsorbed on the surface or in the electrical double layer at the metal electrode surface. Since these processes are important for electrocatalysis, photoelectrocatalysis, photoelectrochemical energy conversion, energy storage, photography, electrophotography, and photobiological processes, surface Raman methods of investigation should have technological as well as fundamental implications. In fact, a key issue in nanoscale chemistry is the ability to activate surface reactions without activating bulk reactions. It is easy to appreciate that optically and electrochemically induced radical and radical ion reactions on surfaces may have important and novel applications in such nanoscale chemistry.

Pimentel [4] pointed out that, in general, our knowledge of the chemical dynamics of molecules is not commensurate with our knowledge of the analysis, synthesis, and structure elucidation of molecules. This lack of knowledge is unfortunately even more true for the dynamics of molecular reactions at surfaces. For optically induced surface processes, fundamental questions have only recently been addressed such as the nature of the initial excitation process, the extent to which the interaction of the surface and the adsorbate molecule affects the excited state and photoproduct dynamics, and how the energy of the reaction finds its way into the reaction channel [5]. What is becoming apparent is that many adsorbate-localized excitation mechanisms involve transfer of charge between the adsorbate and metal surface [5]. The high resolution spectra obtainable with the SERS technique coupled with pulsed laser methods should certainly provide an experimental tool to probe the dynamics of molecular transformations on metal surfaces.

1.B. SURFACE ENHANCED RAMAN SCATTERING

Surface Enhanced Raman Scattering (SERS) is a Raman spectroscopic method for obtaining a high resolution Raman vibrational spectrum of molecules on metal surfaces. The method is an *in situ* probe for the metal/solution, metal/solid, and metal/gas interfaces, including the metal/ultra high vacuum (UHV) interface. In most experimental arrangements, the phase in contact with the metal must be transparent so that both the excitation and scattered light can pass through it. Because a 'giant' enhancement can be provided by the metal surface, the technique is sensitive enough to detect a monolayer or less coverage, and, indeed, depending on the background signal, coverage of a few percent of a monolayer can be observed. The overall surface enhancement which allows for the detection of monolayer concentrations of an adsorbate on a SERS metal surface is of the order 10^5–10^6 over the normal Raman scattering of an isolated molecule. This overall enhancement depends on two mechanisms: (i) enhancement of the electromagnetic (EM) field of the exciting light [6,7] and (ii) enhancement of the Raman scattering cross-section (polarizibility) by metal/absorbate resonance Raman processes [8].

The EM enhancement, which is by far the larger of the two mechanisms, is generated by the coupling of the EM field of the exciting light to the conduction band electrons of the metal, the so-called localized plasmon modes. It is observed for small metal structures on the size order of 10–100 nm. Thus, a method for producing these surface roughness features is required to obtain the EM enhancement. In this article, we will mainly deal with the metal/solution interface in an electrochemical environment where a SERS active, roughened surface can easily be produced by an oxidation-reduction cycle, ORC, on a electrode made of Ag or Au. The EM theory [9] shows that because of their metallic dielectric constants the coinage metals, Cu, Ag, and Au give the largest EM enhancements. For the metal/gas and metal/UHV interfaces vapor deposited metal island films also give large surface EM enhancements.

The enhancement of the Raman scattering cross-section (polarizibility) by a metal/adsorbate resonance Raman processes is caused by the adsorption of molecules on the metal surface which allows the mixing of metal and molecular quantum states. The molecular states are broadened by this interaction and the exciting light can cause resonant charge transfer (CT) transitions between the molecular states and metal states [10]. This excitation process is analogous to the intermolecular photoinduced CT which can take place for charge transfer complexes in solution and leads to resonance Raman scattering. In the metal/adsorbate system, both metal-to-molecule and molecule-to-metal photoinduced CT can take place [11]. The intensity of this adsorption induced resonance Raman process depends on electronic transition moments of the CT processes, on the metal density of states, on Franck–Condon factors, and on Herzberg–Teller coupling matrix elements between metal and molecular states [10]. The magnitude of the enhancement factor due to the Herzberg–Teller resonance Raman process, sometimes called the chemical enhancement, is from ~ 30 to $\sim 10^2$. Since the EM and CT resonance processes are multiplicative, they are both necessary to achieve the 10^5–10^6 surface enhancements observed in SERS.

Another possibility for Raman scattering at a surface is surface enhanced resonance Raman scattering, SERRS, the case in which a pure molecular resonance Raman process occurs for a 'colored' molecule at a metal surface and the scattering is also enhanced by the EM effect. For both SERS and SERRS, the 'giant' enhancement is only for molecules near the metal surface (within molecular dimensions). The EM enhancement is more long range than the CT enhancement since the former falls off more slowly than the latter which can only occur for molecules directly on the surface. Both enhancement mechanisms are necessary to observe Raman scattering for molecules at a metal/solution interface, i.e., to exceed the background (water) scattering. Because only molecules adsorbed directly on the surface give enough sensitivity for observation, SERS is a purely surface sensitive detection method. SERRS becomes a purely surface sensitive method for very dilute solutions where bulk scattering for molecules is too weak to be observed.

In this article we will treat the experimental arrangement for obtaining Time-Resolved SERS (TRSERS) spectra with linear diode array detection. Both electrode potential induced and photoinduced TRSERS pump-probe methods will be considered. We will also discuss the steady spectra of radical ions and TRSERS spectra of systems where radical ions undergo chemical reaction such as reaction with the solvent or intramolecular isomerization. It should be appreciated that in all chemical systems which can be investigated by surface Raman techniques, there is a part of the molecule, such as lone pairs on an N atom or a charged group, which can chemisorb or physisorb to the metal surface.

2. Experimental Techniques

The experimental methods for investigating *steady state* SERS spectra of radical ions has been discussed previously in review papers [12,13] and we will limit the treatment here to TRSERS methods which are of the most importance for investigating radical ions on surfaces. These methods take advantage of the multiplexing and signal averaging afforded by optical multichannel analyzer (OMA) detection, coupled with the enormous enhancement in signal intensity provided by SERS. Two OMA based TRSERS methods are available: (1) a single shot method where a series of full surface Raman intensity vs. wavelength spectra are recorded during the time course of a radical ion producing perturbation, and (2) a pump-probe ensemble averaging method which affords the fastest time resolution for obtaining a full SERS spectra at some fixed time delay from an initiating pulse.

2.A. SINGLE SHOT TRSERS METHOD

Figure 1 shows the diagram of the experimental set-up that we have used for a single shot method. An electrochemical perturbation is applied via a waveform generator/potentiostat system which is used to generate an electrode potential ramp or potential pulse. The potentiostat controls the working electrode potential in a conventional three-electrode electrochemical cell so that it follows the potential-time dependence of the waveform generator. A trigger generator is used to trigger simultaneously an electronic shutter, the waveform generator, and the OMA. The Argon ion laser is focused as a spot of $\sim 80\,\mu m$ diameter on the Ag or Au electrode, usually roughened with a single electrochemical oxidation–reduction cycle, ORC. This ORC depends on the electrode material and electrolyte composition. For example, on Ag in the 0.1 M KCl, the ORC we most often use involves stepping the potential from -0.2 V vs. to $+0.2$ V for 1–2 s and then back. (All voltages cited in this article will be vs. the saturated calomel electrode reference (SCE) unless otherwise specified.) At positive potentials, a AgCl deposit builds up on the electrode which is stripped to form rough Ag deposits when the potential returns to the initial reducing potential. In the absence of an anion in the electrolyte which precipitates Ag^+, the potential is stepped to $+0.5$ V for 1–5s and then back to the initial reducing potential. On Ag only one ORC is necessary to produce surface features which give the giant enhancement. A similar procedure is used for the ORC on Au, but for given experimental solution conditions, the oxidizing voltage, the dwell time for the anodic step, and the number of cycles, must be determined experimentally.

The scattering geometry is usually at a 90° collection angle with a fast collection lens, like an $f/1.2$, 5 cm focal length lens, used to collect the scattered light. The monochromator, in our system a Spex 1877 (resolution $2\,cm^{-1}$), is most conveniently a triple monochromator in the subtractive dispersion mode so that the Raleigh scattered light is removed. The fastest response nongated diode array

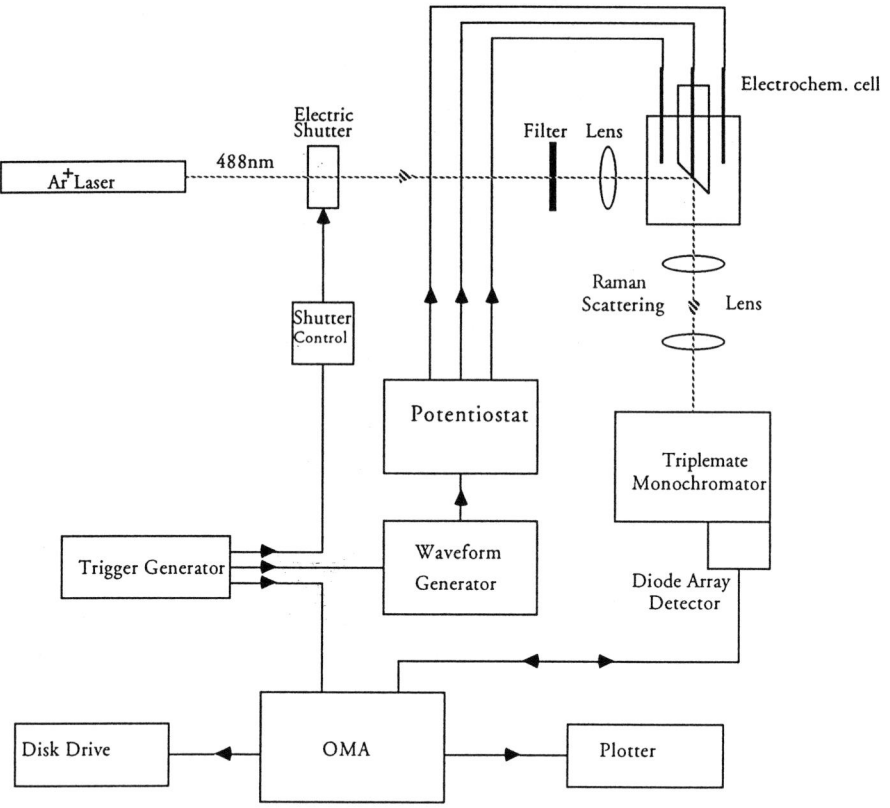

Fig. 1. Time-resolved surface enhanced Raman scattering (TRSERS) instrumental set-up with diode array detector for transient measurements with a triggered shutter. A CW laser (e.g., Ar^+) beam is applied to the working (center) electrode of the electrochemical cell with simultaneous application of a voltage waveform. An entire SERS spectrum is obtained within the dwell time of the detector and this spectral acquisition is repeated under optical multichannel control (OMA) several times during the duration of one or more cycles of the waveform.

detector and OMA system available is capable of a complete scan in a few ms for single shot recording. In most of our single shot studies, spectra were excited with a Spectra Physics Ar^+ model 164 laser and recorded with the Spex 1877 Triplemate spectrometer using an intensified Tracor Nothern photodiode linear array (TN-1223-21) and a model TN-1710 Tracor Northern optical multichannel analyzer (OMA). A grating of 1800 groves per mm gave a spectral coverage of about 900 cm^{-1} with a band pass of around 2 cm^{-1} at 19,436 cm^{-1}. The electrode potential was controlled with an EG&G PARC model 173 potentiostat and EG&G PARC model 179 waveform generator. As explained, the experiment is initiated by simultaneously triggering the electronic shutter, the waveform generator, and the OMA. A potential ramp of a given number of cycles can be applied to the electrode and up 16 spectra measured by the OMA at various set times on this potential waveform. The response time of this system was limited by a 7 ms response time of the electronically controlled shutter and a 10 ms minimum dwell

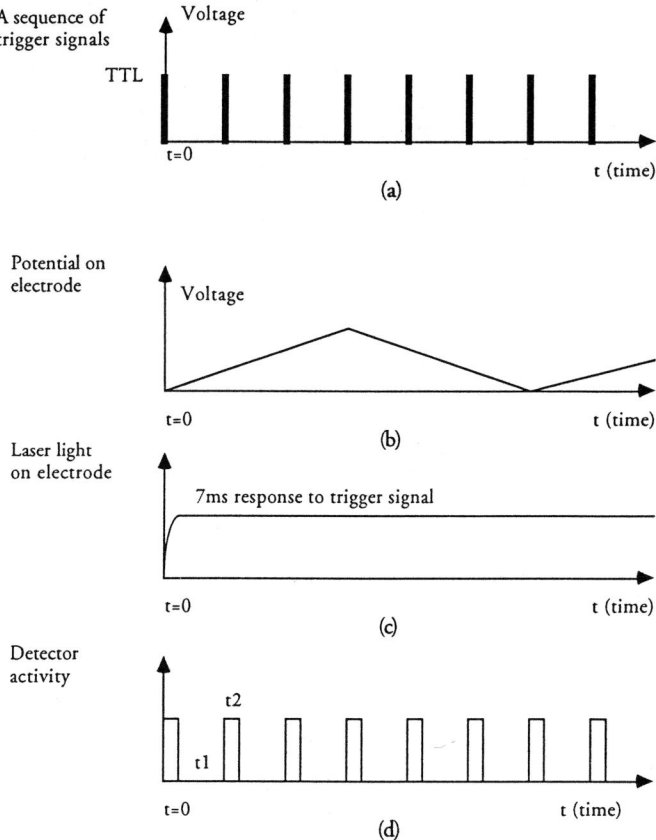

Fig. 2. Timing sequence for TRSERS measurements with a voltage ramp waveform in the instrumental set-up of Figure 1. (a) Sequence of TTL pulses from the trigger generator. (b) Triangular voltage ramp applied to the working electrode. Depending on the OMA used up to around 40 spectra can be obtained on one or more cycles of the waveform. (c) Schematic of the rise time of the electronic shutter which controls the rise time of the light on the working electrode. (d) Control of the diode array detector activity by the OMA. The delay time between acquisition of each spectrum is t_1 and the time the detector is active (dwell time) is t_2.

time for recording of a single spectrum. Figure 2 shows the sequence of trigger signals used for TRSERS on voltage ramp scanning, i.e. for cyclic voltammetry. The delay time between each spectrum, called $t1$, and the detector dwell time or gate width, $t2$, are indicated in Figure 2.

2.B. PUMP-PROBE TRSERS METHOD

After an electrode has been properly prepared for SERS enhancement by means of an oxidation–reduction cycle (ORC), a chemical reaction may be triggered either by a short laser pulse or a potential step. A segment of the surface enhanced spectrum can be obtained by a multichannel analyzer (OMA) attached to a triple monochromator. Time dependence is obtained by using a gated OMA at a prede-

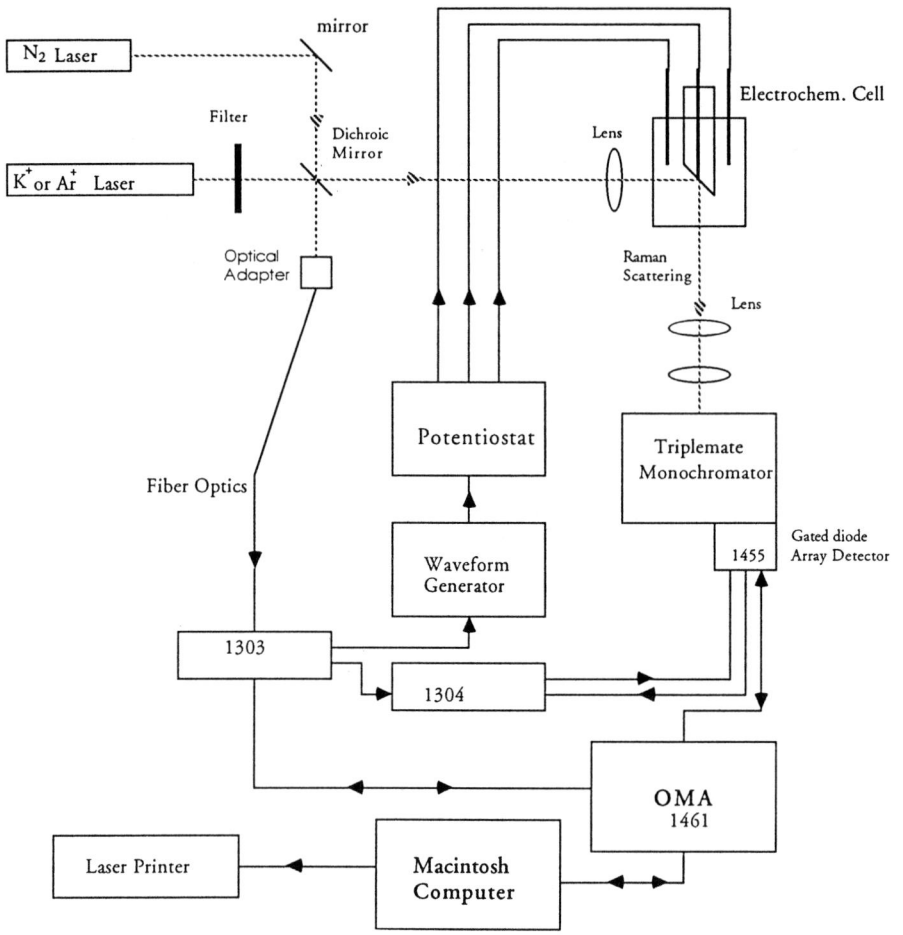

Fig. 3. TRSERS instrumental set-up for pump-probe measurements. Details discussed in text.

termined delay time following the initiation. The width of the gate determines the temporal resolution of the experiment. If reversible the process may be repeated many times to afford good signal averaging. Raman spectra are obtained as a function of delay time to provide a picture of the time development of the various species appearing on the electrode surface.

The apparatus in use in our laboratory for experiments down to the nanosecond regime is shown in Figure 3. An initiation pulse is provided by a 10 ns. pulsewidth [up to 200 Hz repetition rate] with a nitrogen laser pumping a tunable dye laser. Dyes may be selected to provide excitation wavelengths from 370 through 900 nm. It is important to emphasize that too much power in a pulse can either desorb an adsorbate or damage the surface. This is the reason we have used a pulsed nitrogen laser which has a maximum power of ~1 mJ/pulse and a relatively high repetition rate as compared with the higher power but lower repetition rate Nd YAG laser. A small portion of the pulse passes through a beam splitter and through a fiber optic to trigger the detection apparatus. The remainder is focused onto our electrochemical cell, starting the reaction. A CW probe beam is provided either by a Krypton ion laser for a Au electrode (Spectra Physics model 2020) or an Argon ion laser for a Ag electrode (Spectra Physics Model 164) which is also focused on the sample electrode. Alignment of the two lasers in the same region of the electrode surface may be aided by use of a dichroic mirror which transmits visible ion laser light and reflects the ultraviolet pulsed laser. Raman shifted light is collected by an achromatic lens and focused on to the entrance slit of a SPEX triplemate monochromator.

The potential which has been applied to the electrode is controlled by means of a waveform generator (EG&G PARC Model 179) driving a potentiostat (EG&G PARC Model 173). For fast electrode potential initiated experiments at a microelectrode, the potentiostat is not used and a two electrode cell is substituted. During an experimental cycle, an oxidation reduction cycle (ORC) may be programmed for application on the electrode on command by the triggering pulse. The ORC procedure cleans the electrode of any reactant products and activates the surface so that each time the pump-probe cycle starts from nearly the same conditions for the metal/solution interfacial system. Alternatively, if it is necessary to continually refresh the sample at the electrode, it may be replaced by a rotating disc electrode.

Multichannel optical detection in this pump-probe system is provided by a EG&G PARC Model 1455 gated diode array and model 1461 OMA controller which gives a Raman spectrum over a range of 500 cm^{-1}. The EG&G PARC Model 1303/1304 pulser/pulse amplifier system allows one to gate the detector to within 75 ns of the exciting pulse, while substituting the model 1302 enables one to probe the region of 25 to 140 ns. Figure 4 shows the response time sequence for the OMA system. Between 2000 and 20 000 pulses have been averaged (experimental time several minutes to 30 min) and are usually sufficient to develop spectra with good signal-to-noise ratios with highly intense SERS scatterers. A sample calculation shows that for a gate width, $t2$, of 25 ns, time averaging for 1 h allows a signal-to-noise ratio of well over 10 in several cases. On the other hand, in practice, we have found that for a good SERS scatterers and an excitation pulse intensity of ~1 mJ/pulse, a $t2$ of 200 ns is necessary to obtain a SERS spectrum with very high signal-to-noise ratio for 20 000 pulses. By varying the delay times successively, one can obtain high resolution time dependence of the SERS spectrum of a chemical reaction on an electrode surface.

2.C. CHARACTERIZATION OF OBSERVED INTERMEDIATES

Once time dependent spectra are obtained by the above procedures, additional procedures may be needed to determine the nature of any observed transients. In many cases it is useful to take difference spectra, i.e. the difference between the observed time dependent spectrum and that of the zero time delay (reactant) spectrum or the long time delay (product, if different from reactant) spectrum. This is because the predominant species on the surface is usually not the transient of interest, and the spectral lines of the reactants or products tend to obscure the transient spectra. The resulting difference spectra are often much clearer and easier to interpret. Due to the fact that many of the intermediates will simply be

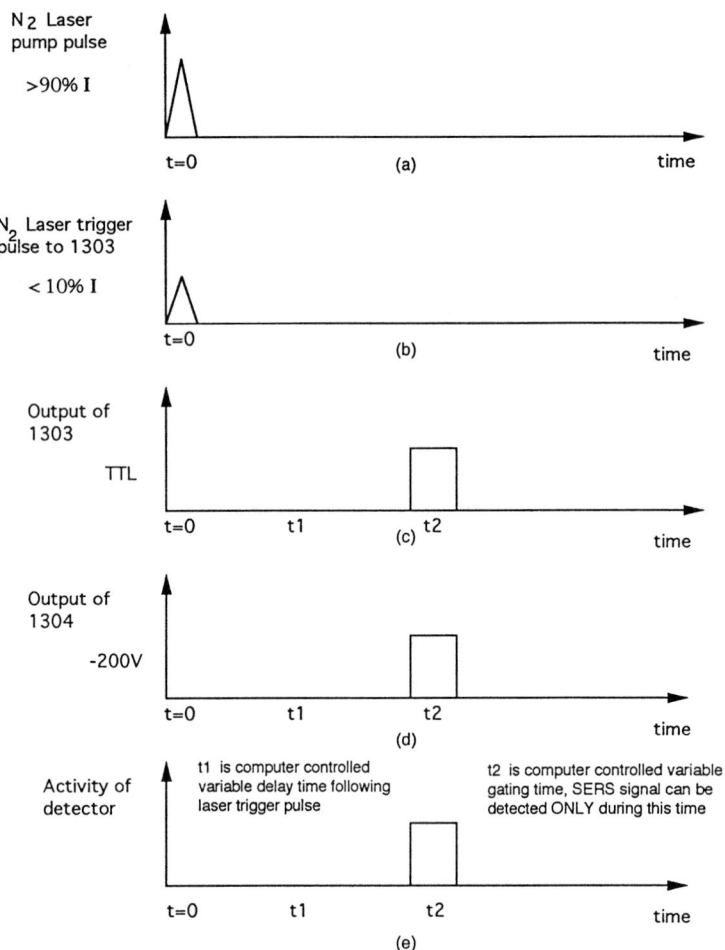

Fig. 4. Timing sequence for the pump-probe instrumental set-up of Figure 3. (a) The part of the pump pulse which is reflected off the dichroic mirror and directed towards the working electrode. (b) The part of the pump pulse which passes through the optical adaptor and triggers the PARC model 1303 pulse gating generator. (c) Output of the model 1303 which provides a TTL gating pulse of width t_2 and delay time t_1. This output triggers the PARC model 1304. The voltage waveform generator can also be triggered by the model 1303 or a steady state voltage can be applied to the electrode. The CW probe laser output in this experiment is always applied to the working electrode. (d) Output of the PARC model 1304 pulse amplifier showing that the gating pulse has been amplified to −200 V for control of the PARC gated diode array detector model 1455. The PARC OMA model 1461 controls the delay time t_1 and the gate width t_2 through the model 1303. (e) Delay time and dwell time of the detector. The OMA 1461 allows many repetitions of this timing sequence at fixed delay time, t_1, for ensemble averaging. Variation of t_1 then allows a series of TRSERS spectra to be recorded as a function of time. True time resolution occurs when t_2 is much shorter than t_1.

cation or anion derivatives of the reactant, the transient spectra will not differ significantly from the reactant, with typically small spectral shifts. Use of difference spectra aids in assignment of the spectral lines since the shifted lines most likely correspond to a similar normal mode to that of the reactant. We have developed capabilities to calculate normal modes, and to optimize the resulting force constants by applying a simplex optimization to a standard normal mode computer program.

Using a normal mode calculation, the SERS spectrum of the sample can be compared with its normal Raman spectrum. It is also possible to investigate how the surface affects the normal vibrational modes by introducing surface metal atoms into the structural model of the surface/molecule scattering center. We have modeled the SERS system [14] with Ag–ligand force constants for surface–molecule bonding interactions and included nonbonding interactions with a Urey–Bradley force field. Determining the change in force constant needed to reproduce the transient spectrum should help characterize the intermediate. For example, in our study of flavin mononucleotide [9], discussed below, we could identify which rings were most strongly effected by photoionization and thereby localize the region of the molecule from which the electron was ejected. Normal coordinate analysis also enabled us to infer that at least one of the photoproducts is an enolic type isomer of the reactant.

It is also not always obvious whether a cation or an anion is initially formed as a photoproduct. Since both molecular HOMOs and LUMOs may in principle be photoenergetically accessible to the metal Fermi level, there is no a priori reason to eliminate one or the other possibility. However, according to the charge transfer theory of surface enhanced Raman spectroscopy [5], these two cases may be distinguished by measuring the potential at which the intensity of a Raman line reaches a maximum, V_{max}, as a function of excitation wavelength, ω_L. This theory is based on a resonant CT excitation of the Raman transition, and since a resonant CT process is necessary for photoproduction of a radical ion species, similar processes are involved. For metal-to-molecule electron transfer (into a LUMO) we expect a positive slope for V_{max} versus ω_L while for molecule-to-metal transfer (from a HOMO) we expect a negative slope. This has been observed experimentally for SERS in several laboratories [15,16]. The same effect should be expected for the appearance of (at least the primary) radical ion transient intermediates if one measures the potential at which the intermediate appears as a function of initiating wavelength. This suggests a technique of studying intermediate appearance potentials using tunable dye laser excitation. These studies may be supplemented by photocurrent studies in which the direction of a detected photocurrent can also indicate the direction of photoinduced charge transfer. However, such studies are not by themselves definitive because there may be other interfering sources of photocurrent, while the voltage maximum studies are specific to the mechanism being studied.

3. Photo and Potential Induced Radical Ion Formation at Metal Electrode Surfaces

The work we discuss can be divided into two broad categories, depending primarily on the mode of initiation of the reaction. A reaction can be initiated either with a short laser pulse or with a discrete potential step on the electrode. Photoinitiated reactions are limited only by the pulse time of the exciting laser, which has been as short as a 10 ns pulse of a N_2 laser in our studies. Such processes presumably involve charge transfer from one of the filled metal conduction bands to a low lying excited state of the adsorbed reactant, forming a radical anion or the opposite mechanism of charge transfer from a filled molecular orbital to an unfilled metal state forming a cation. The mechanism of direct excitation of a molecular transition is somewhat less likely to be observed due to the rapid quenching of molecular excited states when molecules are adsorbed on metal electrode surfaces [17]. The lack of observed fluorescence from adsorbed molecules in SERS testifies to the rapid channels for non-radiative deexcitation available on a metal electrode surface. Presumably the radical cation (or anion) is either formed in its ground state, or decays to it quickly through the above-mentioned rapid quenching of excited states. Further, it is likely that although the initial excitation is a vertical transition, a rapid relaxation of either the radical geometry to its equilibrium position and/or solvent relaxation and isomerization takes place rapidly as well. The consequence is at least some barrier to reverse of the charge transfer, producing a relatively long-lived cation (or anion) at the electrode surface.

The second method of reaction initiation involves electrolysis by a discrete potential perturbation. This method is likely to be much more general, providing a greater variety of possible reactions for study. However, on normal electrodes the RC time constant limits the initiation rate of these reactions to the order of μs or ms. In order to examine still faster reactions it would be necessary to develop this method on microelectrodes, which have much shorter time constants.

The importance of the giant sensitivity of the SERS process for these investigations should not be underestimated since fast time resolved studies of surface phenomena have been impeded by the low sensitivity of most optical techniques [18]. Furthermore, Raman spectroscopy provides sufficient resolution to ensure unambiguous identification of the observed species.

It should be pointed out that the subject of photochemistry of adsorbed molecules at the metal-gas interface has also developed in the last few years. Among the photodissociation reactions studied have been rupture of the carbon–halide bond in methyl halides investigated by Cowin et al. [19–21] and White et al.[22,23], photodissociation in HCl investigated by Polanyi and coworkers [24], and photofragmentation of other molecules such a ketene (CH_2O) [25] and phosgene (Cl_2CO) [26]. These studies have involved well defined metal surfaces such as Pt(111), Ni(111), and Ag(111); however, they were, in fact, preceded by the photofragmentation studies of Goncher and Harris of azabenzenes and other aromatic compounds on roughened Ag [27,30]. These latter studies on roughened Ag and predictions of enhanced photochemistry from theoretical calculations [29,30] based on the SERS electromagnetic (EM) enhancement effect suggest the possibility of photoinduced reactions at the metal-solution interface for electrochemically roughened Ag and Au electrodes. It is clear from the discussions of Nitzan and Brus [30,31] that enhanced photochemistry involving excited states of molecules at metal electrodes is possible only if the rate of surface enhancement of the optical process leading to energy accumulation in the molecule is greater than the rate of excited state energy quenching by energy transfer to the metal surface. However, there must be a chemical de-excitation channel which is also faster than this electronic surface quenching process. One such process we will emphasize is a photoinduced charge transfer followed by an isomerization process.

Molecular excited states are quenched at metal surfaces depending on the distance from the surface [31]. This nonradiative energy transfer is due to the field coupling of the excited molecular dipole to the metal surface, and the electrodynamic theory for flat surfaces has been discussed by Chance, Prock, and Silbey [32,33], Persson [34], Persson and Lang [35], Metiu and Gadzuk [36], and Morawitz and Philpott [37]. Energy transfer to spherical and spheroidal metal structures has also been considered by Rupin [38] and Gersten and Nitzan [39], respectively. Another quenching mechanism is via charge transfer [18,40]. It is quite clear that excited state lifetimes are highly damped by roughened metal surfaces and are in the subpicosecond time domain [18]. The enhanced photochemistry on SERS activated metal surfaces considered by Nitzan and Brus [30,31] involves the mechanism of direct optical excitation of the adsorbate–metal complex. This type of mechanism was invoked for the photofragmentation process found by Harris et al. [28,29] on SERS roughened Ag. Such a process involves direct excitation to a repulsive energy surface which competes favorably with these damping mechanisms. We have proposed [41] that a photoinduced charge transfer process is a more likely mechanism for photochemistry found at a Ag metal electrode in an electrochemical environment.

3.A. PHOTOEXCITATION MECHANISMS

We will discuss photoexcitation mechanisms for an adsorbed neutral molecule at a metal electrode with a simple two level model. Figure 5 shows qualitative diagrams for various events which can occur on photoexcitation localized on a molecule which is electrochemically stable, i.e., its highest occupied molecular orbital, HOMO, level is far below the Fermi level EF of the electrode and its lowest unoccupied level, LUMO, is much above EF. The electronic energy distribution of the HOMO and LUMO levels of the neutral species are not broadened much by interaction with the metal surface and are depicted as horizontal lines in Figure 5a. Upon excitation by an exciting beam of light, an excited state can form which may be quickly quenched, Figure 5a, by electronic energy transfer to the metal surface, i.e., dipole field coupling of molecule and metal. If this process dominates all other channels of reaction then there would be no photochemistry, only nonradiative decay. This field excitation process can be thought of as an excitation of an electron–hole, eh, pair in the metal by the oscillating field at frequency ω of the molecular excitation. The rate of this field coupling [18] depends on the dipole transition moment μ_{ij} and a dielectric function $G(\omega, d)$ where d is the distance from the surface, i.e.,

$$\text{rate}_{NR} = (1/\tau)_{NR} = |\mu_{ij}|^2 G(\omega, d) = \frac{|\mu_{ij}|^2}{4\hbar d^3} \text{im} \frac{[\epsilon(\omega) - 1]}{[\epsilon(\omega) + 1]}$$

where $\epsilon(\omega)$ is the dielectric function of the metal which depends on angular frequency ω. Since $G \sim d^{-3}$ the rate distance-dependence has a long range character. This nonradiative decay mechanism could compete with a photochemical reaction channel and if faster would quench the photochemistry. Such an equation is probably only valid at distances of $> \sim 10\,\text{Å}$ from the surface and thus only

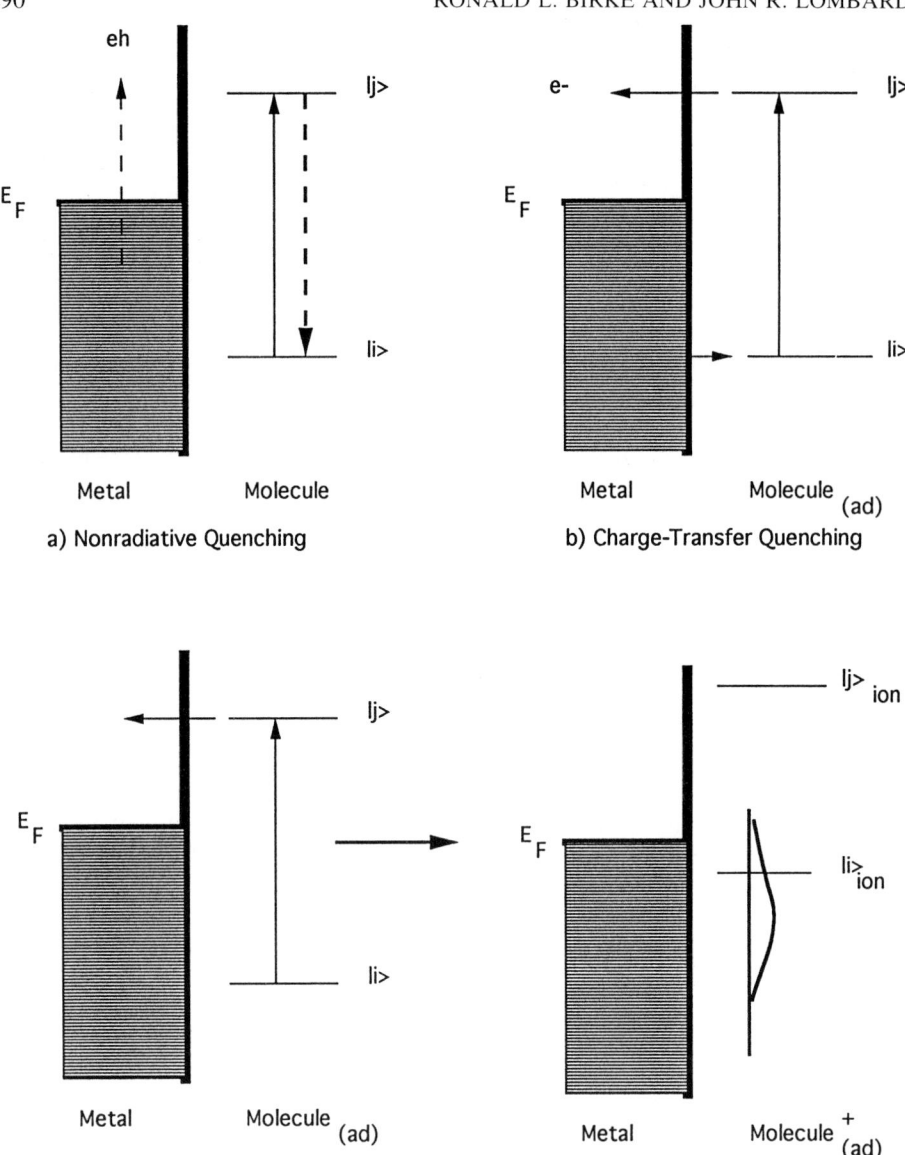

Fig. 5. Photoprocesses which can occur for a molecule adsorbed on a metal surface. (a) Nonradiative quenching of an excited state from energy transfer by dipole coupling between molecule and metal. Excitation of electron–hole (eh) pairs in the metal can be thought of as causing the quenching. (b) Charge transfer quenching involving electron injection from the adsorbed molecule to unoccupied states above the Fermi energy and return from states below the Fermi energy. (c) A photoionization process in which a stable radical cation is formed because some of the ions created have half-vacant HOMO levels above the Fermi energy.

roughly valid for physisorbed systems; however, approximate corrections for separation distances of even chemisorbed molecules can be made which shows that the calculated lifetimes are in reasonable agreement with experiments [18].

Another possibility for quenching is a CT mechanism shown pictorially in Figure 5b. In this case an excited electron of the adsorbate can tunnel to the unoccupied metal orbitals above the Fermi level. The decay of the excited state can be represented by a Fermi Golden Rule formula (first order time dependent perturbation theory) and depends on the square of the tunneling matrix element $|V_k|^2$ and the density, $\rho(E_k)$, of unoccupied metal states at the energy of the excitation, $E_k = (E_i + \hbar\omega)$, i.e.,

$$\text{rate}_{CT} = (1/\tau)_{CT} = \frac{2\pi}{\hbar} \langle |V_k|^2 \rangle \rho(E_k)$$

Now $\langle |V_k|^2 \rangle \sim \exp(-\beta d)$ where β is a tunneling barrier factor and so the rate for charge transfer quenching falls off rapidly with distance, d, from the surface. Excited states of *adsorbed* molecules that decay by this resonant tunneling mechanism have lifetimes in the range of 10^{-14}–10^{-15} s. This CT mechanism produces a radical cation in a nonequilibrium state, $(R^+)_{NE}$, since the initially formed excited state R^* is via a vertical transition. As this $(R^+)_{NE}$ state begins to relax, it may be reduced back to the neutral molecule R by reverse CT as depicted schematically in Figure 5b. The probability for this CT reaction is unity since now there is a large negative free energy difference between the Fermi energy level and the R^+ HOMO energy level. A comparison of rate_{NR} and rate_{CT} for adsorbed molecules (at a metal/gas interface) shows that the charge transfer quenching may be faster-than nonradiative energy transfer. When the HOMO energy level of the cation radical is below the Fermi energy, the time for resonant electron hopping from the conduction band to the R^+ HOMO energy level is 10^{-15}–10^{-16} s [42]. The onset energy $\hbar\omega$ of the vertical CT excitation is approximately given by $\Phi + e_0(\psi_M - \psi_A) - I + (e_0^2/4d)$, where Φ is the work function of the metal electrode in the presence of the adsorbate, ψ_M and ψ_A are the outer electrostatic potentials of the electrode and adsorbate, respectively, I is the ionization potential of the adsorbate, and the last term, which depends on distance d, is an approximation for the image charge interaction of the adsorbate ion with the surface. Outer potentials at interfaces are purely electrostatic potentials of phases which do not include image effects. Since the applied potential controls ψ_M, the onset energy for photoinduced CT will depend on applied potential.

These quenching mechanisms have in the past been considered to prevent the occurrence of photochemistry for adsorbed molecules at an electrode/adsorbate interface [43]. However, a third possibility, Figure 5c, is that once the radical ion is formed, a chemical process can occur which provides a faster (or at least competitive) channel for relaxation than the quenching mechanisms. Such a chemical process shown by potential energy curves in Figure 6 could involve nuclear tunneling, i.e. a photoisomerization step. The isomerized radical cation that is produced from the adsorbed species in a vertical transition interacts with the metal surface plasmon electrons and an electronic image potential, $E_{im} = e_0^2/4d$, results which will raise the radical ion HOMO orbital level towards the Fermi level. In

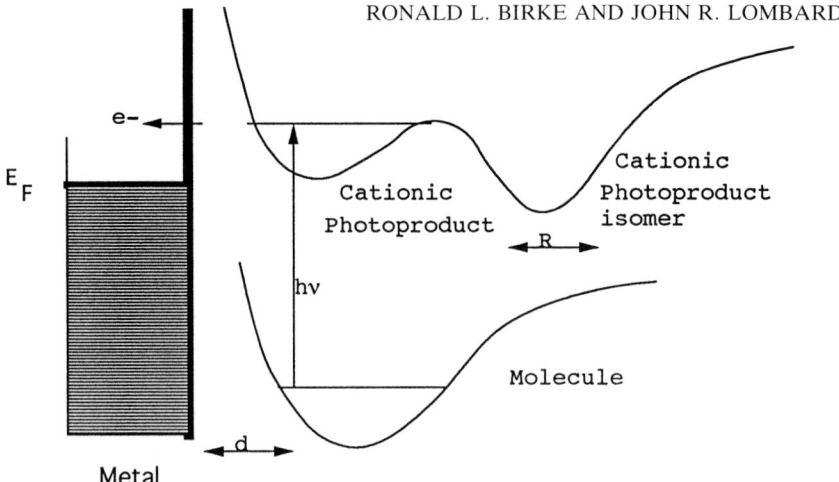

Fig. 6. Schematic energy vs. coordinate curves for the formation of a radical ion which undergoes a chemical reaction following electron injection to the metal. The laser excitation energy is high enough so that radical ions which are created by electron tunneling to the metal can cross over a potential energy barrier representing an isomeric form of the radical ion. The coordinate on the lower curve, d, represents the distance to the electrode, while the coordinate on the upper curve, R, represents the reaction coordinate for formation of the cationic photoproduct isomer.

addition, the adsorbed ion can interact with fast bulk solvent modes, E'_{sol}, which also raises this energy level. Thus the vertical ionization energy, I_{ver}, would be $I_{ver} = I - E_{im} - E'_{sol}$, where I is the ionization energy of the neutral molecule in vacuum. Furthermore, now the radical ion HOMO level is broadened by the electronic interaction with the surface. This leads to a Lorentzian type of distribution of these levels [44]. The result, Figure 5c, may be some states (adsorbed radical cations) in which the singly occupied HOMO level is slightly above the Fermi level. Even if the distribution of singly occupied radical ion HOMO levels is below the Fermi energy level, there may still be a barrier for the reverse CT reaction. These conditions would result in a long lived radical cation species which could be observed by the pump-probe SERS method. If one were able to able to observe SERS spectra of a relaxing radical cation in fs to ps times, the reorganization of internal modes and external solvent modes might be observed. Such experiments do not seem to be feasible at present because of sensitivity problems. At slower times in the ns range, where we have made actual measurements, the decay of photoproducts (e.g., isomerization) and the cation radical itself can be observed in a pump-probe SERS spectra.

3.B. ELECTROCHEMICAL RADICAL ION FORMATION

The electrochemical formation of radical ions from a neutral molecule, R, depends on the relative energy of the metal Fermi level with respect to the radical ion HOMO or LUMO electronic redox level. If an acceptor (LUMO) level is at or below the metal Fermi level, metal-to-molecule CT is possible, or if a donor (HOMO) level is at or above the metal Fermi level, molecule-to-metal CT is possible. The Golden Rule shows that nonradiative electron transfer always occurs between electronic levels which are at the same energy, and since a metal has a manifold of donor or acceptor levels, all that is necessary for CT at the metal/adsorbate interface is that the acceptor level be below or the donor level be above the Fermi level. Henglein [3] has discussed these processes: $R + e^- \rightarrow R^-$ and $R \rightarrow R^+ + e^-$ for solution soluble species. Here, we also consider adsorbed species or those which are close enough to the electrode surface to show SERS enhancement. When the HOMO level of R is far below the metal Fermi level and the LUMO level is substantially above the metal Fermi level, the molecule is electrochemically stable and transitions of electrons into the LUMO level or out of the HOMO cannot take place. If the electrode potential is moved in a negative direction, it is equivalent to raising the metal Fermi level to facilitate transfer of an electron to the LUMO to form R^-; on the other hand, if it is moved in the positive direction, it is equivalent to lowering the metal Fermi level to facilitate transfer of an electron from the HOMO to form R^+. These processes require much less electrode potential (energy) at the metal/solution interface than in the gas phase because of ionic solvation and adsorbate surface interactions. The latter interactions include the image potential of the ion and specific surface binding. These forces pull the singly occupied LUMO level of R^- down towards the metal Fermi level and raise the singly occupied HOMO level of R^+ up towards the metal Fermi level [3].

Thus, many organic and inorganic radical cations and anions can be formed at electrode surfaces in the *ca.* +1.5 V range available for noble metals, like Pt or Au, or in the *ca.* −1.5 V range available for metals with high overpotential to H^+ ion reduction, like Hg or Ag. Since Au and Ag are SERS active metals, the possibility of using applied potential perturbation for TRSERS measurements to investigate chemical processes following radical ion formation are quite general.

Adsorbed species are interesting it that it may be easier to form radical ions electrochemically from the adsorbed state than from the solution soluble state. For an adsorbed species, there is an electronic interaction with the surface, Δ, which must be considered when treating the possibility of charge transfer. This interaction energy broadens the distribution of electronic levels of the adsorbate giving a Lorenzian distribution [42] so that some levels are shifted toward the Fermi level making CT possible at lower applied potentials. On the other hand, the interaction of the adsorbed species with the solvent is different for an adsorbed ion than in the bulk. The adsorbed species loses about one-half of its bulk solvation sphere resulting in about one-half of the solvation energy. This loss may be compensated by the specific adsorption energy, E_{ad}, due to specific chemical interactions when the ion adsorbs and by the image potential of the adsorbed ion at the surface. On balance radical ion formation should be easier for adsorbed species than for solution soluble species. This can be observed in positive(negative) shifts in electrode potential for reduction(oxidation) on the current-potential curve for systems where the electrochemistry of the adsorbate dominates.

In order to calculate the energy required for electrochemical formation of radical ions, it would be necessary to have values for gas phase ionization potentials and electron affinities, radical ion solvation energies, adsorption energies, and the Fermi energy, E_F, of the metal electrode. For an isolated metal E_F is equivalent to the electrochemical potential and is the energy gained in moving an electron from vacuum into the Fermi level of the metal, and thus is the negative of the metal work function (a positive quantity). When there is an applied electrostatic potential on the metal, then $E_F = -(\Phi + e_0\phi_M)$, where Φ is the work function for the metal in vacuum, e_0 is the charge on the unit electron, and ϕ_M is the inner or Galvani potential on the metal. The Galvani potential includes a surface potential, χ, so that $\phi = \psi + \chi$. If we consider the inner potential drop, $\phi_M - \phi_A$, at the electrode surface, then the energy $[\Phi + e_0(\phi_M - \phi_A)]]$ controls whether electron transfer is possible. In practice it is not possible to measure the inner potentials, ϕ, and an applied potential measured against a reference electrode is used in describing experimental systems. Nonetheless using an estimate of the reference electrode potential vs. vacuum and other energy terms in the thermodynamic cycle, it should be possible to estimate whether there is enough energy available for a particular radical ion formation to take place in the permissible voltage window of the electrode.

The above estimate is a thermodynamic calculation. Whether an electrode current flows, indicating the formation of particular radical ion, depends on the kinetics of the reaction. The theory for such a CT reaction at an electrode surface is complicated and still in a state development. Consider a two state system, say R/R^-, for $R + e^- + R^-$ formation. Here, the CT process involves energy of state 1, ϵ_1, of the vacant LUMO on R with the electron in the Fermi level of the metal and the energy of state 2, ϵ_2, for the occupied LUMO on R^-. These energies depend on atomic positions which are fluctuating as a function of time because of the vibrations of the molecule and the solvent and thus ϵ is a function of time. The electron transfer from the Fermi level to the LUMO is a vertical Franck–Condon transition. This means that since the nuclear coordinates do not change during CT, the moment R^- is formed, it will be in a nonequilibrium state which then must relax. The rate constant for electron transfer can be expressed in general in terms of a Fermi Golden Rule expression:

$$k = \frac{2\pi}{\hbar}\langle|V_{12}|^2\rangle\delta(\epsilon_2 - \epsilon_1)$$

where the delta function $\delta(\epsilon_2 - \epsilon_1)$ indicates that state-to-state transition occurs at equal energies for vacant and singly occupied LUMO. Evaluation of the electron hopping matrix element, $|V_{12}|$, is in general complicated because it is difficult to model the electron–vibrational coupling. A complete theory should include the relaxation of both internal vibrational modes of the radical ion and the external solvent modes.

For the case where only external solvent modes are considered, several well-known treatments [45] associated with the names of Marcus–Hush and Levich–Dogonadze–Kuznestov give the microscopic rate constant for electron transfer as

$$k = \frac{|V_{12}|^2}{\hbar}\sqrt{\frac{\pi}{k_B T \lambda_0}}\exp\left[-\frac{(\Delta G^\circ + \lambda_0)^2}{4k_B T \lambda_0}\right]$$

where $|V_{12}|$ is the interaction matrix element, λ_0 the external solvent mode reorganization energy, k_B the Boltzman constant, T the temperature in Kelvins, and ΔG° is the standard electronic free energy change in the reaction at a fixed reaction distance. To obtain the macroscopic rate constant, the proper sum over distance and energy density and occupation probabilities of the electron in the donor and acceptor states (metal and molecule) is necessary. The resulting expression is still quadratic in applied electrode potential, E_a, since $e_0 E_a$ is in the ΔG^0 term of the exponential. In most experimental studies, the data is fit to the phenomenological Butler–Volmer rate constant expression which is linear in applied potential, E_a:

$$k = k^0 \exp[-\alpha n F[E_a - E^0]/RT]$$

where k^0 is called the standard heterogeneous rate constant, α the charge transfer coefficient, E^0 is the standard redox potential, and nF/R is the number of electrons transferred times the Faraday constant, F, divided by the gas constant, R. This is the rate constant for reduction; whereas, for oxidation α is replaced by $\beta = 1 - \alpha$. If the assumption is made that $\Delta G^0/4\lambda \ll 1$, then the theoretical rate constant reduces to the form of the phenomenological rate constant with $\alpha = 1/2$.

TRSERS studies of the electrochemically formed radical ions are limited in time response (slower than charging time of the double layer), and so it is chemical reactions following the formation of a fully relaxed radical ion that will be followed. TRSERS results can be used to identify species involved in CT reactions at electrode and this information used to formulate a mechanism of the electrode process which can be used to numerically simulate current–voltage curves. Curve fitting then allows the electrochemical parameters, k^0, α, and E^0 to be found from current–voltage curves. Two typical reaction possibilities for radical ions are reaction with the solvent and dimerization(polymerization) reactions. Examples of these types of reactions will be given in the next section. In most cases, it appears as if the radical ions desorb, react at the metal/solution interface, and the product then adsorbs and is observed by TRSERS.

4. SERS and TRSERS for Various Radical Ion Systems

4.A. METHYLVIOLOGEN AND OTHER ALKYLVIOLOGENS

These 1,1'-dialkyl-4,4'bipyridinium compounds have been extensively studied at electrode surfaces because of their use as electron mediators [46], electrochromic display materials [47,48] surface adsorbates for modified electrodes [49], and redox reagents in photoelectrolysis systems [50]. The resonance and surface Raman spectroscopy of the dications methylviologen (MV^{++}) and heptylviologen (HV^{++}) has been broadly studied by many research groups [51–57]. We have shown [53] that the SERS spectrum of three redox forms of methyl viologen, MV^{++}, $MV^{+\cdot}$, and MV^0, can be observed on a roughened Ag electrode as the potential is moved from -0.1 V to -1.1 V. The cationic forms are most likely adsorbed to the Ag surface by an electrostatic mechanism involving an ion pair with the counter anion. The strongly adsorbed form of the monoradical cation, $MV^{+\cdot}$, has a SERS spectra on roughened Ag at -0.8 V that differs from the species observed on polished Pt. The spectrum obtained on smooth Pt is nearly identical [53,56,57] to the RR spectrum of the solution soluble species which has an intense blue-purple color. In fact, the surface Raman spectrum of MV^+ on roughened Ag contains bands from a strongly adsorbed form and from the RR active species which is easily removed by washing the electrode under potential control [53] The bands from the strongly adsorbed species are shifted, broader, and do not contain overtone and combination bands which is consistent with our theory of SERS based on Herzberg–Teller scattering [10]. Thus there are two types of MV^+ radical cations called type I and type II by Cotton *et al.* [57]. Type I is not strongly adsorbed and shows only the EM effect of SERRS while type II is strongly adsorbed and shows the adsorption induced resonance (chemical) enhancement effect [10] of SERS. The actual adsorption geometry of the viologen radical cations on Ag is not precisely known. The counter anion strongly affects the nature of the spectrum, and for type II species there are indications that the molecule is adsorbed with its aromatic rings parallel to the surface [56].

SERS experiments show that the adsorbed radical cations of viologens are irreversibly adsorbed from aqueous solutions on Ag after an ORC in the presence of the viologen and are a monomeric species [53,56,57]. Under some conditions, a dimer or polymer film of $MV^{+\cdot}$ can form on metal surfaces, but this was not observed in our experiments on roughened Ag. On the other hand, the neutral MV^0 does form a polymer film on a Ag electrode surface in aqueous solution. The study of the methylviologen as a function of electrode potential illustrates the effect on vibrational frequencies of adding consecutive electrons to a molecule. According to our assignments, the inter ring, C4—C4', stretching vibration is observed at 1292 cm^{-1} in the SERS of MV^{++}, at 1350 cm^{-1} in the SERS of $MV^{+\cdot}$, and at 1380 m^{-1} in the SERS of $MV^{0\cdot}$. The LUMO of MV^{++} should contain electron density in the C—C bridging bond in analogy to biphenyl where MO calculations show considerable contribution of electron density at this bond [58]. The upward shift of the observed frequency of this vibration when an electron is added to the LUMO is consistent with increased electron density at the bridging carbons. The smaller shift when the LUMO is filled can be rationalized by the effect of electron–electron repulsion which causes less electron density build-up at the bridging bond when the second electron is added.

Very recently time-resolved RRS and SERRS studies [59] have been made of film growth on Ag electrodes involving the monocation radical of heptylviologen and other viologen analogs, $RV^{+\cdot}$, containing bulky R groups. For these species, an insoluble purple-blue film grows on the surface by reaction with the counter ion giving rise to the mechanism of an electrochromic display device since voltage reversal causes the film and color to disappear. In these investigations, RRS can be used on smooth Ag since the only RRS active species in the system is the radical cation which is produced directly at the electrode surface. Surface Raman (SERS and SERRS) results from the radical cation viologen species adsorbed directly on a roughened Ag electrode. In these studies, a potential dependent pump-probe method was used with a dye laser excited by a Q-switched Nd:YAG laser with about a 15 ns pulse width and a maximum repetition rate of 15 Hz. The potential was repeatedly stepped from -0.2 V to -0.65 V for t_1 ms and back to -0.2 V for t_2 ms, where $t_2 \approx (3-4)t_1$. Measurements were made by accumulating thousands of laser pulses fired after the rising edge of the step to -0.65 V. At potentials > -0.65 V, the $HV^{+\cdot}$ radical monomer is formed and reacts at the surface; whereas, when the system is returned to -0.2 V, HV^{++} is reformed reversibly, allowing the repetitive cycle to be used.

The time dependent measurements were made over a time range of 0.0 ms to 25.0 ms with the fastest coming at 0.25 ms with 532-nm excitation. Both on smooth (RRS) and roughened Ag electrodes (SERRS), growing doublet bands for the $HV^{+\cdot}$ radical appeared with time. These spectra show two doublets, one with a band at 1530 cm^{-1} and a shoulder at ca. 1510 cm^{-1} and another with a band at 1352 cm^{-1} and a shoulder at 1340 cm^{-1}. The bands at 1530 and 1352 cm^{-1} indicate the presence of the radical monomer or its salt and the shoulder bands at 1510 and 1340 cm^{-1} indicate formation of a radical dimer film at the surface of both the smooth and roughened Ag electrodes. Comparing the intensity vs. time$^{1/2}$ plots with electrode current vs. time$^{1/2}$ plots, it is concluded [58] that at first an instantaneous nucleation process occurs, then three-dimensional growth of nuclei take place, finally followed by two consecutive types of diffusion-controlled film growth processes which have different rates. At both smooth and roughened Ag, the dimer is converted to a radical salt after a film of at least 5 monolayer forms.

4.B. p-NITROBENZOIC ACID

We have found that p-nitrobenzoic acid, PNBA, in an alkaline medium is photo-electrochemically reduced on roughened Ag electrodes [41],. In the presence of the laser excitation, a time dependency in the spectra is observed even at potentials positive to the electrochemical reduction wave [41]. This behavior was attributed to a photoinduced reduction process. Even the straight electroreduction of arylnitro compounds is a complicated process which has received much attention in the literature [60–63]. These studies indicate that reduction of an arylnitro compound (Ar—NO$_2$), in our case p-nitrobenzoate (Ar = —OOC—C$_6$H$_4$—NO$_2$), should produce p-nitrosobenzoate (Ar—NO) as an intermediate which, however, does not accumulate but forms p-hydroxylaminobenzoic acid (Ar—NHOH) on most metal electrode surfaces. In alkaline medium, the intermediate products such as hydroxylamine and nitroso compounds can react chemically to produce an azoxy compound, i.e., Ar—NHOH + Ar—NO → Ar—N(→O)=N—Ar + H$_2$O, which may be further reduced to azo and hydrazo compounds. In all of these compounds, the benzoate group can bind to the Ag surface through both oxygens of the carboxylate group, and indeed, the SERS spectra show a broad band at 1395 cm^{-1} for the symmetrical COO stretching vibration. Since TRSERS techniques allow monolayer sensitivity and the possibility of identifying the structure of transient intermediates and products, we applied a single shot TRSERS method to elucidate the mechanism of electrochemical and photoelectrochemical reduction of p-nitrobenzoic acid (PNBA) on a Ag electrode in basic solution [64, 65].

The single shot TRSERS method employed a triggered shutter and sequential triggering of the OMA on a triangular potential sweep (CV) [65] or a potential step [64], as discussed in the experimental section of this article. Rapid scan OMA SERS spectra (10 ms to 100 ms dwell time) were obtained at various times on CV sweeps or on a potential step. Other investigations of the electroreduction of nitrocompounds have been carried out by time-resolved SERS on slower time scales in order to identify products or intermediates [66, 67]; however, only the long-lived or stable intermediates were detected since the time scale was in seconds. Our *in situ* SERS spectra during the CV experiment show that p-nitrosobenzoate, hydroxylamine, and azoxy compounds appear in the reduction process and are stable in certain regions of potential. It appeared that essentially the same electrode mechanism takes place on a smooth electrode as on a SERS activated (roughened) electrode [65].

By using a step potential waveform, we were able to observe TRSERS spectra of a transient intermediate species which we attribute to the p-nitrosobenzoate radical anion. A potential step to -0.9 V was applied to the ORC roughened Ag electrode for a period of 200 ms to reduce the p-nitrobenzoate to the hydroxylamine compound, and the potential was stepped back to -0.2 V where the hydroxylamine was oxidized to the nitroso compound. Full TRSERS spectra were recorded with 10 ms dwell time by an optical multichannel analyzer (OMA) at various times after the potential step returns to -0.2 V (Figure 7). Transient bands were found at 966, 1233 and 1580 cm^{-1}. These bands appear slightly earlier than those bands for the final products and then disappear about 70 ms later. The experimental curves of the intensities of the transient 1233 and 1580 cm^{-1} bands versus time are shown in Figure 8. These transient bands were suggested to be the radical anion intermediate, Ar—NO$^{-\cdot}$, produced during the electrochemical oxidation of the hydroxylamine compound:

$$\text{Ar—NHOH} + 2\text{OH}^- \rightarrow \text{Ar—NO}^{-\cdot} + 2\text{H}_2\text{O} + e^-$$

$$\text{Ar—NO}^{-\cdot} \rightarrow \text{Ar—NO} + e^-$$

The lifetime of this species on the Ag electrode surface is about 70 ms. ESR studies also provide evidence for the existence of nitroso radical anion but the lifetime of this free radical in solution is much longer, about two minutes [68]. The shorter lifetime in our experiment is understandable since the species is not free in solution but is situated on the electrode surface or in the electrical double layer at the Ag surface. The 1580 cm^{-1} band of the transient species may be assigned as the benzene ring stretching mode. The resonance effect of the free radical with benzene ring would tend to stabilize the radical and thus cause a shift of the ring stretch band from 1600 to 1580 cm^{-1}. Furthermore, the transient band at 1233 cm^{-1} may be related to the carbon-nitrogen stretching vibration since the resonance effect favors the double bond structures and the carbon-nitrogen single bond vibration band is located at about 1145 cm^{-1}.

On the basis of the TRSERS results which show spectral lines which can be assigned to the nitro, nitroso, nitroso radical anion, hydroxylamine, and azoxy compounds, we were able to propose an overall electrode process for the electrochemical mechanism of reduction on Ag:

$$\text{Ar—NO}_2 + \text{H}_2\text{O} + e^- \xrightarrow{E_1^0, k_1^0} \text{Ar—NO}_2\text{H} + \text{OH}^- \qquad (1)$$

$$\text{Ar—NO}_2\text{H} + \text{H}_2\text{O} + e^- \xrightarrow{E_2^0, k_2^0} \text{Ar—NO}_2\text{H} + \text{OH}^- \qquad (2)$$

$$\text{Ar—NO}_2\text{H}_2 \xrightarrow{k_{c1}} \text{Ar—NO} + \text{H}_2\text{O} \qquad (3)$$

INVESTIGATION OF RADICAL IONS WITH TRSERS

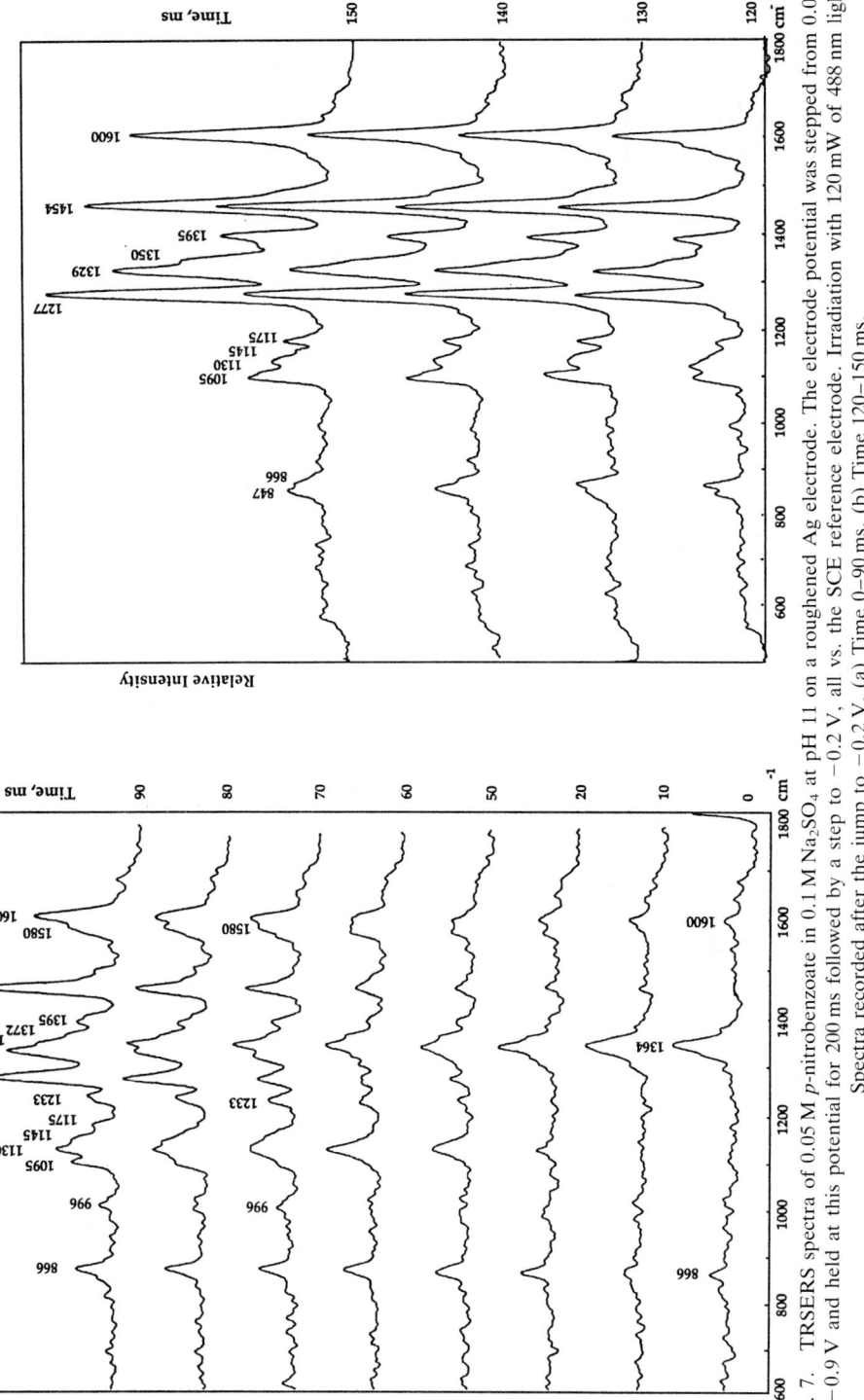

Fig. 7. TRSERS spectra of 0.05 M p-nitrobenzoate in 0.1 M Na_2SO_4 at pH 11 on a roughened Ag electrode. The electrode potential was stepped from 0.0 V to −0.9 V and held at this potential for 200 ms followed by a step to −0.2 V, all vs. the SCE reference electrode. Irradiation with 120 mW of 488 nm light. Spectra recorded after the jump to −0.2 V. (a) Time 0–90 ms, (b) Time 120–150 ms.

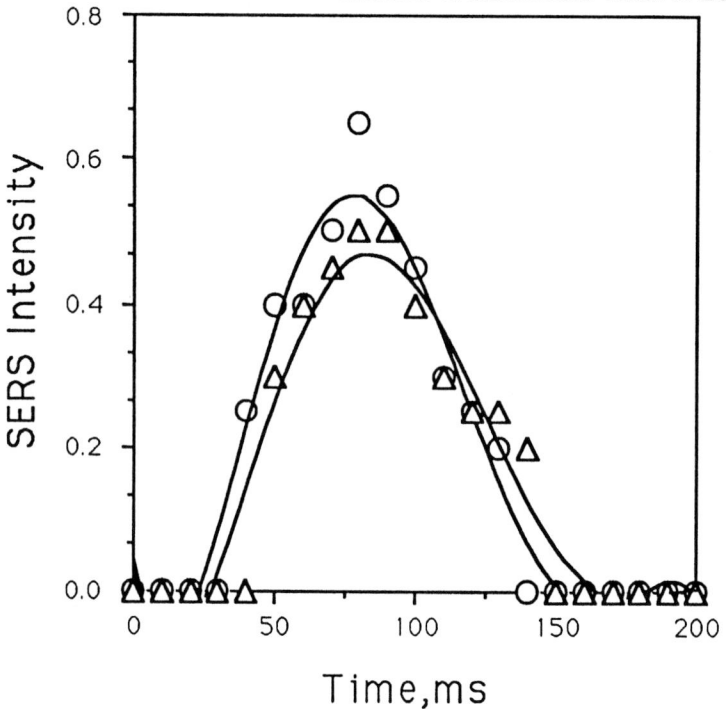

Fig. 8. Time dependence of SERS intensity for transient bands in the potential step experiment of Figure 7. Bands plotted: (○) 1233 cm^{-1}, (△) 1580 cm^{-1}.

$$\text{Ar—NO} + e^- \xrightarrow{E_3^0, k_3^0} \text{Ar—NO}^- \quad (4)$$

$$\text{Ar—NO}^- + e^- + 2\text{H}_2\text{O} \xrightarrow{E_4^0, k_4^0} \text{Ar—NHOH} + 2\text{OH}^- \quad (5)$$

$$\text{Ar—NO} + \text{Ar—NHOH} \xrightarrow{k_{c2}} \text{Ar—N(O)=N—Ar} + \text{H}_2\text{O} \quad (6)$$

The mechanism was fitted by digital simulation to the cyclic voltammetry curves over three orders of magnitude in scan rate and the following parameters found: $E_1^0 = E_2^0 = -0.27$ V, $k_1^0 = k_2^0 = 1.08 \times 10^{-5}$ cm s^{-1}, $\alpha_1 = \alpha_2 = 0.5$; which represents in a stepwise manner an electrochemically irreversible two electron, two proton reduction for the first step in the mechanism. The fast loss of a water molecule occurs in chemical reaction (3), $k_{c1} = 500$ s^{-1}, which is only a lower bound for the rate of this reaction. The formation of the transient nitroso radical anion is given in reaction (4) with electrochemical parameters $E_3^0 = -0.34$ V, $k_3^0 = 0.0600$ cm s^{-1}, $\alpha_3 = 0.63$; however, this species is reduced in reaction (5) to the hydroxylamine at slightly more positive potentials and with a relatively fast heterogeneous rate constant. The electrochemical parameters for this step are $E_4^0 = -0.32$ V, $k_4^0 = 0.0165$ cm s^{-1}, $\alpha_4 = 0.63$. Thus, this radical anion is electrochemically unstable on the reduction sweep of the CV curve and can only be observed as a transient species. Finally, the shape of the CV curve is only correctly simulated by adding a chemical coupling reaction for the formation of the azoxy compound with rate constant, $k_{c2} = 4.5 \text{ s}^{-1}$. Good agreement was obtained between the simulated and experimental curves [65].

4.C. p-SUBSTITUTED PYRIDINE COMPOUNDS

Many *para* —X substituted pyridine (C_6H_5X) compounds are known which are electrochemically active in the negative potential range (-0.9 V to -1.2 V vs. SCE) on Ag. We have found [69] using TRSERS studies that when —X is a cyano group (—CN), aldehyde group (—CHO), or hydroxymethyl (—CH_2OH) group that a radical can form which then dimerizes to form products. These mechanisms are complicated because further reduction usually takes at the electrode surface. For 4-cyanopyridine (4-CNPy), a dimerization reaction occurs when the parent compound is both adsorbed on the surface and in the bulk solution at relatively high concentrations. (At low bulk concentrations, the molecule is reduced at Ag to pyridine and cyanide.) With 0.020 M 4-CNPy in the bulk in 0.1 M KCl electrolyte, if the electrode potential is stepped or scanned to -1.2 V, a dimerization reaction occurs. This reaction has been studied with TRSERS in a pump-probe mode with a voltage step from -0.4 V to -1.2 V as the pump and a 488 nm Ar$^+$ CW laser as the probe with a gated detection on a linear diode array detector [69]. Each spectrum was averaged from 5000 shots and recorded over 0.0 to 5.0 ms with the first spectrum recorded at 0.2 ms. The time-resolved SERS results (Figure 9) show that, at the beginning, the relative intensity of the 4-CNPy bands at 1450, 1262 and 970 cm^{-1} bands decreases, but the 847 cm^{-1} band increases. No new bands were found in the first 2.0 ms. After about 2.5 ms, two new bands appear at 1416 and 1471 cm^{-1} and the original 1520 cm^{-1} band gradually shifts to 1530 cm^{-1}. Although no transient intermediate bands were found in the 10 ms time window, the decrease of the intensity of three initial bands in the first 2.0 ms can be attributed to the formation of 4-CNPY radical anion by one electron reduction. A careful use of difference spectra in the μs region might reveal the radical anion spectrum. The final spectrum represents a new species generated by the electrode reactions. The bands at 1416 and 1471 cm^{-1} are assigned to an azo (—N=N—) stretching mode formed by the reaction of the 4-CNPy radical anion with the parent species followed by further reduction in the scheme:

$$PyCN + e^- \rightarrow Py\text{—}CN^-$$
$$Py\text{—}CN^- + Py\text{—}CN + e^- + 2H^+$$
$$\rightarrow Py\text{—}CH\text{=}N\text{—}N\text{=}CH\text{—}Py \quad \text{(azin species)}$$
$$Py\text{—}CH\text{=}N\text{—}N\text{=}CH\text{—}Py + 2e^- + 2H^+$$
$$\rightarrow Py\text{—}CH_2\text{—}N\text{=}N\text{—}CH_2\text{—}Py \quad \text{(azo species)}$$

Similar TRSERS voltage pump-laser probe experiments for a 0.01 M solution of 4-pyridine carboxyaldehyde (4-CHO—Py) in neutral 0.1 M KCL indicate [69] that at -0.8 V a radical species is formed in the reactions

$$Py\text{—}CHO + e^- \rightarrow Py\text{—}CHO^-$$
$$Py\text{—}CHO^- + H_2O \rightarrow Py\text{—}CHOH^{\cdot}$$

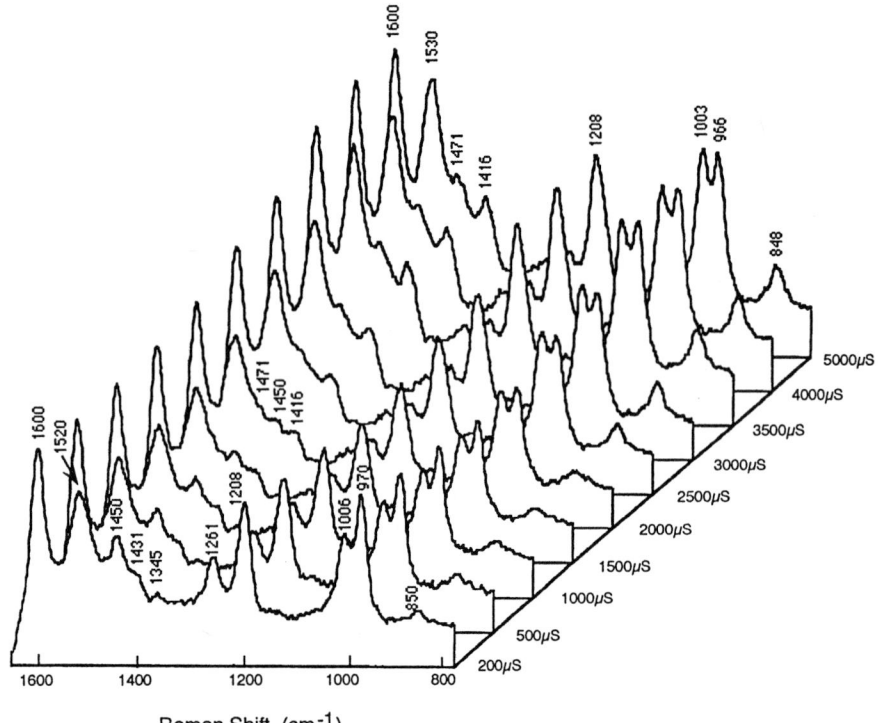

Fig. 9. TRSERS spectra obtained during the electroreduction of 20.0 mM 4-cyanopyridine (4-CNPY) in 0.1 M KCl solution. A potential step is applied to the ORC roughned Ag electrode. Initial potential: -0.4 V; final potential: -1.2 V. Time scale from front to back: (1) 200 μs; (2) 500 μs; (3) 1000 μs; (4) 1500 μs; (5) 2000 μs; (6) 2500 μs; (7) 3000 μs; (8) 3500 μs; (9) 4000 μs; (10) 5000 μs.

The reduction is initiated by the negative potential pulse to -0.8 V in the time window from 100 μs to 30 ms. Up to 1.0 ms, no new bands are found in the TRSERS but several original bands decay, located at 1571, 1270, 1150, 1092 and 1059 cm^{-1}. A transient spectrum is detected during the time window 1.0 to 10.0 ms. Several new bands are found in the transient spectrum, which are at 1384, 1356, 1330 and 1276 cm^{-1}. A product spectrum appears after 10 ms and its intensity is much stronger than that of both the reactant and the intermediate. This product spectrum has two very strong bands at 1592 and 1208 cm^{-1}, six strong bands at 1554, 1532, 1494, 1476, 1322 and 1004 cm^{-1}, four medium strong bands at 1424, 1248, 878 and 842 cm-1 and three weak bands at 1636, 1090 and 1056 cm^{-1}. The change in spectrum in the first millisecond is probably due to the one electron addition to 4-CHO—Py with the formation of a neutral radical, Py—CHOH˙ through hydration. A transient spectrum is detected during the time window 1.0 to 10.0 ms. The most likely reaction path is the formation of a pinacol (1,2-bishydroxy-1,2-bis(4-pyridyl)ethane) species, Py—CH(OH)—CH(OH)—Py, by dimerization of the aldehyde radical, Py—CH(OH)˙

$$2\text{Py—CH(OH)}\cdot \rightarrow \text{Py—CH(OH)—CH(OH)—Py}$$

Further reduction of this intermediate in the reaction

$$Py-CH(OH)-CH(OH)-Py + 4e^- + 4H^+$$
$$\rightarrow Py-CH_2-CH_2-Py + 2 H_2O$$

gives the product, 1,2-bis(4-pyridyl)ethane (BPE), $Py-CH_2-CH_2-Py$. The SERS spectra of BPE, which is commercially available, were measured at both pH 7 and 3. The authentic BPE spectrum at -0.8 V at pH 7 is almost identical to the product spectrum of reduction of 4-CHO—-Py except for the presence of a band at 1636 cm^{-1}. The 1636 cm^{-1} band could be identified as coming from the existence of a 4-PyCH$_2$OH species which is thought to form in a side reaction from the protonated form of the initial radical

$$Py-CH(OH)^{\cdot} + H^+ = Py-CH_2OH^{\cdot+}$$
$$Py-CH_2OH^{\cdot+} + e^- \rightarrow Py-CH_2OH$$

during the reduction. Therefore, the product spectrum observed after 10 ms is a combination of the spectra of 4-PyCH$_2$OH and BPE, with later the major product.

The total reduction process can be written as follows,

<1.0 ms	$Py-CHO + e^- + H_2O \rightarrow Py-CHOH^{\cdot} + OH^-$	(1)
1.0–10 ms	$2Py-CHOH^{\cdot} \rightarrow Py-CH(OH)CH(OH)-Py$	(2)
>10 ms	$Py-CH(OH)-CH(OH)-Py + 2e^- + 2H^+$	(3)
	$\rightarrow Py-CH_2-CH_2-Py + 2 H_2O$	

The electroreduction process of the 4-hydroxymethylpyridine (Py—CH$_2$OH) compound can be followed by time-resolved SERS in the same way [69] as the aldehyde and shows a similar radical coupling indicated in Scheme 1 below:

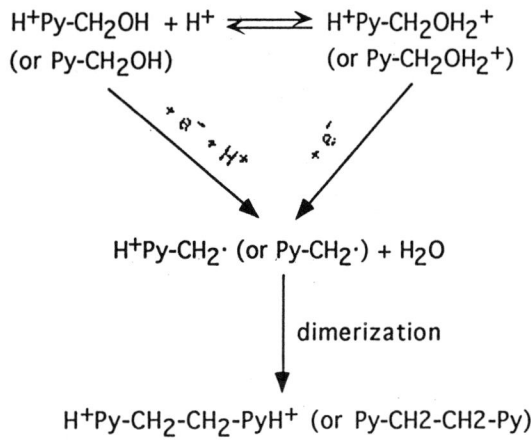

Scheme 1.

The product of this process is again 1,2-bis(4-pyridyl)ethane. Here the possibility of the existence of various protonated species are indicated. The SERS spectra show that in neutral medium there is an equilibrium between the neutral and singly protonated forms. The extent to which the radical coupling reactions occur in the diffusion layer at the electrode surface or on the surface has not been determined.

4.D. FLAVIN MONONUCLEOTIDE, FMN

Flavin mononucleotide (FMN) is a coenzyme which plays an important role in redox enzymes and is well known to have interesting excited state chemistry [70]. It has also been used as a reagent to produce a highly oxidizing excited state redox species in a photogalvanic cell which utilizes solar energy to oxidizes organic material at the illuminated electrode and reduces H^+ to hydrogen gas at the dark electroder [71]. We have made such a cell with FMN adsorbed on the surface of roughened Ag and demonstrated that it works without FMN in the bulk solution [72]. This means that the *adsorbed FMN* must be directly involved in the photochemistry. Our electrochemistry experiments show that only one monolayer of FMN is on the electrode surface.

The direct photoinduced charge transfer from adsorbed flavin mononucleotide (FMN) to a Ag electrode has been observed by a ns pulse laser pump-CW laser probe TRSERS system described in the experimental section. These results [1] indicate two photoproducts in FMN adsorbed on Ag surfaces, after illumination with a 10 ns laser pulse (337 nm. N_2 laser). The first product is obtained within the laser pulse duration plus an instrumental delay of 75 ns, and decays to the second photoproduct by 775 ns. Over still longer times ($>\mu$s) the second photoproduct reverts to the initial species FMN, allowing continuous repetition of the experiment.

Spectra were taken as displayed in Figure 10, which show the surface enhanced Raman spectrum of FMN at -0.4 V in the region of 800 to 1640 cm^{-1}, using the 488 nm Ar^+ Laser line as probe beam. The gate width is 200 ns, and each spectrum represents an average over 10 000 pulses. The lowest curve (a) is the spectrum with the 337 nm light blocked, and since it is identical to the SERS spectrum of FMN itself, indicates that under these conditions no reaction takes place. The middle spectrum (b) is obtained when 337 nm light is allowed to fall on the electrode with a 75 ns delay. To clearly highlight the effect of the UV laser, we show in curve (c) the difference between (b) and (a) (multiplied by a factor of 4). The resultant spectrum shows slight, but significant differences from that of FMN. The most significant shifts involve the 1572 cm^{-1}, the 1532 cm^{-1}, 1344 cm^{-1} and the 1089 cm^{-1} lines which shift to 1576 cm^{-1} ($+4$ cm^{-1}), 1528 cm^{-1} (-4 cm^{-1}) 1358 cm^{-1} ($+14$ cm^{-1}) and 1093 cm^{-1} ($+4$ cm^{-1}) respectively. In addition, two new lines appear, one at 1548 cm^{-1} and another as a shoulder at 1600 cm^{-1}.

Difference spectra at longer times show that by *ca.* 1500 ns the difference spectrum decays to zero, indicating reversion to the starting material FMN. Further by the time 775 ns has elapsed several new changes are evident. The line at 1093 cm^{-1} shifts still further to 1096 cm^{-1} and the 1257 cm^{-1} line shifts to 1252 cm^{-1}. The line at 1269 cm^{-1} disappears and a new, probably related line appears at 1321 cm^{-1}. Most of the other lines remain unchanged. From the above observed results, it is clear that two distinct photoproducts are produced: one,

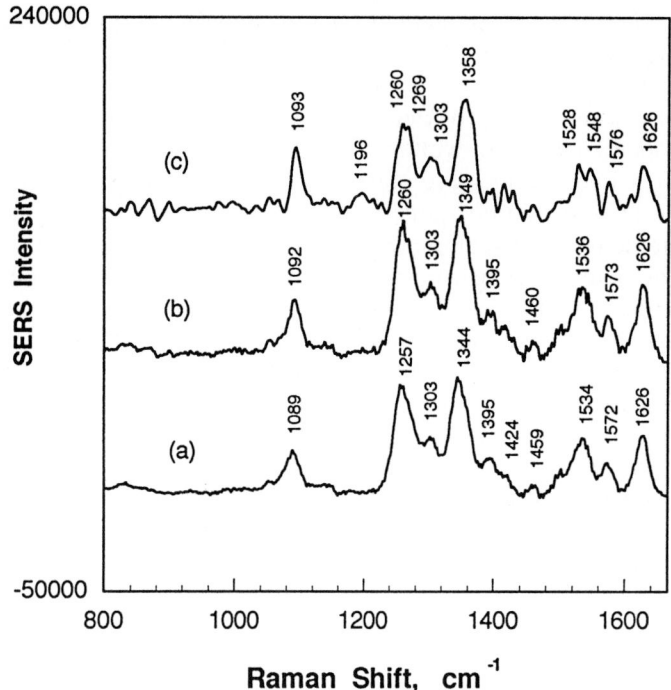

Fig. 10. TRSERS pump-probe spectra of flavin mononucleotide (FMN) on a roughened Ag electrode. Pump beam 337 nm, N_2 laser light with *ca.* 15 ns pulse width. Probe beam Ar^+ CW laser 488 nm light. Dwell time 200 ns. (a) probe beam only. (b) Pump-probe experiment with 75 ns delay time. (c) Difference of curve a and curve b (×4).

photoproduct **I**, after the 75 ns delay time and the other, photoproduct **II**, after the 775 ns delay time.

One likely interpretation of the above results is that a radical ion of FMN is produced at the electrode surface via charge transfer either to or from the Fermi level of the metal electrode. One technique for the testing of this phenomenon, and determination of the direction of charge transfer, is to measure the potential threshold at which spectral changes take place as a function of the wavelength of exciting light. We have carried out such studies [1] in FMN. Using several CW laser lines from either an Argon ion or dye laser, we have varied the potential from −0.4 V vs SCE, where FMN is stable on the surface in the presence of 488 nm light, in a more positive direction until a product spectrum just appears. In these steady state experiments, photoproduct **II** appears first as the excitation energy is increased. Photoproduct **I** is observed in these experiments at 364 nm and 350 nm, but not at lower energy. When the threshold potential is plotted against photon energy of the exciting light, a straight line is observed with a negative slope (−0.74). These results indicate that photoproduct **II** is a cation derivative of FMN, i.e. that charge is transferred during the process from the molecule to the metal at or near the Fermi level. The intercept of this line, which falls at 1.7 V, represents the potential (at zero photon energy) at which electron transfer would take place.

Further evidence of the interrelationship of the photo and electrochemical nature of the observations are provided by photovoltage and photocurrent measurements. The open circuit photovoltage observed as a function of excitation wavelength shows a decided increase starting at wavelengths shorter than 514 nm, indicating the need for an energetic threshold near about 2.4 eV. Similar results are evident in the photocurrent measurements. The dark photocurrent is near zero throughout the potential range, while with modulated 488 nm laser light impinging on the electrode, a sizable modulated photocurrent develops for potentials more positive than −0.2 V, once again indicating a threshold near 2.4 eV. Since the photocurrent increases as the potential is increased in a positive direction, the charge transfer process is oxidative, i.e. a cationic species is produced.

In order to understand these results we must first discuss the Raman spectrum in terms of the normal modes each line represents. FMN is composed of an isoalloxazine ring to which a triglycerol-phosphate linkage is attached. All of the Raman frequencies observed may be assigned to the isoalloxazine, which itself is composed of three rings, labeled **I**, **II**, and **III**, respectively, shown below.

According to our normal coordinate analysis, the normal vibrations may be viewed as arising from one or another of the rings. All the spectral lines associated exclusively with ring **I** are essentially unshifted throughout the reaction. These include lines at 740 cm^{-1} (ν_{28}), 803 cm^{-1} (ν_{27}), 833 cm^{-1} (ν_{26}), 863 cm^{-1} (ν_{25}), 1303 cm^{-1} (ν_{16}), 1459 cm^{-1} (ν_{11}), 1532 cm^{-1} (ν_9), and 1626 cm^{-1} (ν_6). We take this to mean that most of the changes which take place are in rings **II** and **III**. All the lines which involve these two rings show some degree of change during the course of the experiment. Further analysis show that those lines which change in forming photoproduct I, namely 555 cm^{-1} (ν_{32}), 994 cm^{-1} (ν_{24}), 1089 cm^{-1} (ν_{22}), 1344 cm^{-1} (ν_{15}), 1572 cm^{-1} (ν_8) almost all (except 667 cm^{-1} (ν_{29})) involve ring **II**, while those changing from photoproduct I to photoproduct II, namely 667 cm^{-1} (ν_{29}), 1257 cm^{-1} (ν_{18}) and 1269 cm^{-1} (ν_{17}) tend to involve ring **III**, although some slight changes in the ring **II** lines may also be observed. The vibrational bands ν_{22} and ν_{15} (at 1089 cm^{-1} and 1344 cm^{-1}) belonging to ring **II** stretching vibrational modes observed in TRSERS of adsorbed FMN at 75 ns have been shifted to higher frequency from the steady spectrum. Also the band at 1269 cm^{-1} in the 75 ns spectrum has been shifted from 1257 cm^{-1} in the steady state spectrum and is attributed to a pure ring **III** stretching mode. We take these shifts to higher wavenumber in the 75 ns spectrum to indicate an increase in bond order or aromaticity of the three member isoalloxazine ring at ring **II** and ring **III** of the photoproduct **I** intermediate.

In order to figure out the chemistry involved in the photoproduct transitions, band assignments [72] for steady state SERS spectral bands of (3N-H)FMN and (3N-D)FMN were made from our own normal mode calculations [72] in comparison with the experimental results from resonance Raman, isotopic substitution resonance Raman, and the normal coordinate calculations of M. Abe [73]. The most significant changes in the SERS bands on going from (3N-H)FMN to (3N-D)FMN are the following: the band found at 1424 cm^{-1} in the steady state SERS spectrum, assigned to a vibrational mode of the ring III N_3—H bending coupled with C_2=O stretching, disappears in (3N-D)FMN and (3N-D)LF (lumiflavin) steady state SERS spectra. Two new bands, 1170 m^{-1} and 948 cm^{-1}, were observed in (3N-D)FMN or (3N-D)LF SERS spectra, and these can be assigned to normal modes at 1174 cm^{-1} and 932 cm^{-1} for the N_3—D bending vibrations. Several other bands shift to higher wavenumber upon deuteration.

The band at 1420 cm^{-1} which is assigned to a N_3—H bending mode disappears at 75 ns in the TRSERS spectrum and a new band at 1196 cm^{-1} appears at this time. This 1196 cm^{-1} band is assigned in our normal mode calculation to a vibration involving the C—O—H bending and stretching modes. The 1196 cm^{-1} band disappears at 775 ns while the 1420 cm^{-1} band reappears at 775 ns. These vibrational band changes at the various delay times in TRSERS spectra indicate that a proton transfer occurs. The first proton transfer is from N_3—H to an adjacent C=O bond of ring **III** to form a C—O—H bond (enolization) within the first 75 ns after the laser pulse. The changes in 775 ns spectrum where the 1196 cm^{-1} band disappears and 1420 cm^{-1} band appears indicate a de-enolization process for photoproduct **I** going to photoproduct **II**. Also at 775 ns the band at 1269 cm^{-1}, which is attributed to a ring **III** stretching mode, shifts to 1252 cm^{-1} a new band at 1321 cm^{-1} appears, and the band at 1610 cm^{-1} disappears. These band shifts indicate that photoproduct **II** loses bond order or aromaticity in its ring **III** at longer time. Resonance structures show that the enol form of the cation radical formed by loss of an electron from a nitrogen atom in ring **III** will have increased aromaticity over the neutral FMN and that de-enolization lowers this aromaticity. Thus the spectral interpretation is quite consistent with cation radical formation.

Additional TRSERS spectra have been investigated with the deuterated species (3N-D)FMN at a 75 ns time delay with gate width of 500 ns. The experimental conditions were the same as those in the TRSERS of (3N-H)FMN. The band at 1172 cm^{-1} which is attributed to N_3—D bending mode disappears at the 75 ns delay time. New bands for the C—O—D stretching and bending modes were observed at 926 cm^{-1} and 900 cm^{-1}[21]. Also, similar bands shifts as found in the TRSERS of (3N-H)FMN corresponding to ring **II** and ring **III** vibrations have been observed. This experiment is additional evidence that FMN undergoes a photo-enolization process to form photoproduct **I** at 75 ns which decays after *ca.* 775 ns at the electrode surface to photoproduct **II**. The overall process is illustrated in Scheme 2.

All of our experiments were carried out at neutral pH, where proton transfer from the solvent is unlikely. We thus interpret these results as stemming from a photoinduced molecule-to-metal charge transfer, forming a flavin radical cation.

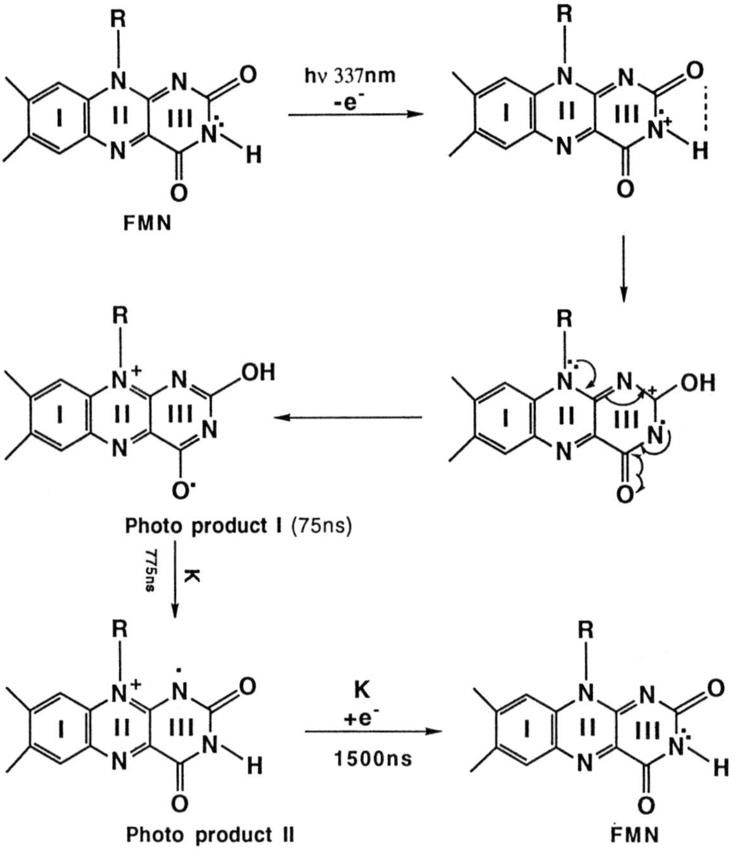

Scheme 2.

Such a species has been identified in the one electron oxidation of a flavin triplet species [74, 75]. The energy needed for ionization is considerably below that for the liquid phase ionization potential if the excited electron tunnels to the low lying empty Fermi levels of the metal, and indeed we can measure a threshold in the visible region for this process as a function of Fermi level. On the other hand, UV irradiation at 253.7 nm has been required to produce flavin radical cations in solution [76]. Thus, for an adsorbed molecule on a metal surface where the metal substrate and adsorbate wave function overlap, we infer that an excitation to an electronic level above the Fermi level can lead to electron tunneling to the metal by resonant charge transfer. It must be assumed that the reverse process is quenched by a relaxation mechanism, involving intramolecular enolization, solvent reorganization, and radical ion surface interactions, leading to a photoproduct **I**. To our knowledge, these studies are the first report of SERS observation of nanosecond time scale kinetics. All the experimental evidence indicates direct photoexcitation of an adsorbate to a radical cation on a metal electrode surface.

Acknowledgements

The authors are indebted to the PSC-BHE Research Award Program of the City

University of New York, the National Institutes of Health MBRS program and the Chemistry Section of NSF for financial assistance for unpublished work discussed in this article. We also acknowledge the work of our students, C. Shi, W. Zhang, and A. Vivoni, whose unpublished results appear in this article.

References

1. C. Shi, W. Zhang, R. L. and J. R. Lombardi: *J. Phys. Chem.* **96**, 10093 (1992).
2. J. R. Birk and S. P. Perone: *Anal. Chem.* **40**, 466 (1986).
3. A. Henglein: Pulse Radiolysis and Polarography: Electrode Reactions of Short-lived Free Radicals, in A. J. Bard (ed.), *Electroanalytical Chemistry*, Vol. 9, Marcel Dekker, Inc., New York (1976), pp. 163–244.
4. G. C. Pimental (ed.): *Opportunities in Chemistry*, National Academy Press, Washington, S. 86 ff. (1985).
5. R. R. Cavanagh, D. S. King, J. C. Stephenson, and T. F. Heinz: *J. Phys. Chem.* **97**, 786 (1993).
6. T. E. Furtak: 'Optical and Electronic Resonances: "The Underlying Sources of Surface Enhanced Raman Spectroscopy"' in B. Garetz and J. R. Lombardi (eds.), *Advances in Laser Spectroscopy*, Vol. 2, John Wiley and Son, Chichester (1984), pp. 175–205.
7. R. K. Chang and B. L. Laube: 'Surface-Enhanced Raman Scattering and Nonlinear Optics Applied to Electrochemistry', in *CRC Critical Reviews in Solid State and Material Science*, Vol. 12, pp. 1–73, CRC Press, Inc., Boca Raton (1984).
8. R. L. Birke and J. R. Lombardi: 'Surface-Enhanced Raman Scatterings', in R. J. Gale (ed.), *Spectroelectrochemistry: Theory and Practice*, Plenum Press, New York (1988), pp. 263–348.
9. E. J. Zeman and G. C. Schatz: *J. Phys. Chem.* **91**, 634 (1987).
10. J. R. Lombardi, R. L. Birke, T. Lu, and J. Xu: *J. Chem. Phys.* **84**, 4174 (1986).
11. J. R. Lombardi, R. L. Birke, L. A. Sanchez, I. Bernard, and S. C. Sun: *Chem. Phys. Lett.* **104**, 240 (1984).
12. R. P. Van Duyne: 'Laser Excitation of Raman Scattering from Adsorbed Molecules on Electrodes Surfaces', in C. B. Moore (ed.), *Chemical and Biochemical Applications of Lasers*, Vol. IV, Academic Press Inc., New York (1979), pp. 101–185.
13. R. L. Birke, T.-H. Lu, and J. R. Lombardi: 'Surface-Enhanced Raman Spectroscopy', in R. Varmi and J. R. Selman (eds.), *Techniques for the Characterization of Electrodes and Electrochemical Processes*, John Wiley and Son, Inc., New York (1991), pp. 211–277.
14. A. Vivoni, J. R. Lombardi, and R. L. Birke: unpublished results (1993).
15. J. Billman and A. Otto: *Solid State Commun.* **44**, 105 (1982).
16. T. E. Furtak and S. H. Macomber: *Chem. Phys. Lett.* **95**, 328 (1983).
17. Ph. Avouris and B. N. J. Persson: *J. Phys. Chem.* **88**, 837 (1984).
18. J. Bokor: *Science* **246**, 1130 (1989).
19. E. P. March, F. L. Tabares, M. R. Schneider, and J. P. Cowin: *J. Vac. Sci. Technol.* **A5**, 519 (1987).
20. E. P. Marsh, M. R. Schneider, T. L. Gilton, F. L. Tabares, W. Meier, and J. P. Cowin: *Phys. Rev. Lett.* **60**, 2551 (1988).
21. E. P. Marsh, T. L. Gilton, W. Meier, M. R. Schneider, and J. P. Cowin: *Phys. Rev. Lett.* **61**, 2725 (1988).
22. S. A. Costello, B. Roop, Z.-M. Liu, and J. M. White: *J. Phys. Chem.* **92**, 1019 (1988).
23. Z. M. Liu, S. A. Costello, B. Roop, S. R. Coon, S. Akhter, and J. M. White: *J. Phys. Chem.* **93**, 7681 (1989).
24. C. C. Cho, B. A. Collings, R. E. Hammer, J. C. Polanyi, C. D. Stanners, J. H. Wang, and G.-P. Xu: *J. Phys. Chem.* **93**, 7761 (1989).
25. B. Roop, S. A. Costello, C. M. Greenlief, and J. M. White: *Chem. Phys. Lett.* **143**, 38 (1988).
26. X. L. Zhou and J. M. White: *J. Phys. Chem.* **94**, 2643 (1990).
27. G. M. Goncher and C. A. Harris: *J. Chem. Phys.* **77**, 3767 (1982).
28. G. M. Goncher, C. A. Parsons, and C. A. Harris: *J. Phys. Chem.* **88**, 4200 (1984).
29. A. Nitzan and L. E. Brus: *J. Chem. Phys.* **74**, 5321 (1981).

30. A. Nitzan and L. E. Brus: *J. Chem. Phys.* **75**, 2205 (1981).
31. K. H. Drexhage, H. Kuhn, and F. P. Schafer: *Ber. Bunsenges. Phys. Chem.* **22**, 329 (1968).
32. R. R. Chance, A. Prock, and R. Silbey: *J. Chem. Phys.* **60**, 2744 (1974).
33. R. R. Chance, A. Prock, and R. Silbey: *Adv. Chem. Phys.* **37**, 1 (1978).
34. B. J. N. Persson: *J. Phys. C* **11**, 4251 (1978).
35. B. J. N. Persson and N. D. Lang: *Phys. Rev. B* **26**, 5409 (1982).
36. H. Metiu and J. W. Gadzuk: *J. Chem. Phys.* **74**, 2641 (1981).
37. H. Morawitz and M. R. Philpot: *Phys Rev. B* **10**, 4863 (1974).
38. R. Rupin: *J. Chem. Phys.* **76**, 1681 (1982).
39. J. Gersten and A. Nitzan: *J. Chem. Phys.* **74**, 1139 (1981).
40. Ph. Avouris and B. N. J. Persson: *J. Chem. Phys.* **79**, 5156 (1983).
41. S. Sun, R. L. Birke, J. R. Lombardi, K. P. Leung, and A. Z. Genack: *J. Phys. Chem.* **92**, 5965 (1988).
42. P. Avouris and R. E. Walkup: 'Fundamental Mechanisms of Desorption and Fragmentation Induced by Electronic Transitions at Surfaces' in H. Stock, G. T. Babcock, and C. B. Moore (eds.), *Annual Reviews of Physical Chemistry*, Vol. 40, Annual Reviews Inc., Palo Alto (1989), p. 173.
43. R. Meming and G. Kursten: *Ber. Bunsenges. Physik. Chem.* **76**, 4 (1972).
44. W. Schmickler: *J. Electroanal. Chem.* **100**, 533 (1979).
45. W. L. Reynolds and R. W. Lumry: *Mechanisms of Electron Transfer*, The Ronald Press Co., New York (1966), pp. 149–154.
46. L. N. Mackey and T. Kuwana: *Bioelectrochem. Bioenerg.* **3**, 596 (1976).
47. C. J. Schoot, J. J. Ponjee, H. T. van Dam, R. A. Van Doorn, and P. T. Bolwijn: *Appl. Phys. Lett.* **23**, 64 (1973); H. T. van Dam and J. J. Ponjee: *J. Electrochem. Soc.* **121**, 1555 (1974).
48. R. Jasinski: *J. Electrochem. Soc.* **124**, 533 (1979).
49. R. C. Ciesliski and N. R. Armstrong: *J. Electrochem. Soc.* **127**, 2605 (1980).
50. F. F. Fan, B. Reichman, and A. Bard: *J. Am. Chem. Soc.* **102**, 1488 (1980).
51. A. Regis and J. Corset: *J. Chim. Phys.* **78**, 687 (1981).
52. C. A. Melendres, P. C. Lee, and D. Meisel: *J. Electrochem. Soc.* **130**, 1523 (1983).
53. T. Lu, R. L. Birke, and J. R. Lombardi: *Langmuir* **2**, 305 (1986).
54. Q. Feng and T. M. Cotton: *J. Phys. Chem.* **90**, 983 (1986).
55. A. Yasuda, H. Kondo, M. Itbashi, and J. Seto: *Electroanal. Chem.* **210**, 265 (1986).
56. T. Lu and T. M. Cotton: *J. Phys. Chem.* **91**, 5978 (1987).
57. Q. Feng, W. Yue, and T. M. Cotton: *J. Phys. Chem.* **94**, 2082 (1990).
58. M. J. S. Dewar and N. Trinajstic: *Collec. Czech. Chem. Commun.* **35**, 3136 (1970).
59. Y. Misono, K. Shibasaki, N. Yamasara, Y. Mineo, and K. Itoh: *J. Phys. Chem.* **97**, 6054 (1993).
60. G. L. McIntire, D. M. Chiappardi, R. L. Casselberry, and H. N. Blount: *J. Chem. Phys.* **86**, 2632 (1982).
61. A. J. Fry: in S. Patai (ed.), *The Chemistry of Amino, Nitroso, Nitro Compounds and their Derivatives*, Part I, John Wiley and Son, New York (1982), p. 320.
62. I. Rubinstein: *J. Electroanal. Chem.* **183**, 379 (1985).
63. W. J. Albery, B. A. Coles, and A. M. Couper: *J. Electroanal. Chem.* **65**, 901 (1975).
64. C. Shi, W. Zhang, R. L. Birke, and J. R. Lombardi: *J. Phys. Chem.* **94**, 4767 (1990).
65. C. Shi, W. Zhang, R. L. Birke, and J. R. Lombardi: *J. Phys. Chem.* **95**, 6276 (1991).
66. P. Gao, D. Gosztola, and M. J. Weaver: *J. Phys. Chem.* **92**, 7122 (1988).
67. P. Gao, D. Gosztola, and M. J. Weaver: *Anal. Chim. Acta* **212**, 201 (1988).
68. C. Nishihara and M. Kaise: *J. Electroanal. Chem.* **149**, 287 (1983).
69. C. Shi, W. Zhang, R. L. Birke, and J. R. Lombardi: unpublished results (1993).
70. P. F. Heelis: *Chem. Soc. Rev.* **11**, 15 (1982).
71. J. Yamase: *Photochem. Photobiol.* **34**, 11 (1981).
72. W. Zhang, R. L. Birke, and J. R. Lombardi: unpublished results (1993).
73. M. Abe: in R. J. H. Clark and R. E. Hester (eds.), *Spectroscopy of Biological Systems*, Ch. 7, John Wiley & Sons Ltd (1986) and references cited therein.
74. P. F. Heelis, B. J. Parsons, B. Thomas, and G. O. Phillips: *J. Chem. Soc. Chem. Commun.* 954 (1985).
75. P. F. Heelis, B. J. Parsons, G. O. Phillips, and A. J. Swallow: *J. Phys. Chem.* **90**, 6833 (1986).
76. N. Getoff, S. Solar, and D. B. McCormick: *Science* **201**, 616 (1978).

INDEX

ab initio calculations, 111
ab initio Hartree–Fock (HF)ΔSCF calculations, 62
ab initio MO CI calculations, 68
　the XPS core spectra of NiCO, NiN$_2$, PdCO, Pd$_2$CO, PdN$_2$ and NiNH$_3$, 63, 68
accelerators, 259, 260, 263
acetone ketyl radical, 234, 239
acetylene, 135, 137, 188
acid-catalytic reaction, 188
acid sites, 39, 40, 42, 45, 57
activated alumina, 92
activated zeolites, 119
adsorbed molecules, 148, 164, 176
adsorption energy, 74
Ag cations, 194
Ag$^+$, 194, 196
Ag$_6^{n+}$, 196
Ag^0, 194, 195, 196
Ag$_3^0$, 194, 195, 197
Ag$_5^0$, 194
Ag$_2^+$, 194, 195
Ag$_6^+$, 195
Ag$_3^{2+}$, 194, 195, 196
Ag$_4^{3+}$, 195
Ag$_6^{n+}$, 195
Ag$_6$-M$^+$A, 195
Ag$_1$-NaA, 195, 196
^{27}Al, 92, 101, 108
Al^{+3} (cus) site, 102
alkali, 147, 148, 151, 152, 153, 156, 158, 159, 161, 162, 163, 176
alkali cationic clusters, 189
alkali metal particles, 192
alkali metal vapours, 190
alkaline earth metals, 147
alkene adsorption on H-zeolites, 128
alkenes, 188, 192
alkynes, 187, 188, 192
^{27}Al nuclear modulation, 192
allyl radical, 265
alumina, 201, 202, 219, 223, 224
Anderson model, 61
angle resolved near edge X-ray, 64
aniline, 183
anion radicals, 104

antraquinone anion radical, 244
aromatic molecules, 119
aromatic radicals, 91
aromatic radical cations, 125, 127
aromatics, 188
Auger spectrum, 64, 77, 81, 82, 84
autoionization, 64, 77, 78, 81, 83, 84
autoionization spectra of Ar adsorbed on a number of substrates; graphite, Pt(111), Cu(100), Au(110) and Ag(110), 80
autoreduction, 194
avoided crossings, 262, 264
avoided-level-crossing, 263, 264, 268

^{11}B, 101
backbonding peak, 82
back-donation, 64, 74, 76
back-electron transfer, 244
band structure effect, 62
benzene, 179, 183, 184, 185, 187, 188, 194, 197, 201, 202, 213, 214, 215, 216, 217, 218, 219, 221, 222, 223, 265, 268, 269, 270, 271, 272, 273, 274
benzene cation radicals, 185, 187, 188
benzene dimeric cations, 185
benzene radical cation, 127
benzonaphthothiophene, 110
benzoquinone, 40, 41, 42, 46, 47, 48, 49, 245
benzothiophene, 110
β-proton hyperfine coupling, 98
bicyclo[3.3.0]oct-1-ene, 134
bicyclo[3.3.0]oct-2-ene, 134
bimodal cavity, 234
binomial intensity, 241, 242, 245
biphenyl radical cation, 127, 128
Bloch equations, 231
Boltzmann equilibrium, 236, 237
bonding mechanism, 65
Born–Harber cycles, 74
BQ anion, 245
branched alkanes, 182
bridged-loop-gap, 233
Brønsted (acid) site, 125, 187
butene, 128, 129, 188, 189

C, 181
C$_{60}$, 196

^{13}C, 101
CaO, 201, 202, 203, 209, 210, 211, 212, 213
catalysis, 3, 33
cationic Na clusters, 189
cation-radical, 201, 202, 213, 214, 215, 216, 217, 218, 219, 222
cavities, 233, 234
$C_6D_6^+$, 217
$C_6H_6^+$, 213, 214, 215, 217
charge delocalization, 64, 78, 80, 84
charge separation, 202, 203, 204, 205, 207, 209
charge transfer, 208, 215, 216, 218, 222
charge transfer reactions, 190
chemical decay, 236, 239
chemical kinetics, 231, 236, 238
chemical reduction, 196
Chemically Induced Dynamic Electron Polarization (CIDEP), 230, 236
CH$_2$OH, 34, 35
C$_2$H$_4$, 32, 35, 36
C$_2$H$_6$, 36
CH$_3$OH, 34, 35
CIDEP, 229, 230, 232, 236, 237, 238, 239,242, 243, 245, 246
CIDEP mechanisms, 229, 231, 236
CI method, 63, 66, 67, 73, 76
Cini–Sawatzky formula, 77
cis- and *trans-*decalin radical cations, 182
cluster approaches, 63
cluster model, 63, 65, 81
cluster model of the adsorption site, 63
clusters, 179, 180, 189, 191, 192, 193, 194, 195, 196
^{13}C-NMR, 188
CNDO (complete neglect of differential overlap), 62
complete breakdown of the one-electron picture, 70
conc. H$_2$SO$_4$, 92
constant ionic (final) state spectroscopy (CISS) spectrum, 83
continuous-wave EPR, 89, 90
CoO–MgO, 15, 16, 17
cooperative core-hole screening mechanism, 64
cooperative screening mechanism, 76
core hole screening, 64, 65, 83, 84
core hole XPS spectra of CO(N$_2$) on a Ni metal surface, 62
core-valence valence (CVV) Auger spectrum, 77
coronene radical, 96
CTAC, 245, 248, 249, 250, 251
CT (charge transfer), 61

CT screening, 62, 68, 69, 71, 78, 81
Cu, 7, 8, 9, 25, 26, 29
Cu^{2+}, 7, 8, 23, 24, 25, 26, 27, 29
cycloalkenes, 188
cyclododecene, 134
cyclohexadienyl radical, 185, 264, 265, 267, 268, 269, 271, 272, 273, 274
cyclohexane radical cations, 182
cyclooctadiene, 134
cyclooctene, 134, 135
cyclopentene, 131, 132, 133, 134, 139, 141
CYCLOPS, 233, 235

d^1, 7
d^5, 21
d^9, 5, 17
deadtime, 232, 234, 236, 241
9-decalinyl radical, 122
de-excitation electron spectroscopy (DES), 64
delay time, 235, 239, 242, 247, 249, 250, 253
delocalization, ionization, 78
deprotonation, 182
DES Spectra of Adsorbates, 77
diagrammatic two particle-hole Tamm–Dancoff approximation, 63
dibenzothiophene, 110
dienes, 188
diffusion, 39, 48, 49, 50, 51, 53, 54, 55, 56, 57, 58, 259, 265, 271
diffusion coefficient, 39, 48, 49, 50, 51, 53, 54, 55, 56
diffusion measurements, 48
dimethylacetylene, 125, 135
3,3-dimethylbut-1-ene, 131
2,5-dimethylhexa-2,4-diene (DMHD) radical cation, 129
1,3- and 1,4-dimethylcyclohexene, 135
dinaphthothiophene, 110
dinitrobenzene radicals, 106
dinitrobenzene, 104
dipole-dipole interactions, 101, 237
dispersion, 240
donor-acceptor, 218
doublet–quartet mixing, 240
doublet radicals, 237, 240, 241
doublet-triplet encounter, 241
DQ, 241–254
durene, 128
duroquinone, 241
dynamic averaging, 184
dynamics, 184, 180

ECA, 73, 74

electrochemical radical ion formation, 292
electron, 201, 202, 203, 204, 205, 206, 207, 209, 213, 215, 217, 219, 220, 224
electron capture, 120
electron centres, 204, 206, 208, 212, 213
electron-gain centres, 126
electron loss centres, 120
electron magnetic resonance, 89
electron-nuclear double resonance (ENDOR), 89
electron-nuclear hyperfine interactions, 90
electron paramagnetic resonance (EPR), 148, 259
electron-rich MgO, 148, 164, 166, 168, 176
electron spin echo envelope modulation (ESEEM), 102
electron spin echo (ESE), 89, 91, 231
electron spins, 234
electron spin–spin dipolar interaction, 252
electron transfer, 147, 148, 166, 176
electron transfer reactions, 180, 189, 190, 196
electronic Zeeman interaction, 90
ENDOR, 39, 40, 41, 42, 43, 44, 57, 184, 185
ENDOR spectroscopy, 91
enhanced absorption, 237, 242, 252
EPR, 89, 201, 202, 205, 207, 208, 209, 210, 211, 212, 213, 214, 215, 216, 218, 219, 220, 222, 223, 224, 229, 230, 231, 232, 233, 234, 235, 236, 237, 238, 239, 240, 241, 243, 244, 245, 246, 247, 248, 249, 250, 251, 252, 253, 254, 255
equilibrium cascade, 129
equivalent core approximation (ECA), 71
ESE, 39, 40, 43, 48, 49, 50, 51, 52, 53, 54, 55, 56, 57
ESEEM, 89, 192
ESE–T method, 54, 55, 56, 57, 58
ESE-tomography, 52
ESR, 39, 40, 42, 43, 46, 47, 48, 49, 52, 58, 89, 179, 180, 181, 185, 186, 187, 188, 189, 190, 191, 192, 193, 194, 195, 196, 197
ethane, 180, 181, 182
ethyl radical, 271
'exact cancellation', 106
exchanged zeolites, 189, 190, 192, 194
exchange interaction, 238, 240, 244, 248
excitation energies, 64, 71, 72, 73, 74, 75, 76, 84
excited singlet state, 237
experimental and theoretical XPS core hole spectra of CO/Ni(100) (NiCO), N_2/Ni(100) (NiN$_2$) and PdCO, 65
extramolecular screening, 61

F centre, 151, 155
faujasite, 93

FID, 231, 232, 233, 234, 236, 240, 243, 244, 248, 252, 254
field-sweep two-pulse ESE, 192
final state rule, 75, 84
fine structure, 26
flavin mononucleotide, 287, 304, 305
fluorinated benzenes, 179, 185
Fourier transformation
Fourier Transform Electron Paramagnetic Resonance (FT-EPR), 229
F-pair CIDEP, 239
F-pair spin polarization, 239
free radicals, 229, 230, 231, 232, 233, 235, 236, 234, 238, 239, 240, 241, 244, 245
free induction decay (FID), 231, 262
freon matrices, 120, 124, 128, 131, 135, 137, 138
frequency domain spectrum, 234
FT-ESEEM, 191
FT-NMR, 231, 232, 234
fullerenes, 196, 275
furans, 104

γ-alumina, 93
g-anisotropy, 95
geminate radical pairs, 239
giant satellite, 61, 70, 76
5 eV giant shakeup satellite, 70
Green's function method, 63
g-strain broadening, 95
g value, 229, 238, 239, 254
GVBCI (generalized valence bond configuration interaction), 62
gyromagnetic ratio, 232
γ-irradiation, 195
γ-radiolysis, 182

^{1}H, 101
^{2}H, 101
halocarbon matrix, 179, 182, 184, 185
heterogeneous catalysis, 119, 120, 259, 272, 275
heterogeneous medium, 254
hexa-1,5-diene, 135, 137
hexafluorobenzene, 185
hexamethylbenzene, 124, 128, 135
hexamethylethane (HME), 182
heteroatoms, 104
H-mordenite, 119, 125, 127, 128, 129, 131, 132, 134, 135, 137, 139, 140, 141, 183, 189
hole centres, 202, 203, 204, 205, 208, 209, 210, 211, 212, 213, 224
homogeneous electron transfer, 254
Houndry-M46, 95
Hückel Molecular Orbital, 111

hydrogen abstraction reaction, 241
hydrogen isotope, 260
hydroxylated surface, 271
hydroxymethyl radical, 34
hyperfine anisotropy, 264
hyperfine coupling constants, 271, 274
hyperfine interactions, 229, 237, 238, 239
hyperfine structure, 5, 7, 15, 18, 19
HY zeolite, 184
H-ZSM-5, 188, 189

inequivalent sites, 274
inert-gas matrices, 179
inorganic materials, 147
intersystem crossing (isc), 237
ionic clusters, 180, 189, 190, 191, 192, 195
ionic solids, 147
ionization potential, 179, 185, 196
ion–molecule reaction, 182
ion pair, 25
ion-radical, 201, 202, 204, 206, 208, 209, 211, 212, 213, 216, 217, 219, 220, 221, 222, 224
impregnation, 191, 192

Jahn–Teller distortion, 96, 123, 184, 188

K_3^{2+}, 191, 192
K_4^{3+}, 191, 192
kinetic measurements, 230
Koopman's theorem (KT) state, 68

Larmor frequency, 232
laser pulse, 231, 234, 235, 237, 242, 243, 247
linear, 182
linear model molecules, 62
linear prediction-singular value decomposition (LP-SVD), 236
linewidth alternation, 132
local metal configuration, 64, 68, 69, 70, 83
local metal, local ligand and metal-ligand CT excitations, 63
local metal $s - d$ population changes, 68
loop-gap, 233
loop-gap resonator, 233, 234
Lorentzian line, 234
LP-SVD analysis, 244

M_4^{3+}, 191
magnesium oxide, 201, 202, 203, 204
magnetization, 231, 232, 233, 240
m-difluorobenzene, 185
m-dinitrobenzene, 104

MgO, 201–213, 223
MgO surface, 204, 205, 207, 208, 209
metal oxides, 147, 179, 181, 182, 188
metal particles, 180, 192, 197
methane, 179, 180, 181
methane radical cation, 181
methanol oxidation, 3, 34, 35, 36
2-methylbut-1-ene, 131
2-methylbut-2-ene, 131
methylcyclohexane, 141
methylcyclohexane radical cations, 183
3-methylpentane, 141
^{25}Mg, 92, 108
micelle, 238, 246, 247, 248, 249, 250, 251
microwave field, 230, 231, 232, 233
microwave pulse, 231, 232, 234, 235, 240, 242, 243, 244
$\pi/2$ microwave pulse, 231
2 mm-band ESR, 39, 39, 40, 46, 57
Mo, 3, 7, 8, 9, 10, 11, 14, 18, 19, 20, 25, 34, 35, 36
Mo^{5+}, 18
molecular orbital (MO) approach, 62
Mn, 21
Mn^{2+}, 21
Mobility of alkene radical cations in zeolites, 139
monoalkali cationic centres, 156
mordenite, ZSM-5, X, Y, 120
multi-frequency EPR approach, 90
muonium, 260
muon spin resonance, 259, 262, 264

Na_2^+, 193
Na_3^{2+}, 193
Na_4^{3+}, 189, 190, 191, 192, 193
Na_4^{4+}, 193
Na_5^{4+}, 191
Na_6^{5+}, 189, 191, 192
$Na_n^{(n-1)+}$, 189
Na-Ω-5, 123, 182, 183
Na-Y, 188, 191, 192, 193
n-$C_6H_{14}^+$, 182
n-$C_8H_{18}^+$, 182
neutral organic radicals, 274
NH$_3$ adsorbed on Ni metal surface, 71
Ni, 3, 30, 37
NiCO, 61, 62, 63, 65, 66, 68, 69, 70, 71, 72, 75, 76, 77
NiN$_2$, 62, 63, 65, 66, 67, 68, 69, 71, 72, 73, 75, 76, 77
NiPF$_3$, 63
nitroaromatics, 104

nitrobenzene, 106, 201, 219, 220, 222, 223, 224
nitroxyl radicals, 220, 221, 224
NO, 189
NO^+, 189
NO_2, 189
NO adsorption, 17
N_2O adsorption, 35, 36
norbornadiene, 124, 188
normal Auger decay and the CT core hole screening process, 81
nuclear spin state, 237, 238, 239, 243, 250
nuclear quadrupole interaction, 106
nuclear quadrupole resonance (NQR), 92

^{17}O, 92, 101, 108
O^-, 34, 36
O_2^-, 9, 15, 16, 17
O_2 adsorption, 26
9-octalin, 119, 132, 140
9-octalin radical cation, 132, 137
oligomerization, 188
one-electron donor and acceptor sites, 93
organic radical cations, 179, 185
organic radicals and radical ions, 180
'orientation selection', 112
O_2^- –radical, 189
oscillator strength, 64, 72, 74
out-of-phase, 240
out-of-phase signal, 244
O $1s$ DES spectra of CO/H/Ni, 82
O $1s$ DES spectra of the CO/Ni(100) system, 82
oxygen anion-radical, 201, 202, 203, 204, 207
oxygen radical, 209

paramagnetic, 229–234, 236, 240, 245, 251, 254
paramagnetic Ag clusters, 195, 196
paramagnetic complexes, 39, 41, 42, 44, 46, 57
paramagnetic metal ions, 179
participant decay, 64, 78, 79, 80, 81
PdCO, 63, 65, 67, 68, 69, 71
penta-1,4-diene, 137, 139
pentafluorobenzene, 185
pentamethylbenzene, 128
perylene (Pe) radicals, 93
phase shift, 244
phase transition, 274
photochemical, 229, 230, 231, 234, 236, 237, 238, 241
photoexcited triplet, 236
photoinduced electron transfer, 229, 241, 243, 246, 251
photoprocesses, 201, 202, 211, 224
photosensitizers, 182

piperylene, 137, 139
polymerization, 188
polynuclear aromatic compounds (PAs), 92
polytetrafluoroethylene, 188
porphyrins, 241
powder ENDOR pattern, 97
probe molecule, 3, 6, 10, 11, 13, 15, 24, 36
propagating radical, 188
propene, 128, 132, 189
proton catalysed, 128, 130, 131, 138, 142
pseudo-rotation, 184
pulsed ENDOR, 89, 91
pulsed-EPR, 231, 232, 240
pulse radiolysis, 193
pulse width, 232
pump-probe TRSERS method, 283
π CT shakeup satellite, 71
π bonding, 64, 65, 68, 74, 77, 78
π relaxation, 68, 69, 70
π satellite, 71

quadricyclane, 124, 188
quasi-in-situ studies, 30
quinone, 39, 40, 41, 42, 45, 46, 48, 57, 241

radical anions, 147, 164, 176
radical cations, 119, 120, 121, 122, 123, 124, 125, 127, 128, 129, 130, 131, 139, 140, 141, 142, 143
radical cation of hexamethyl(Dewar benzene), 124
radical ion formation at metal electrode surfaces, 287
radical ions, 179, 183, 193
radical pair, 217, 218, 219, 229, 238, 239, 240, 244, 246, 248, 249, 251, 254
radicals, 179, 180, 181, 182, 183, 185, 193, 197
radiolysis, 119, 120, 121, 124, 125
Rb_n and Cs_n metal particles, 192
reflectance UV-visible, 188
relaxation processes, 48, 50
resonance frequency, 233, 240
resonantly core excited state, 64
resonant photoemission, resonant Auger emission, autoionization) spectra of adreversed s–$d\sigma$ mechanism, 74, 76, 77
resonator, 233
rotational correlation time, 237, 250, 251
rotational diffusion, 271
RPM CIDEP, 239, 245

SERS and TRSERS for various radical ion systems, 295
^{29}Si, 92, 101

silica, 259, 260, 261, 265, 266, 267, 268, 269, 270, 271, 272, 273, 274
silica-alumina, 92
silica gel, 181, 185, 241, 251, 252, 253, 254
silicalite, 121, 122, 123
silica supported metal catalyst, 260, 271
silicates, 182, 183
silver clusters, 191, 194, 195, 197
silver radicals, 179
simulations, 92
Si-Na clathrates, 196
single shot TRSERS method, 281, 297
singlet-triplet mixing, 238, 239
SnO_2, 18, 19, 20
sodium clusters, 179, 191, 192, 193
sodium dodecylsulfate (SDS), 247
sodium Na_2^+, 193
species partially immobilized on surfaces, 95
spectral width, 232, 233
spin-correlated radical pairs (SCRP), 240, 244
spin decoupling, 25
spin dynamics, 231, 236, 240, 243, 245
spin Hamiltonian, 90
spin label, 259, 260
spin-orbit (SO) coupling, 104
spin polarization, 232, 237, 238, 239, 240, 244, 245, 246, 250
spin relaxation rates, 229
spin selective, 237
'spontaneous' formation, 119
spontaneous generation, 125
static distortion, 184
stimulated-emission, 237
superhyperfine structure, 15, 19, 24, 32
surface, 147, 148, 149, 150, 151, 152, 153, 155, 156, 157, 158, 159, 161, 162, 163, 164, 166, 167, 168, 169, 170, 171, 172, 173, 174, 175, 176
surface enhanced Raman scattering, 277, 279, 282
surface enhanced resonance Raman scattering, 277, 280
surface F_S^+ centres, 151
surfactant molecules, 245, 246, 250

temperature-dependent line shape, 184
temperature-programmed desorption, 188
1,2,4,5-tetrafluorobenzene, 185, 186, 197
tetramethylallene, 123
tetramethylallene radical cation, 188
tetramethylcyclobutadiene (TMCB) radical cation, 135
tetramethylcyclopropane (TMC) radical cation, 122

tetramethylene, 188
tetramethylethylene, 119
thermal equilibrium, 234, 236, 237, 238, 242, 245, 254
thermogravimetric analysis, 188
thiophene (THI), 104, 110
time domain signal, 231
time-domain spectrum, 234
time-resolved SERS, 278, 297, 301, 303
time-resolved studies, 234
TiO_2, 18, 19, 21, 22
toluene, 183, 194
trans-2,3-dimethyloxirane, 122
transmission, 188
transverse magnetization, 231, 232
TREPR, 230, 231, 240, 241, 245, 248, 252
1.3.5-trifluorobenzene, 185
1,1,2-trimethylactivation energies, 271, 272, 273, 274
trimethylbenzene, 183
trimethylbutane (TMB), 182
2,3,3-trimethylbut-1-ene, 130
trinitrobenzene, 104
triplet excited state electron transfer, 246, 252
Triplet mechanism, 236
Triton X-100 (TX100), 246

unpaired spin density, 111
'unscreened' peak, 61
'unscreened' state, 70
'unscreened state' (Koopmans state, KT state), 62
UV irradiation, 181
UV irradiation in boric acid glass, 92

V, 19, 21, 22
V^{4+}, 21, 22
valence photoemission spectra, 63
van der Waals interaction, 260, 271
very high frequencies (VHF, above 35 GHz), 90
Vycor glass, 181, 182, 183

W-band (95 GHz), 95
'well screened' peak, 61

XAS spectrum of the N_2/Ni system, 75
XAS width, 83
X and Y zeolites, 192, 193
XPS core spectra of CO/Ni(100) and N_2/Ni(100), 62
XPS core spectra of CO/H/Ni system, 70
XPS spectra of NO on Ni(100), 72
XPS spectrum of N_2 adsorbed on Ni–Ti alloy, 70
X-ray absorption spectroscopy (XAS), 64

X-ray diffraction, 191, 195
X-ray photoelectron spectroscopy (XPS), 61

Zeeman interaction, 238
zeolite, 3, 4, 9, 21, 23, 24, 25, 26, 36, 39, 40, 41, 42, 43, 44, 45, 46, 48, 49, 57, 179, 180, 182, 183, 185, 187, 188, 189, 190, 191, 192, 193, 194, 195, 196, 197, 198, 199, 259, 273, 275, 276

zeolite 4A, 180, 181, 189, 190
ZnTPP, 241, 242, 243, 244, 245, 246, 251, 252, 253, 254
ZnTPPS, 245, 246, 247, 248, 249, 250, 251, 254
ZSM-5, 182, 185, 187, 188, 189, 215, 218
ZSM-5 zeolite, 182, 185, 186, 187, 188, 201, 202, 213, 215, 218
ZSM-34, 182, 183

TOPICS IN
MOLECULAR ORGANIZATION AND ENGINEERING

Honorary Chief Editor: W. N. Lipscomb, Harvard, U.S.A.
Executive Editor: Jean Maruani, Paris, France

1. J. Maruani (ed.): *Molecules in Physics, Chemistry, and Biology.*
   Vol. 1: General Introduction to Molecular Sciences. 1988
   ISBN 90-277-2596-9
2. J. Maruani (ed.): *Molecules in Physics, Chemistry, and Biology.*
   Vol. 2: Physical Aspects of Molecular Systems. 1988
   ISBN 90-277-2597-1
3. J. Maruani (ed.): *Molecules in Physics, Chemistry, and Biology.*
   Vol. 3: Electronic Structure and Chemical Reactivity. 1989
   ISBN 90-277-2598-5
4. J. Maruani (ed.): *Molecules in Physics, Chemistry, and Biology.*
   Vol. 4: Molecular Phenomena in Biological Sciences. 1989
   ISBN 90-277-2599-3
5. E. Schoffeniels and D. Margineanu: *Molecular Basis and Thermodynamics of Bioelectrogenesis.* 1990   ISBN 0-7923-0975-8
6. A. Lund and M. Shiotani (eds.): *Radical Ionic Systems.* Properties in Condensed Phases. 1991   ISBN 0-7923-0988-X
7. P.I. Lazarev (ed.): *Molecular Electronics.* Materials and Methods. 1991
   ISBN 0-7923-1196-5
8. E. Rizzarelli and T. Theophanides (eds.): *Chemistry and Properties of Biomolecular Systems.* 1991   ISBN 0-7923-1393-3
9. L.A. Montero and Y.G. Smeyers (eds.): *Trends in Applied Theoretical Chemistry.* 1992   ISBN 0-7923-1745-9
10. M.T. Pope and A. Müller (eds.): *Polyoxometalates: From Platonic Solids to Anti-Retroviral Activity.* 1994   ISBN 0-7923-2421-8
11. N. Russo, J. Anastassopoulou and G. Barone (eds.): *Properties and Chemistry of Biomolecular Systems.* 1994   ISBN 0-7923-2666-0
12. Y.G. Smeyers (ed.): *Structure and Dynamics of Non-Rigid Molecular Systems.* 1994   ISBN 0-7923-2774-8
13. A. Lund and C.J. Rhodes (eds.): *Radicals on Surfaces.* 1995
    ISBN 0-7923-3108-7

KLUWER ACADEMIC PUBLISHERS – DORDRECHT / BOSTON / LONDON